Radon Series on Computational and Applied Mathematics 1

Lectures on Advanced Computational Methods in Mechanics

Edited by

Johannes Kraus
Ulrich Langer

Walter de Gruyter · Berlin · New York

Editors

Johannes Kraus
Radon Institute for Computational and Applied Mathematics
Austrian Academy of Sciences
Altenberger Straße 69
A-4040 Linz
Austria
e-mail: johannes.kraus@oeaw.ac.at

Ulrich Langer
Radon Institute for Computational and Applied Mathematics
Austrian Academy of Sciences
Altenberger Straße 69
A-4040 Linz
Austria
e-mail: ulrich.langer@assoc.oeaw.ac.at

Key words:
Partial differential equations, elliptic boundary value problems, evolutionary problems, the Stokes problem, linear elasticity, parameter identification, finite element methods, domain decomposition methods, iterative methods, preconditioning, hierarchical and multilevel splittings, hysteresis, Preisach operator, a posteriori error control, modeling error.

Mathematics Subject Classification 2000:
00B25, 35Q60, 35R30, 47J40, 65F10, 65M15, 65N15, 65N20, 65N22, 65N30, 65N55.

⊗ Printed on acid-free paper which falls within the guidelines of the ANSI to ensure permanence and durability.

Library of Congress Cataloging-in-Publication Data

A CIP catalogue record for this book is available from the Library of Congress.

ISBN 978-3-11-019556-9

Bibliographic information published by the Deutsche Nationalbibliothek

The Deutsche Nationalbibliothek lists this publication in the Deutsche Nationalbibliografie; detailed bibliographic data are available in the Internet at http://dnb.d-nb.de.

Printed in Germany
Cover design: Martin Zech, Bremen.
Printing and binding: Hubert & Co. GmbH & Co. KG, Göttingen.

Preface

This volume is a collection of lectures delivered during the *Special Semester on Computational Mechanics* held at the Johann Radon Institute for Computational and Applied Mathematics (RICAM) of the Austrian Academy of Sciences, Linz, Austria, October 3 – December 16, 2005. *Computational Mechanics* is certainly one of the most developed Computational Sciences. On the one hand, there is a growing complexity of purely mechanical models (complicated material behavior, large deformations, friction in solid and structural mechanics, turbulence modeling in fluid mechanics etc.) which requires a fundamental mathematical analysis as well as new and more efficient numerical methods. The efficient handling of model identification problems, optimal control problems, shape, topology, material optimization etc. is not only a major research topic, but is also of prime importance for practical applications. On the other hand, there are a lot of new applications where mechanical fields are interacting with other fields leading to coupled or multifield problems such as magnetomechanical and piezoelectrical, or Fluid-Solid-Interaction (FSI) problems, which are far less studied in the mechanical and in the mathematical communities. The modeling, the analysis and the solution of this kind of problems require knowledge from different fields and research disciplines.

The aim of this Special Semester was to bring together applied mathematicians with experts working in mechanics and related fields. The Special Semesters was focused on new mathematical and methodological questions which have recently appeared in Computational Mechanics and Computational Mathematics with the aim to increase the potential of solving challenging problems in Mechanics.

The following main topics were treated in different forms like lectures, seminars, tutorials, workshops etc.:

(1) Novel Discretization and Solver Techniques in Mechanics,

(2) Inverse, Control, and Optimization Problems in Mechanics,

(3) Challenging Problems in Mechanics.

We invited leading experts to organize tutorials, lectures, seminars, workshops and panel discussions for young researchers, in particular PhD students and young postdocs. The invited experts were cooperating with local people (local co-organizers) helping them in organizing and coordinating the activities.

Topic (1) on **"Novel Discretization and Solver Techniques in Mechanics"** was organized as follows:

- *Discontinuous Galerkin methods:*
  Organizer (invited): Raytcho Lazarov (TAMU, USA).
  Co-organizer (local): Satyendra Tomar (RICAM, Austria).
  This topic was organized in form of lectures given by Raytcho Lazarov, regular seminars led by Raytcho Lazarov and Satyendra Tomar, and a Miniworkshop.

- *Non-standard mixed FE-techniques:*
  Organizer (invited): Dietrich Braess (University of Bochum, Germany).
  Co-organizers (local): Joachim Schöberl (RICAM, Austria).
  This topic was organized in form of lectures given by Dietrich Braess.
- *hp-techniques in Mechanics:*
  Organizer (local): Alexander Duester (TU Munich, Germany)
  Co-organizer (local): Joachim Schoeberl (RICAM, Austria)
  This topic was organized in form of a block lecture given by Alexander Duester and a Miniworkshop.
- *A posteriori estimates and adaptivity in solid mechanics:*
  Organizer (invited): Sergey Repin (Steklov Institute, St. Petersburg, Russia).
  Co-organizer (local): Jan Valdman (SFB F013, Austria).
  This topic was organized in form of lectures given by Sergey Repin and a Miniworkshop. The material presented in these lectures is summarized in Lecture 4.
- *H-matrix techniques:*
  Organizer (invited): Wolfgang Hackbusch (MPI, Leipzig, Germany).
  Co-organizer (local): David Pusch (RICAM, Austria).
  This topic was organized in form of lectures given by Wolfgang Hackbusch and regular seminars led by Steffen Boerm and Lars Grasedyck.
- *Robust parallel algebraic multigrid and multilevel techniques:*
  Organizer (invited): Svetozar Margenov (BAS, Bulgaria).
  Co-organizer (local): Johannes Kraus (RICAM, Austria).
  This topic was organized in form of lectures given by Svetozar Margenov and regular seminars led by Svetozar Margenov and Johannes Kraus. The material presented in these lectures are summarized in Lecture 2.
- *Domain decomposition methods:*
  Organizer (invited): Sergei Nepomnyaschikh (RAS, Novosibirsk, Russia).
  Co-organizer (local): Sven Beuchler (JKU, Austria).
  This topic was organized in form of block lectures given by Sergei Nepomnyaschikh and a Miniworkshop. The material presented in these block lectures are summarized in Lecture 3.
- *Numerical simulation of coupled fields:*
  Organizer (invited): Manfred Kaltenbacher (University of Erlangen, Germany).
  Co-organizer (local): Marco Discacciati (RICAM, Austria).
  This topic was organized in form of lectures given by Manfred Kaltenbacher and Marco Discacciati. The material presented in these lectures are partly summarized in Lecture 1.

The second topic on **"Inverse, Control, and Optimization Problems in Mechanics"** is closely connected with the first topic since fast numerical methods are needed for solving the forward problem anyway. Topic (2) was organized as follows:

- *Inverse problems and model identification in mechanics:*
  Organizer (invited): Barbara Kaltenbacher (University of Erlangen, Germany).
  Co-organizer (local): Martin Burger (JKU, Austria).

This topic was organized in form of lectures given by Barbara Kaltenbacher, tutorials led by Tom Lahmer and a Miniworkshop. The material presented in these lectures are partly summarized in Lecture 1.

- *Control and optimization problems in mechanics:*
  Organizer (invited): Ekkehard Sachs (University of Trier, Germany).
  Co-organizer (local): Roland Griesse (RICAM, Austria) and Karl Kunisch (University of Graz and RICAM, Austria)
  This topic was organized in form of two workshops and various seminars.
- *Conversion from imperfection-sensitive into imperfection-insensitive elastic structures:*
  Organizer (invited): Herbert Mang (Vienna University of Technology and Austrian Academy of Sciences, Austria).
  Co-organizer (local): Roman Stainko (SFB F013, Austria)
  This topic was organized in form of block lectures given by Herbert Mang.

The **"Challenging Problems in Mechanics"** were presented and discussed at the Challenges Day where we invited scientists who presented mechanical problems which require the development of new or significantly enhanced computational techniques. In addition to this we organized lectures and talks on new challenging models and problems in Mechanics like biomechanical problems, geomechanical problems and problems in Mechantronics.

We refer the reader to

`http://www.ricam.oeaw.ac.at/specsem/sscm/structure/`

for more information about the topics (1)–(3).

The ultimate highlights of this Special Semester were certainly the International Workshop on "Direct and Inverse Field Computations in Mechanics" and the public lecture on "Gottfried Wilhelm Leibniz – far ahead of his time" by Erwin Stein (University of Hannover, Germany).

A complete overview over the major activities, events and contributions of the Special Semester on Computational Mechanics can be found on the home page

`http://www.ricam.oeaw.ac.at/specsem/sscm/`

of the Special Semester.

This book contains four contributions to different topics belonging to the fields (1) and (2) as presented above. In Lecture 1, Barbara and Manfred Kaltenbacher discuss the modeling and iterative identification of hysteresis via Preisach operators in partial differential equations (PDEs). Hysteresis effects play a key role in many physical processes ranging from ferromagnetics to porous media filtration. The efficient and precise numerical modeling of such phenomena is a big challenge for science and an outstanding topic combining researchers working on forward and inverse problems. The authors concentrate on hysteresis effects in piezoelectric (hyperbolic PDE) and magnetic (parabolic PDE) materials. The hysteresis modeling is based on the Preisach operator, which can be considered to be the most used phenomenological hysteresis model in practice. After introducing the defining PDEs for piezoelectricity and magnetics a detailed discussion on the properties of the Preisach operator including memory deletion as well as its efficient computation via Everett functions is provided. Next, the authors invest in the well-posedness of the PDEs and derive efficient numerical schemes for their numerical solution. Having established the framework for the forward problems, the authors discuss the identifiability and develop iterative schemes mainly based on Newton type methods for dealing with the nonlinear inverse problems. Finally, numerical tests demonstrate the applicability of the inverse schemes for the identification of hysteresis in parabolic and hyperbolic PDEs.

Lectures 2 and 3 are devoted to advanced solver techniques for large-scale systems of finite element equations. Multigrid and multilevel solvers are among the most efficient iterative methods, however, their robustness deserves close attention. This issue is investigated by Johannes Kraus and Svetozar Margenov for the class of anisotropic elliptic problems (Lecture 2). The authors discuss various preconditioners based on multilevel extensions of two-level finite element methods, which lead to solvers that are of optimal order of computational complexity with respect to the size (or discretization parameter) of the system. Both, the construction and the analysis of these methods rely on a (recursive) two-level splitting of the underlying finite element space. The key role in the derivation of optimal convergence rate estimates plays the abstract angle between the coarse space and its complementary space, which is defined in virtue of the constant γ in the so-called strengthened Cauchy–Bunyakowski–Schwarz (CBS) inequality. The methods can be of block-matrix factorized form, recursively extended via certain matrix polynomial approximations of the arising Schur complement matrices, or, on additive, i.e., block diagonal form, using stabilizations of the condition number at certain levels. The focus of this survey of algebraic multilevel iteration (AMLI) methods is on their robustness with respect to coefficient and mesh anisotropy. The cases of conforming (Courant) and nonconforming (Crouzeix–Raviart) linear finite elements are considered. Lecture 3 is devoted to Domain Decomposition methods which provide another technology for the development of efficient solvers in complex geometries with possibly unstructured finite element meshes. The author Sergey Nepomnyaschikh is one of the inventors of the so-called Additive Schwarz Method (ASM) which are very suitable for the implementation on massively parallel computers. In his contribution, the author presents two preconditioning techniques,

- the Domain Decomposition (DD) method, and
- the Ficticous Space Method (FSM).

For DD methods, the computational domain is decomposed into several overlapping or nonoverlapping domains with simpler geometrical structure, where multilevel solvers are available. The key role for the fast implementation of nonoverlapping Domain Decomposition preconditioners play the preconditioners on the traces of the subdomains and the development of efficient explicit extension operators. The Ficticous Space Method embeds the computational domain into a larger domain of simpler structure where multilevel preconditioners can be used again. This technique is applicable for systems of linear algebraic equations arising form the discretization of unstructured finite element meshes and is an alternative to Algebraic Multigrid Methods (AMG). The presented material is suitable for a one-semester course on Domain Decomposition methods.

Nowadays, a posteriori error control methods and closely related adaptive methods of computer simulation form one of the most rapidly developing parts of numerical analysis. In Lecture 4, Sergey Repin gives an overview of a posteriori error estimation methods for PDEs and exposes a new functional approach to the derivation of a posteriori estimates. The overview is intended to show the development of a posteriori methods in the historical perspective and explains the underlying mathematical ideas in the most transparent way. It starts with the first methods suggested by Runge, Prager and Synge, and Mikhlin. Further, a posteriori methods developed for iteration methods and monotone operators are considered. An essential part of the paper is concerned with a posteriori error estimation methods for finite element approximations. Here, a reader will find motivations of the main error indication methods used in modern mesh adaptive codes (e.g., explicit and implicit residual methods, dual-weighted residual method, methods based on post-processing of approximate solutions). The subsequent part is devoted to functional a posteriori estimates which were introduced in the publications of the author and his co-workers. Functional estimates contain no mesh-dependent constants and are valid for any conforming approximations of a problem. In the paper, two methods of their derivation are discussed on the paradigm of diffusion type problems, linear elasticity, Stokes problem, variational inequalities and other problems in mathematical physics.

The editors of this collection of lectures and the organizers of Special Semester are exceedingly grateful to Wolfgang Forsthuber, Magdalena Fuchs, Florian Tischler and Annette Weihs for their administrative and technical support. Last but not least we would like to thank the Austrian Academy of Sciences for their support.

Linz, May 2007

Johannes Kraus
Ulrich Langer

Table of Contents

Radon Series Comp. Appl. Math **1**, 1–45

Modelling and iterative identification of hysteresis via Preisach operators in PDEs

Barbara Kaltenbacher and Manfred Kaltenbacher

Key words. Hysteresis, Preisach operator, partial differential equations, parameter identification, piezoelectricity, magnetics

AMS classification. 35R30, 47J40, 35Q60

1 Introduction

Hysteresis is a memory effect that appears in a wide range of time-dependent processes such as ferromagnetism, elastoplasticity, or porous media filtration, to name just a few examples. It is characterized by a lag behind in time of some output in dependence of the input history.

Figure 1.1, for example, shows the curve describing the polarization (and depolarization) P of some ferroelectric material in dependence of the applied electric field E: As E increases from zero to its maximal positive value $E_{\max}$ (virgin curve), the polarization also shows a growing behaviour that lags behind E, though. Then E decreases, and again P follows with some delay. As a consequence, there is a positive remnant

First author: Support by the Austrian Academy of sciences as well as by the DFG under grant KA 1778/1 is gratefully acknowledged.
Second author: Support by the Austrian Academy of sciences is gratefully acknowledged.

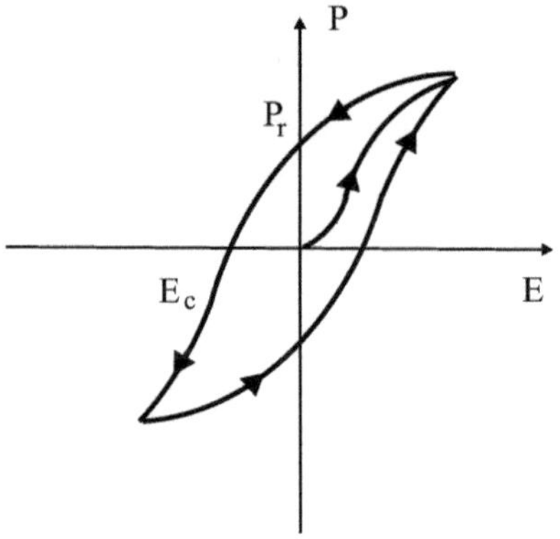

Figure 1.1 Polarization hysteresis for a ferroelectric medium.

polarization P_r for vanishing E that can only be completely removed by further decreasing E until a critical negative value $-E_c$ is reached. After passing this threshold, a polarization in negative direction – so with the same orientation as E – is generated, until a minimal negative value is reached. The returning branch of the hysteresis curve ends at the same point $(E_{\max}, P_{\max})$ where the outgoing branch had started but takes a different path, which results in a gap between these two branches. We write

$$P(t) = \mathcal{P}[E](t)$$

with some hysteresis operator $\mathcal{P}$.

Probably the most simple example of hysteresis is the behaviour of a switch or *relay* (see Figure 1.2, as well as Example 2.1.1 in [3]), that is characterized by two threshold values $\alpha > \beta$. The output value p is either -1 or $+1$ and changes only if the input value e crosses one of the switching thresholds α, β: If, at some time instance t, $e(t)$ increases from below to above α, the relay will switch up to $+1$, if e decreases from above to below β, it will switch down to -1, in all other cases it will keep its value – either plus or minus one, depending on the preceding history. Obviously, there is some ambiguity in case the input only touches one of the thresholds without really crossing it. Neglecting this latter detail for the moment, we just formally define the relay operator $\mathcal{R}_{\beta,\alpha}$ by

$$\mathcal{R}_{\beta,\alpha}[e] = p$$

according to the description above.

A practically important phenomenological hysteresis model that was originally introduced in the context of magnetism but plays a role also in many other hysterestic processes, is given by the *Preisach operator*

$$\mathcal{P}[e](t) = \iint_{\beta \leq \alpha} \wp(\beta, \alpha) \mathcal{R}_{\beta,\alpha}[e](t)\, d(\alpha, \beta), \tag{1.1}$$

which is a weighted superposition of elementary relays. The initial values of the relays

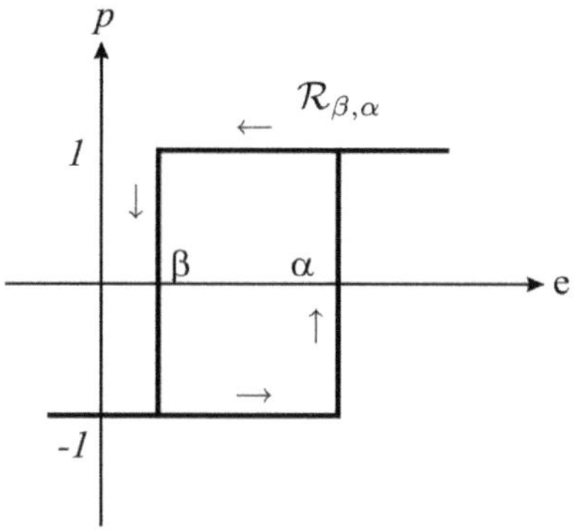

Figure 1.2 Hysteresis of an elementary relay.

$\mathcal{R}_{\beta,\alpha}$ (assigned to some "pre-initial" state e_{-1}) are set to

$$\mathcal{R}_{\beta,\alpha}[e_{-1}] = \begin{cases} -1 & \text{if } \alpha > -\beta \\ +1 & \text{else.} \end{cases} \tag{1.2}$$

This in some sense reflects a neutral (or virginal) initial state. Note that due to the integration with an integrable weight function, the ambiguity of the relay operator for the case of only touching but not crossing thresholds does not contribute essentially (cf. [3]). It will be this Preisach model and its identification within PDEs of hyperbolic and parabolic type that will constitute the main topic of this paper. Determining $\mathcal{P}$ obviously amounts to determining the weight function $\wp$ in (1.1). The domain $\{(\beta,\alpha) \mid \beta \leq \alpha\}$ of $\wp$ is called the *Preisach plane*. Assuming that the $\wp$ is compactly supported and by a possible rescaling, we can restrict our attention to the Preisach unit triangle $\{(\beta,\alpha) \mid -1 \leq \beta \leq \alpha \leq 1\}$ within the Preisach plane, see Fig. 1.3, which shows the Preisach unit triangle with the sets S^+, S^- of up- and down-switched relays at the initial state (1.2).

As will be stated in more detail below, there is a different bivariate function that – along with the independently of $\wp$ valid *Preisach memory* – uniquely determines $\mathcal{P}$, namely the *shape function* (or *Everett function*) on whose identification we will concentrate here. Motivated by its high versatility that, e.g., enables to reproduce minor loops as well, see Figure 1.4, we will concentrate on Preisach operators for hysteresis modelling in this paper.

Proving unique identifiability in nonlinear PDEs and, still more, in PDEs with hysteresis is a challenging task, which will not be treated in this paper. In this context, we wish to refer to a recent Diploma thesis by Jais, supervised by Brokate, in which first identifiability results for parabolic PDEs with hysteresis are presented [9].

While well analyzed numerical methods exist for identifying hysteresis in direct input-output models (cf., e.g., Hoffmann, Sprekels and Visintin [7], Hoffmann and Meyer [6], Kuhnen [20], Reiländer [25], and the references therein), this does not seem to be the case for determining hysteresis operators appearing in partial differential equations from indirect (e.g., boundary) measurements. As a matter of fact,

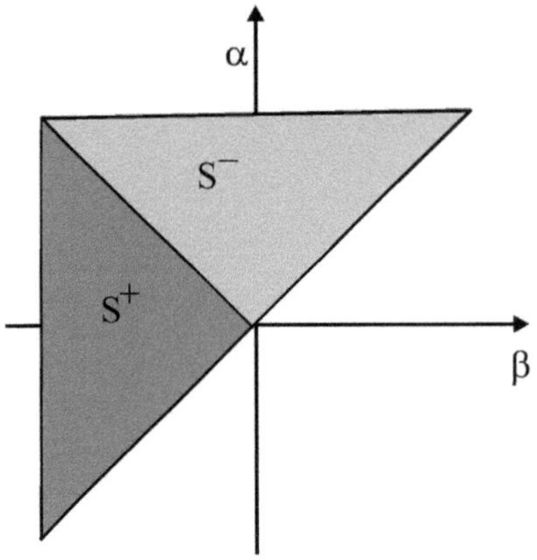

Figure 1.3 Preisach plane.

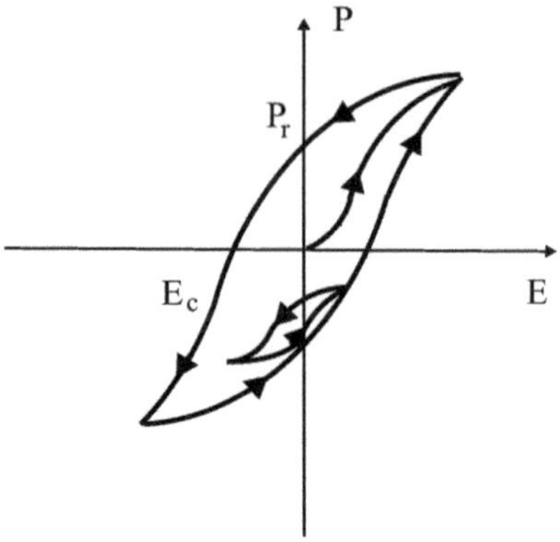

Figure 1.4 Minor loop.

according to the authors' knowledge, no such methods can be found in the literature so far in the context of infinite dimensional hysteresis models such as the Preisach model in PDEs (for an identification procedure for the basically five dimensional Jiles Atherton hysteresis model, we refer to [2]). This was the motivation for deriving and investigating numerical methods for hysteresis identification in this paper. Here we present new methods based on fixed point and Newton type iterations and test their numerical performance.

The paper is organized as follows: The next section introduces two application examples, namely piezoelectricity and magnetics, leading to hysteresis in hyperbolic and parabolic PDEs, respectively. From these we derive two model problems that the discussion of the paper concentrates on, where the main emphasis is put on the first, i.e., the hyperbolic one. In Section 3 we provide some definitions and results on hysteresis operators mainly taken from [3] that will be required for our purpose. Section 4 provides some well-posedness results for the forward problem of solving the PDE with hysteresis, as well as iterative procedures for their numerical solution. In Section 5, we derive and investigate a number of iterative procedures for hysteresis identification in these PDEs. The paper concludes with some final remarks on open problems and future research tasks.

2 Two examples

2.1 Piezoelectricity: Hysteresis in a hyperbolic PDE

The piezoelectric effect is a direct coupling between the electric and the mechanical behaviour of certain materials. In the spatially fully three dimensional case it can be modelled by a system of coupled PDEs for the vector of mechanical displacements $\vec{d}$

and the electric potential ϕ. In the linear case of small excitations this system reads as

$$\begin{aligned} \rho\vec{d}_{tt} &= \mathrm{DIV}\left(\mathbf{c}^E\mathrm{DIV}^T\vec{d}+\mathbf{e}^T\mathrm{grad}\phi\right) \\ 0 &= -\mathrm{div}\left(\mathbf{e}\mathrm{DIV}^T\vec{d}-\varepsilon^S\mathrm{grad}\phi\right), \end{aligned} \tag{2.1}$$

where ρ is the mass density, $\mathbf{c}^E$ the elastic tensor, $\mathbf{e}$ the piezoelectric coupling tensor, and ε^S the dielectric tensor. These PDEs can be derived by an application of Newton's law

$$\rho\vec{d}_{tt} = \mathrm{DIV}\vec{\sigma} \tag{2.2}$$

on the mechanical side and Gauss' law

$$0 = \mathrm{div}\vec{D} \tag{2.3}$$

on the electrical side, with the constitutive relations (in Voigt notation)

$$\begin{aligned} \vec{\sigma} &= \mathbf{c}^E\vec{S}-\mathbf{e}^T\vec{E} \\ \vec{D} &= \mathbf{e}\vec{S}+\varepsilon^S\vec{E} \end{aligned} \tag{2.4}$$

for the mechanical stress $\vec{\sigma}$ and strain $\vec{S} = \mathrm{DIV}^T\vec{d}$, as well as the electric flux density $\vec{D}$ and field $\vec{E} = -\mathrm{grad}\phi$, respectively.

For large electric fields $\vec{E}$, the constitutive relations become nonlinear and hysteretic, due to switching processes occurring in the medium, i.e., (2.4) is replaced by

$$\begin{aligned} \vec{\sigma} &= \mathbf{c}^E(\vec{S}-\vec{S}^i)-\mathbf{e}^T\vec{E} \\ \vec{D} &= \mathbf{e}(\vec{S}-\vec{S}^i)+\varepsilon^S\vec{E}+\vec{P}^i, \end{aligned} \tag{2.5}$$

with $\vec{S}^i$, $\vec{P}^i$ denoting the irreversible parts of the mechanical strain and electric polarization, respectively, cf., e.g., [15].

In the here considered setting of excitation by a large electric field $E = -\mathrm{grad}\phi$, hysteresis affects polarization and strain in the sense that

$$\vec{P}^i = \vec{P}^i[\vec{E}] = \vec{P}^i[-\mathrm{grad}\phi], \qquad \vec{S}^i = \vec{S}^i[\vec{E}] = \vec{S}^i[-\mathrm{grad}\phi]$$

and will therefore contained in terms in the right hand side of the hysteretic version of (2.1): A hysteresis operator acts on first spatial derivatives of the field quantity ϕ and is acted on by a first order spatial differential operator, cf. (2.2), (2.3). Hence, there is an obvious correspondence to a wave equation with hysteresis

$$\rho u_{tt} = \mathcal{P}[u_x]_x, \tag{2.6}$$

which we will use as a model problem in the sequel, see Sections 4.1, 5.1.

More concretely, we give an exemplary situation, where hysteresis identification in the piezoelectric system can be reduced to hysteresis identification in (2.6), with just those boundary conditions that will be considered for this model problem later on. To do so, we concentrate on the spatially one-dimensional case, where the PDEs read as

$$\begin{aligned} \rho d_{tt} &= (\mathrm{c}d_x+\mathrm{e}\phi_x-\mathrm{c}S^i)_x \\ 0 &= (\mathrm{e}d_x-\varepsilon\phi_x-\mathrm{e}S^i+P^i)_x \end{aligned} \tag{2.7}$$

with d the mechanical displacement, ϕ the electric potential, c the elasticity coefficient, ε the dielectric coefficient, and e the piezoelectric coupling coefficient. Appropriate boundary conditions in the situation of $x = 0$ and $x = L$ representing electrodes attached to opposite sides of the piezoelectric transducer are

$$\begin{aligned}
&(\mathrm{c}d_x + \mathrm{e}\phi_x - \mathrm{c}S^i)(0,t) = \sigma_0(t) \quad (\text{or } d(0,t) = 0)\,, \\
&(\mathrm{c}d_x + \mathrm{e}\phi_x - \mathrm{c}S^i)(L,t) = \sigma_L(t)\,, \\
&(\mathrm{e}d_x - \varepsilon\phi_x - \mathrm{e}S^i + P^i)(0,t) = (\mathrm{e}d_x - \varepsilon\phi_x - \mathrm{e}S^i + P^i)(L,t) = -\tfrac{q}{A}\,,
\end{aligned} \tag{2.8}$$

which corresponds to prescribed boundary stresses σ_0, σ_L, (or a clamped left hand boundary), as well as measured charge q at an electrode, where A is the electrode area. Excitation is done via an input voltage at one of the electrodes.

Modelling of hysteresis in piezoelectric systems is a highly complex task and a field of active research (cf., e.g., [15], [16], [21], [23], [27]). We here restrict ourselves to a simplified but to some extent plausible situation that can be seen as a phenomenological model regarding hysteresis effects purely as consequence of switching processes induced by polarization changes. Thus, we assume the following:

- The hysteresis mainly affects polarization, i.e.,

  $$P^i = \tilde{\mathcal{P}}[E]$$

  with a hysteresis operator $\tilde{\mathcal{P}}$.

- The irreversible mechanical strains are just a function of polarization

  $$S^i = \psi(|P^i|)\,, \qquad \text{e.g., } S^i = \frac{S^i_{\max}}{P^i_{\max}}|P^i|\,,$$

  with a known function ψ. Note that dependence on the absolute value of P rather than on P itself is natural in view of the fact that in 1-d, switching strains depend on the modulus of the polarization, but not on its orientation.

- In view of the fact that in absence of polarization, there is no piezoelectric coupling, and that a changing orientation of polarization reverses the sign of the coupling coefficient, the coefficient e is written as a (known) function of the polarization

  $$\mathrm{e} = \varphi(P)\,, \qquad \text{e.g., } \mathrm{e} = \frac{\mathrm{e}_{\max}}{P^i_{\max}}P^i\,.$$

Here $S^i_{\max}$, $P^i_{\max}$, $\mathrm{e}_{\max}$ are the maximal absolute values of the irreversible strain, the irreversible polarization and the coupling coefficient. $S^i_{\max}$, $P^i_{\max}$ can be obtained directly from large signal measurements, the coefficients c, ε, and $\mathrm{e}_{\max}$ can be determined from small signal measurements, cf. [8], [13].

With these replacements (2.7) becomes

$$\begin{aligned}
\rho d_{tt} &= (\mathrm{c}d_x + \varphi(\tilde{\mathcal{P}}[-\phi_x])\phi_x - \mathrm{c}\psi(|\tilde{\mathcal{P}}[-\phi_x]|))_x \\
0 &= (\varphi(\tilde{\mathcal{P}}[-\phi_x])d_x - \varepsilon\phi_x + \tilde{\mathcal{P}}[-\phi_x] - \varphi(\tilde{\mathcal{P}}[-\phi_x])\psi(|\tilde{\mathcal{P}}[-\phi_x]|))_x\,.
\end{aligned} \tag{2.9}$$

By the boundary conditions (2.8), the second line of (2.9) yields

$$d_x = \frac{\varepsilon\phi_x - \tilde{\mathcal{P}}[-\phi_x] + \varphi(\tilde{\mathcal{P}}[-\phi_x])\psi(|\tilde{\mathcal{P}}[-\phi_x]|) - \frac{q}{A}}{\varphi(\tilde{\mathcal{P}}[-\phi_x])} =: \mathcal{W}[-\phi_x]$$

so that assuming the hysteresis operator $\mathcal{W}$ to be invertible, we can insert $-\phi_x = \mathcal{W}^{-1}[d_x]$ into the first equation of (2.9) to arrive at

$$\rho d_{tt} - \mathcal{P}[d_x]_x = 0$$

with boundary conditions

$$\begin{aligned} \mathcal{P}[d_x(0,\cdot)] &= \sigma_0 \quad (\text{or } d(0,t) = 0)\,, \\ \mathcal{P}[d_x(L,\cdot)] &= \sigma_L\,, \end{aligned}$$

where

$$\mathcal{P}[e] = \mathrm{c}\Big(e - \psi(|\tilde{\mathcal{P}} \circ \mathcal{W}^{-1}[e]|)\Big) - \varphi(\tilde{\mathcal{P}} \circ \mathcal{W}^{-1}[e])\, \mathcal{W}^{-1}[e]\,.$$

Under reasonable assumptions, $\mathcal{P}$ will be a hysteresis operator in the sense described below, cf. Sections 2.10 and 2.11 in [3]. A task of interest is to identify $\mathcal{P}$ from additional measurements of the displacement

$$d(L,t) = f_L(t)\,.$$

Note that the model problem (2.6) also appears to be related to hysteresis in mechanical problems, i.e., elasto-plasticity, where, one often works with Prandtl–Ishlinskii hysteresis operators. These can be seen as special instances of Preisach operators, see Example 2.1.6 in [3].

2.2 Magnetics: Hysteresis in a parabolic PDE

Ferromagnetic hysteresis is one of the fields that actually initiated a large part of the existing research on the mathematical theory of hysteresis. In this context we especially wish to refer to the monographs by Krasnoselskii and Pokrovskii [17], as well as Mayrgoyz [22]. For some advanced hysteresis models in magnetics, see, e.g., [28].

For measuring the nonlinear relation between the magnetic flux density $\vec{B}$ and the magnetic field intensity $\vec{H}$, the usual experimental setup consists of a coil wound around a material probe, which is supplied with an impressed current I. The magnetic flux Φ through the coil computes as

$$\Phi = \int_{A_c} \vec{B} \cdot \vec{n}\, d\Gamma\,,$$

where A_c denotes the cross sectional area of the coil, is measured, cf., e.g., [12], [14]. A mathematical model of this setting is given by a combination of part of Maxwell's equations together with constitutive equations. Ampére's law together with the fact that in this context the change $\vec{D}_t$ of the electric flux density is negligible, as well as Faraday's law give the equation

$$\text{curl}\,\frac{1}{\gamma}\text{curl}\,\vec{H} = \text{curl}\,\frac{1}{\gamma}J_i - \vec{B}_t \text{ in } \Omega \times [0,T]\,,$$

where γ is the electric conductivity, and J_i the impressed current density according to

$$J_i(x,t) = \begin{cases} \frac{I(t)}{|A_c|}\vec{e}_J(x) & \text{for } x \text{ in the coil region } \Omega_c \subset \Omega \\ 0 & \text{else.} \end{cases}$$

The region $\Omega \subseteq \mathrm{I\!R}^3$ includes the coil, the probe as well as the surrounding air and is supposed to be sufficiently large so that basically no magnetic field lines leave Ω

$$\vec{H} \cdot \vec{n} = 0 \text{ on } \partial\Omega \times [0,T]\,.$$

The magnetic flux density $\vec{B}$ is solenoidal, which is expressed by

$$\text{div}\vec{B} = 0\,, \tag{2.10}$$

and the constitutive relation between $\vec{B}$ and $\vec{H}$ is given via the magnetization $\vec{M}$ in

$$\vec{B} = \mu_0\vec{H} + \vec{M}\,,$$

where $\vec{M}$ depends on $\vec{H}$ in a hysteretic way

$$\vec{M} = \boldsymbol{\mathcal{P}}[\vec{H}]\,.$$

Here, μ_0 denotes the magnetic permeability in air, and $\boldsymbol{\mathcal{P}}$ a vectoral hysteresis operator. For simplicity, we assume here that $\boldsymbol{\mathcal{P}}$ acts in a component-wise decoupled way

$$(\boldsymbol{\mathcal{P}}[\vec{H}])_i = \mathcal{P}_i[H_i] \qquad i \in \{1,2,3\} \tag{2.11}$$

with scalar hysteresis operators $\mathcal{P}_i$ and H_i the components of $\vec{H}$, see [11] for the treatment of a more general vectorial hysteresis model.

With $\gamma = 1$, this leads to the model problem

$$\mu_0\vec{u}_t + \boldsymbol{\mathcal{P}}[\vec{u}]_t + \text{curl curl}\,\vec{u} = f\,, \tag{2.12}$$

see Sections 4.2, 5.2, and cf., e.g., [2], [10] for an alternative formulation via a magnetic vector potential.

3 Hysteresis operators

We would like to start this section with pointing out three characteristic features of hysteresis operators in general, and especially of Preisach operators (1.1), that will play a role in the following:

Firstly, the output $p(t)$ at some time t depends on the present as well as past states of the input $e(t)$, but not on the future (Volterra property).

Secondly, it is rate independent, i.e., the values that the output attains are independent of the speed of the input in the sense that for any continuous monotonically increasing transformation κ of the time interval $[0, T]$ with $\kappa(0) = 0$, $\kappa(T) = T$, and all input functions e, there holds

$$\mathcal{P}[e \circ \kappa] = \mathcal{P}[e] \circ \kappa \tag{3.1}$$

As a consequence, given a piecewise monotone continuous input e, the output is (up to the speed in which it is traversed) uniquely determined by the local extrema of the input only, i.e., the values of e at instances where e changes its monotonicity behaviour from decreasing to increasing or vice versa. Accordingly, there is a one-to-one correspondence between hysteresis operators mapping piecewise monotone functions $t \mapsto e(t)$ onto piecewise monotone functions, and hysteresis operators understood as mapping between sets of (alternating) strings $(e_0, e_1, \ldots, e_N)$ consisting of finitely many real values (see Section 2.2 in [3]). Note that this rate independence distinguishes hysteresis e.g. in the context of elastoplasticity from other important physical phenomena like creep or viscoelasticity.

The third important characteristic of hysteresis is that it typically does not keep the whole input history in mind but forgets certain passages in the past. I.e., there is a certain deletion in memory and it is quite important to take this into account also when doing computations: in a finite element simulation of a system with hysteresis, each element has its own history, so in order to keep memory consumption in an admissible range it is essential to delete past values that are not required any more.

There are many other aspects of hysteresis like energy dissipation or fatigue analysis, that we will not go into more detail here, since we will not directly make use of them. The interested reader is pointed to the literature cited, e.g., in [3].

After this intuitive introduction, we will now give some mathematically more precise definitions and results that are taken from the monograph [3]:

Rate independent functionals and hysteresis operators

Let, for some interval $[0, T] \subset \mathbb{R}$, $\mathrm{Map}([0, T])$ denote the set of all functions from $[0, T]$ to $\mathbb{R}$. We denote by $M_{\mathrm{pm}}([0, T])$ the set of all *piecewise monotone* functions in $\mathrm{Map}([0, T])$, i.e., of all functions e that have a so-called monotonicity partition $0 = t_0 < t_1 < t_2 < \ldots < t_N = T$ such that e is monotone on each of the subintervals t_i, t_{i+1}, and by $C_{\mathrm{pm}}([0, T])$ the intersection of $M_{\mathrm{pm}}([0, T])$ with $C([0, T])$.

A functional $\mathcal{H} : C_{\mathrm{pm}}([0, T]) \to \mathbb{R}$ is called *rate independent* if for all monotonically increasing transformations κ of the time interval $[0, T]$ with $\kappa(0) = 0$, $\kappa(T) = T$, and for all $e \in C_{\mathrm{pm}}([0, T])$, there holds

$$\mathcal{H}[e \circ \kappa] = \mathcal{H}[e]\,.$$

To each piecewise monotone function e, we can assign the sequence of local minima and maxima, i.e., the values of e at the standard monotonicity partition given by $t_0 = 0$ and

$$t_{i+1} := \max\{t \in [t_i, T] \mid e \text{ is monotone in } [t_i, t]\}\,.$$

We denote this restriction operator by $\rho : M_{\text{pm}}([0,T]) \to S_A$.

On the other hand we can assign to each alternating real-valued string, i.e., each element of

$$S_A = \{(e_0, e_1, \ldots, e_N) \in S \mid (e_{i+1} - e_i)(e_{i-1} - e_i) < 0\}$$

where

$$S = \{(e_0, e_1, \ldots, e_N) \mid N \in \mathbb{N}, \ e_i \in \mathbb{R}\},$$

a function $e \in C_{\text{pm}}([0,T])$ by doing linear interpolation between the values e_i. This gives an extension operator $\pi : S_A \to C_{\text{pm}}([0,T])$ that is extendable to all of S in a straightforward manner. In the context of rate independent functionals, this mutual correspondence between piecewise monotone functionals and alternating strings is obviously important. Actually, deletion rules can very well be formulated in terms of strings, as will be shown below. Moreover by

$$\mathcal{H}[e] := \mathcal{H}[\pi \circ \rho[e]]$$

one can extend rate independent functionals to all of M_{pm}.

Definition 3.1 (Definition 2.2.8 in [3]) An operator $\mathcal{P} : M_{\text{pm}}([0,T]) \to \text{Map}([0,T])$ is called a *hysteresis operator* if and only if

$$\mathcal{P}[e](t) = \mathcal{H}[e|_{[0,t]}]$$

holds for some rate independent functional $\mathcal{H}$. Here, $e|_{[0,t]}$ denotes the truncation of e at t

$$e|_{[0,t]}(\tau) = \begin{cases} e(\tau) & \text{if } 0 \le \tau \le t \\ e(t) & \text{if } t \le \tau \le T. \end{cases}$$

Note that this definition implies not only rate independence (3.1) but also the Volterra property. By inserting $t = T$ we immediately see that for any hysteresis operator $\mathcal{P}$, the generating rate independent functional $\mathcal{H}$ is uniquely determined by

$$\mathcal{P}[e](T) = \mathcal{H}[e|_{[0,T]}] = \mathcal{H}[e],$$

which motivates the notation *generating functional* or *final value mapping* $\mathcal{P}_f := \mathcal{H}$. Via the above mentioned correspondence between piecewise monotone functions and strings, one can define hysteresis operators also for strings $s \in S$ via

$$\tilde{\mathcal{P}}[s] = \mathcal{P}[\pi(s)].$$

Since no confusion will occur, we will omit the tilde in the sequel.

Relays and consequently also their weighted linear superpositions are obviously rate independent, cf. Example 2.2.12 in [3]. As a matter of fact, it can be rigorously proven that the Preisach operator (1.1) is a hysteresis operator in the sense of Definition 3.1, if $\wp \in L^1_{\text{loc}}(\mathbb{R} \times \mathbb{R})$, see Definition 2.4.6 in [3].

If we think of a continuum in space being subject to a process hysteretic in time, this means that at each space point the field quantity under consideration evolves according to the action of a hysteresis operator. In principle, each space point could have a different initial state. To simplify exposition and since in fact for the application we have in mind, this seems to be the reasonable, we confine ourselves to the initial state being the same in all space points, namely equal to the neutral initial state described in (1.2). Hence we deal with so-called $\mathcal{P}_0$ operators according to the notation of [3] (note that the parametrization of curves in the Preisach plane used here is slightly different from the one in [3]).

Memory deletion

Deletion, i.e., the way in which hysteresis operators forget, can be described by appropriate orderings on the set S of strings, together with the above mentioned correspondence to piecewise monotone input functions.

Definition 3.2 (Definition 2.7.1 in [3]) Let $\preceq$ be an ordering (i.e., a reflexive, antisymmetric, and transitive relation) on S. We say that a hysteresis operator forgets according to $\preceq$, if

$$s' \preceq s \Rightarrow \mathcal{P}_f(s) = \mathcal{P}_f(s') \quad \forall s, s' \in S\,.$$

Due to this implication, strings can be reduced according to certain rules. With the notation

$$[[e, e']] := [\min\{e, e'\}, \max\{e, e'\}]$$

the relevant deletion rules for Preisach operators with neutral initial state (1.2) can be written as follows (for an illustration see Figure 3.1):

- Monotone deletion rule: only the local maxima and minima of the input are relevant.

$$\begin{aligned}&(e_0, \ldots, e_N) \mapsto (e_0, \ldots, e_{i-1}, e_{i+1}, \ldots, e_N)\\ &\text{if } e_i \in [[e_{i-1}, e_{i+1}]]\,.\end{aligned} \tag{3.2}$$

- Madelung rule: inner minor loops are forgotten.

$$\begin{aligned}&(e_0, \ldots, e_N) \mapsto (e_0, \ldots, e_{i-1}, e_{i+2}, \ldots, e_N)\\ &\text{if } [[e_i, e_{i+1}]] \subset [[e_{i-1}, e_{i+2}]] \wedge e_i \notin [[e_{i-1}, e_{i+1}]] \wedge e_{i+1} \notin [[e_i, e_{i+2}]]\,.\end{aligned} \tag{3.3}$$

- Wipe out: previous absolutely smaller local maxima (minima) are erased from memory by subsequent absolutely larger local maxima (minima).

$$\begin{aligned}&(e_0, \ldots, e_N) \mapsto (e_1, \ldots, e_N)\\ &\text{if } e_0 \in [[e_1, e_2]]\,.\end{aligned} \tag{3.4}$$

- Initial deletion: a maximum (minimum) is also forgotten if it is followed by an minimum (maximum) with sufficiently large modulus.

$$(e_0, \ldots, e_N) \mapsto (e_1, \ldots, e_N) \quad \text{if } |e_0| \leq |e_1|\,. \tag{3.5}$$

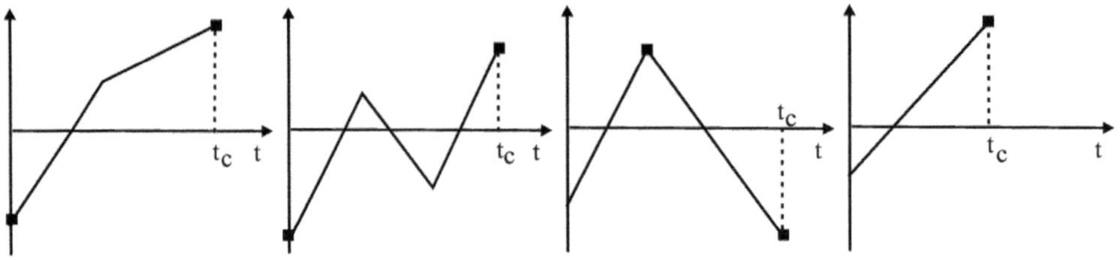

Figure 3.1 Illustration of deletion rules (3.2)–(3.5). Here the filled boxes mark the dominant input values, i.e., those sufficing to compute output values after time t_c.

It can be shown that irreducible strings for this Preisach ordering with neutral initial state are given by the set

$$S_0 = \{s \in S \mid s = (e_0, \ldots, e_N) \text{ is fading and } |e_0| > |e_1|\}$$

where

$$s = (e_0, \ldots, e_N) \text{ is fading } \Leftrightarrow$$
$$\Big(s \in S_A \text{ and } |e_0 - e_1| > |e_1 - e_2| > |e_2 - e_3| > \ldots > |e_{N-1} - e_N|\Big),$$

see Proposition 2.7.4 in [3].

Considering an arbitrary input string, the rules above have to be applied repeatedly to generate an irreducible string with the same output value, which could lead to a considerable computational effort. When computing the hysteretic evolution of some output function by a time stepping scheme, we have to update the input string and apply deletion in each time step. Fortunately, in that situation this reduction can be done within one iteration per time step without having to recursively apply rules (3.2)–(3.5). To see this, note that we augment an irreducible input string from the previous time step by appending one value, namely the input value at the current time step. The new string can then be made irreducible as follows (see Figure 3.2 for an illustration)

Lemma 3.3 *Let, for some* $N \in \mathbb{N}_0$, $s = (e_0, \ldots, e_N) \in S_0$, $e_{N+1} \in \mathbb{R}$, *and denote by* $\tilde{s} = (e_0, \ldots, e_N, e_{N+1})$ *the string obtained from* s *by appending* e_{N+1}. *Then, for*

$$s_{\text{new}} = \begin{cases} (e_0, \ldots, e_{\max_\mathcal{I}}, e_{N+1}) & \text{if } \mathcal{I} \neq \emptyset \\ (e_{N+1}) & \text{if } \mathcal{I} = \emptyset\,, \end{cases}$$

where

$$\mathcal{I} = \{i \in \{1, \ldots, N\} \mid e_{N+1} \in [e_{i-1}, e_i]\}\,,$$

we have

$$s_{\text{new}} \in S_0 \text{ and } s_{\text{new}} \preceq \tilde{s}\,.$$

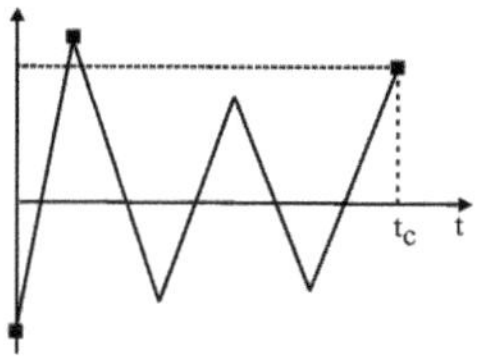

Figure 3.2 Illustration of Lemma 3.3.

Proof. It is obvious that s_{new} is irreducible. To show that it can be obtained from $\tilde{s}$ by applying the deletion (3.2)–(3.5), we make use of the following recursion:

DELETION:
IF $\max_{\mathcal{I}} = N$, i.e., $e_{N+1} \in [[e_{N-1}, e_N]]$, hence $s_{\text{new}} = \tilde{s}$; STOP.
ELSE i.e., $e_{N+1} \notin [e_{N-1}, e_N]$
    IF $e_N \in [[e_{N-1}, e_{N+1}]]$ we have $(e_0, \ldots, e_{N-1}, e_{N+1}) \preceq \tilde{s}$ by (3.2), hence
        SET $\tilde{s} = (e_0, \ldots, e_{N-1}, e_{N+1})$, $N = N - 1$, GOTO DELETION
    IF $e_N \notin [[e_{N-1}, e_{N+1}]]$, then $e_{N+1} \notin [[e_{N-1}, e_N]]$ and the fact that $(e_0, \ldots, e_N)$ is fading imply that (3.3) is applicable, hence
        SET $\tilde{s} = (e_0, \ldots, e_{N-2}, e_{N+1})$, $N = N - 2$, GOTO DELETION

Due to the finite length of the string, the algorithm has to stop after finitely many steps at $\max_{\mathcal{I}} = N$. To see applicability of (3.3) in the second subcase $e_N \notin [[e_{N-1}, e_{N+1}]]$, note that by $s \in S_0$ and the fact that $\tilde{s}$ is alternating, it suffices to show $[[e_{N-1}, e_N]] \subset [[e_{N-2}, e_{N+1}]]$, i.e., $e_{N-1}, e_N \in [[e_{N-2}, e_{N+1}]]$:

In case $e_{N+1} \geq e_N$ (hence $e_{N-1} \geq e_N$, $e_{N-1} \geq e_{N-2}$) this follows from

$$e_{N+1} \geq e_N = \underbrace{e_N - e_{N-1}}_{\geq -|e_{N-1}-e_{N-2}| = e_{N-2}-e_{N-1}} -(e_{N-2} - e_{N-1}) + e_{N-2} \geq e_{N-2}$$

and

$$e_{N-2} \leq e_{N-1} \leq e_{N+1}$$

where for the last inequality we have used the fact that $e_N \leq e_{N+1}$ and $e_{N+1} \notin [[e_{N-1}, e_N]]$.

If $e_{N+1} \leq e_N$ (hence $e_{N-1} \leq e_N$, $e_{N-1} \leq e_{N-2}$) we get $[[e_{N-1}, e_N]] \subset [[e_{N-2}, e_{N+1}]]$ from

$$e_{N+1} \leq e_N = e_N - e_{N-1} + \underbrace{e_{N-1} - e_{N-2}}_{\leq -|e_N - e_{N-1}| = -(e_N - e_{N-1})} + e_{N-2} \leq e_{N-2}$$

and

$$e_{N-2} \geq e_{N-1} \geq e_{N+1}\,;$$

the last inequality follows from the fact that $e_N \geq e_{N+1}$ and $e_{N+1} \notin [[e_{N-1}, e_N]]$. □

Thus, to apply a Preisach operator to some input function e, it suffices to form the string of all local minima and maxima, reduce this string according to the deletion rules above (see Lemma 2.6.4 in [3] for the question concerning the order of application of deletion, as well as Lemma 3.3 above for efficient deletion in the context of time stepping) until it is irreducible and then apply the hysteresis operator to an element of S_0. The latter can be done efficiently by so-called shape or Everett functions:

Everett functions

The possible shapes of the hysteresis loops for some given operator $\mathcal{P}$ are given by the values of its level functions l_N defined by

$$l_N(x) = \mathcal{P}_f(e_0, \ldots, e_{N-1}, x) \tag{3.6}$$

on irreducible strings $s = (e_0, \ldots, e_{N-1}, x)$. For Preisach operators (1.1) it is readily checked that if $s = (e_0, \ldots, e_{N-1}, e_N) \in S_0$, then, starting at $(e_0, \ldots, e_{N-1})$ and letting x go from e_{N-1} to e_N, we just switch the relays with threshold pairs (β, α) within the triangle

$$\Delta(e_{N-1}, e_N) = \begin{cases} \{(\beta,\alpha) \mid e_{N-1} \le \beta \le \alpha \le e_N\} & \text{if } e_{N-1} \le e_N \\ \{(\beta,\alpha) \mid e_N \le \beta \le \alpha \le e_{N-1}\} & \text{else.} \end{cases}$$

Thus, with the so-called *shape function* or *Everett function* (cf. [4]),

$$h(e_{N-1}, e_N) = 2\,\mathrm{sign}(e_N - e_{N-1}) \iint_{\Delta(e_{N-1}, e_N)} \wp(\beta, \alpha)\, d(\alpha, \beta)\,, \tag{3.7}$$

we obtain

$$\mathcal{P}_f(e_0, \ldots, e_N) = \mathcal{P}_f(e_0, \ldots, e_{N-1}) + h(e_{N-1}, e_N)$$

and, inductively,

$$\mathcal{P}_f(s) = \mathcal{P}_f(e_0) + \sum_{k=1}^{N} h(e_{k-1}, e_k) \qquad \forall s = (e_0, \ldots, e_N) \in \overline{S}_0\,, \tag{3.8}$$

where

$$\overline{S}_0 = \{(e_0, \ldots, e_N) \in S_A \mid |e_0 - e_1| \ge |e_1 - e_2| \ge \ldots \ge |e_{N-1} - e_N| \text{ and } |e_0| \ge |e_1|\}\,.$$

As a matter of fact, among the more general class of hysteresis operators of Preisach type (see Section 2.4 in [3]), the Preisach operators (1.1) can be characterized as those having shape functions h satisfying (3.8), via the identity

$$\wp(\beta, \alpha) = -\frac{1}{2}\partial_1\partial_2 h(\beta, \alpha)\,, \tag{3.9}$$

cf. Theorem 2.9.4 in [3]. From (3.8) we can – even without assuming that h is defined by (3.7) – conclude that h vanishes on the main diagonal

$$h(e, e) = 0 \quad \forall e \in \mathbb{R}\,,$$

since $\mathcal{P}_f(e) = \mathcal{P}_f(e, e) = \mathcal{P}_f(e) + h(e, e)$, and, with $s := (e_1, e_2, e_1, e_2) \in \bar{S}_0$, there holds $\mathcal{P}_f(e_1, e_2) = \mathcal{P}_f(e_1, e_2, e_1, e_2) = \mathcal{P}_f(e_1, e_2) + h(e_2, e_1) + h(e_1, e_2)$, hence

$$h(e_1, e_2) = -h(e_2, e_1) \quad \forall e_1, e_2 \in \mathbb{R}\,. \tag{3.10}$$

Additionally, in many applications, one may assume symmetry

$$\wp(-\alpha, -\beta) = \wp(\beta, \alpha) \tag{3.11}$$

of the weight function and therewith

$$h(-e_2, -e_1) = h(e_1, e_2) \quad \forall e_1, e_2 \in \mathbb{R}\,, \tag{3.12}$$

which we will also do in the following. In that case, it suffices to specify $h := \overline{h}$ in the cone

$$\nabla = \{(e_1, e_2) \in \mathbb{R}^2 \mid e_1 \le 0 \wedge e_1 \le e_2 \le -e_1\} \tag{3.13}$$

and set (cf. Figure 3.3)

$$h(e_1, e_2) = \begin{cases} \overline{h}(e_1, e_2) & e_1 \le 0 \wedge e_1 \le e_2 \le -e_1 & I \\ -\overline{h}(e_2, e_1) & e_2 \le 0 \wedge e_2 \le e_1 \le -e_2 & II \\ -\overline{h}(-e_1, -e_2) & e_1 \ge 0 \wedge -e_1 \le e_2 \le e_1 & III \\ \overline{h}(-e_2, -e_1) & e_2 \le 0 \wedge -e_2 \le e_1 \le e_2\,. & IV \end{cases} \tag{3.14}$$

Moreover, the symmetry (3.11) together with (1.2) implies

$$\mathcal{P}[0] = 0\,. \tag{3.15}$$

Therewith, the final value mapping for strings of length one, as appearing in the first term on the right hand side of (3.8) can be expressed in terms of $\overline{h}$, using the fact that for $e \in \mathbb{R}$, $\mathcal{P}_f(e)$ describes the virgin curve and we start with the initial value (1.2): Starting at (1.2) and moving the input from zero to some $e > 0$, we switch all relays with thresholds in the triangle

$$\{0 \le \alpha \le e, -\alpha \le \beta \le \alpha\}$$

from -1 to $+1$, so that

$$\mathcal{P}_f(e) = 2 \iint_{\{0 \le \alpha \le e, -\alpha \le \beta \le \alpha\}} \wp(\beta, \alpha) \mathcal{R}_{\beta,\alpha}[e](t)\, d(\alpha, \beta)\,.$$

Taking into account the fact that

$$\begin{aligned} h(-e, e) &= 2 \iint_{\{-e \le \beta \le \alpha \le e\}} \wp(\beta, \alpha) \mathcal{R}_{\beta,\alpha}[e](t)\, d(\alpha, \beta) \\ &= 2 \iint_{\{0 \le \alpha \le e, -\alpha \le \beta \le \alpha\}} \wp(\beta, \alpha) \mathcal{R}_{\beta,\alpha}[e](t)\, d(\alpha, \beta) \\ &\quad + 2 \iint_{\{0 \le -\beta \le e, \beta \le \alpha \le -\beta\}} \wp(\beta, \alpha) \mathcal{R}_{\beta,\alpha}[e](t)\, d(\alpha, \beta) \\ &= 4 \iint_{\{0 \le \alpha \le e, -\alpha \le \beta \le \alpha\}} \wp(\beta, \alpha) \mathcal{R}_{\beta,\alpha}[e](t)\, d(\alpha, \beta)\,, \end{aligned}$$

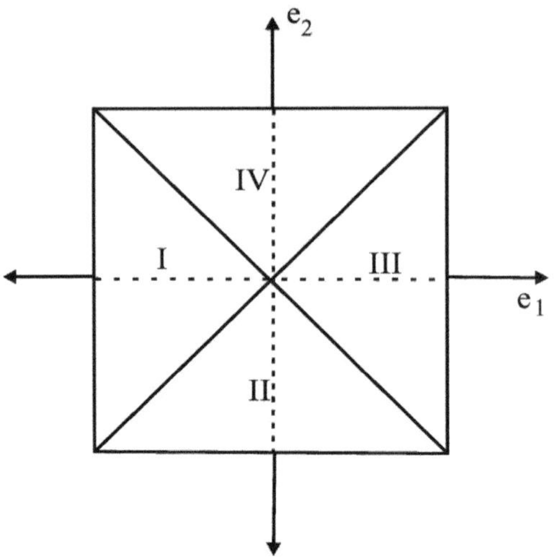

Figure 3.3 Cases in the definition of h in case of symmetry (3.12).

where we have used the symmetry (3.11) in the last equality, we obtain

$$\mathcal{P}_f(e) = \frac{1}{2}h(-e,e) = \frac{1}{2}\overline{h}(-e,e) \quad \text{for } e > 0\,. \tag{3.16}$$

Analogously, one obtains

$$\mathcal{P}_f(e) = \frac{1}{2}h(-e,e) = -\frac{1}{2}\overline{h}(e,-e) \quad \text{for } e < 0\,.$$

Remark 3.4 The already above mentioned fact that a Preisach operator is determined by its shape function h and the Preisach memory has crucial consequences in an efficient implementation of hysteresis simulation: Therewith, each time step can be divided into a first stage of applying deletion rules (3.2), (3.3), (3.4), (3.5) (e.g., by using Lemma 3.3), and a second stage of summing up shape function values according to (3.8). This is especially important in the context of hysteresis in PDEs, where each space point is subject to a hysteretic process.

In the context of hysteresis in PDEs, we require certain monotonicity and convexity properties of the hysteresis operator, according to the following definition:

Definition 3.5 (Definitions 2.2.14 and 2.8.6 in [3])

(i) A hysteresis operator $\mathcal{P}$ is called *piecewise (strictly) increasing* (resp. *piecewise* C^k) if its level functions l_N according to (3.6) are (strictly) increasing functions (resp. piecewise C^k functions) for any $N \in \mathbb{N}_0$ and any $(e_0, \ldots, e_{N-1}) \in S_A$.

(ii) A hysteresis operator $\mathcal{P}$ is called *clockwise (counterclockwise) piecewise convex* on the interval $I \subset \mathbb{R}$ if its level functions l_N according to (3.6) are convex (concave) in $(-\infty, e_{N-1}] \cap I$ and concave (convex) in $[e_{N-1}, \infty)$ for any $N \in \mathbb{N}_0$ and any $(e_0, \ldots, e_{N-1}) \in S_A$.

Hysteresis operators on restricted domains $I \subset \mathbb{R}$ can be defined via strings $s \in S_I$ consisting of values in I only, or continuous functions $e \in C_I([0,T])$ assuming values in I only, see Definition 2.11.2 in [3]. Therewith one can define composition and inverses of hysteresis operators, see Proposition 3.6 below.

Preisach operators (1.1) with weight functions $\wp \in L^1_{\text{loc}}(\mathbb{R} \times \mathbb{R})$ belong to the more general class of *operators of Preisach type*, that have been analyzed in detail in [3] (see Section 2.4 and especially Definition 2.4.6 therein). As a consequence, we can infer the following properties:

Proposition 3.6 *Let $\mathcal{P}$ be a Preisach operator (1.1).*

(i) If

$$\int_0^\infty \sup_{s \in \mathbb{R}} |\wp(s-r, s+r)| \, dr < \infty \,,$$

then $\mathcal{P}$ is Lipschitz continuous on $C([0,T])$.

(ii) If $\wp$ is bounded on bounded subsets of $\mathbb{R}^2$, then $\mathcal{P}$ is Lipschitz continuous on bounded subsets of $C([0,T])$. Moreover, $\mathcal{P}$ maps bounded subsets of $W^{1,p}([0,T])$ into bounded subsets of $W^{1,p}([0,T])$ for each $1 \leq p \leq \infty$.

(iii) If

$$\int_0^y \int_{-y}^y |\wp(s-r, s+r)| \, ds \, dr \leq c_1 |y| + c_2 \quad \forall y \geq 0$$

with some constants c_1, c_2 independent of y, then the linear growth estimate

$$\|\mathcal{P}[e]\|_{L^\infty} \leq c_3 \|e\|_{L^\infty} + c_4$$

holds for all $e \in C([0,T])$, with constants c_3, c_4 independent of e.

(iv) Let $I \subset \mathbb{R}$ be some interval. Every piecewise strictly monotone $\mathcal{P}_0$-operator $\mathcal{P}$ with domain I has an inverse $\mathcal{P}^{-1}$, which is a piecewise strictly monotone $\mathcal{P}_0$-operator with domain $J = \mathcal{P}_f(I)$.

(v) Let $\mathcal{P}$ be a strictly monotone operator of Preisach type with domain $I \subset \mathbb{R}$ an interval. Assume that $\mathcal{P} : C_I([0,T]) \to C([0,T])$ is continuous and that

$$\chi_I(x) = \inf\{|h(e_1, e_2)| \mid |e_1 - e_2| = x \,,\ e_1, e_2 \in I\} > 0 \quad \forall x > 0 \,.$$

Then $\mathcal{P}$ is invertible and its inverse can be extended to a continuous operator $\mathcal{P}^{-1} : C_J([0,T]) \to C_I([0,T])$, where $J = \mathcal{P}_f(I)$.

Proof. Part (i) follows from Proposition 2.4.11 in [3], part (ii) from Corollary 2.4.12 in [3], part (iii) from Remark 2.4.13 in [3] part (iv) from Proposition 2.11.4 in [3], and part (v) from Theorem 2.11.20 in [3]. □

4 The forward problem

4.1 A hyperbolic PDE with hysteresis

Consider the problem

$$\rho u_{tt} - (\mathcal{P}[u_x])_x = 0 \quad x \in (0, L), \quad t \in (0, T)\,, \tag{4.1}$$

with boundary conditions

$$\begin{aligned} \mathcal{P}[u_x(0,t)] &= g_0(t) \\ u(L,t) &= f_L(t) \end{aligned} \qquad t \in (0, T)\,, \tag{4.2}$$

and initial conditions

$$u(x,0) = u_0(x), \quad u_t(x,0) = u_1(x) \quad x \in (0, L)\,, \tag{4.3}$$

for given $u_0 : [0, L] \to \mathrm{I\!R}$, $u_1 : [0, L] \to \mathrm{I\!R}$, $f_L : [0, T] \to \mathrm{I\!R}$, $g_0 : [0, T] \to \mathrm{I\!R}$ satisfying the compatibility conditions

$$u_0(L) = f_L(0)\,, \quad u_1(L) = f_L'(0)\,, \quad u_0'(L) = 0\,, \quad g_0(0) = 0\,.$$

Note that the last two conditions imply $\mathcal{P}[u_0'(L)] = g_0(0)$, provided $\mathcal{P}[0] = 0$, (see (v) in Assumption 4.1 below,) which is automatically satisfied in the symmetric case (3.11), cf. (3.15).

Well-posedness

Solvability of the wave equation with hysteresis has been proven by Krejčí, cf. [18], [19]. We here follow the exposition in [3] to give sufficient conditions for the solvability of (4.1) (see (H1) in Section 3.5), with a slight generalization concerning the boundary conditions:

Assumption 4.1 Let $I \subset \mathrm{I\!R}$ denote some closed interval such that $0 \in \mathrm{int}(I)$. We assume that $\mathcal{P}$ is a hysteresis operator of Preisach type with domain I which has the following properties:

(i) $\mathcal{P} : C_I[0,T] \to C[0,T]$ is locally Lipschitz continuous, piecewise C^2, maps $W_I^{1,1}(0,T)$ into $W^{1,1}(0,T)$ and has linear growth.

(ii) $\mathcal{P}$ is piecewise strictly increasing and satisfies

$$\underline{\eta}|e'(t)| \leq |\mathcal{P}[e]'(t)| \leq \overline{\eta}|e'(t)| \quad \forall e \in W_I^{1,1}(0,T)$$

for some constants $0 < \underline{\eta} < \overline{\eta}$.

(iii) $\mathcal{P}$ is clockwise piecewise convex on I.

(iv) $\mathcal{P}^{-1}$ is continuous on its domain in $C[0,T]$.

(v) $\mathcal{P}[0] = 0$ (compatibility condition).

Note that in general the hysteresis operators depend on space via their initial state. Since we have assumed to always have the same initial state, this is not the case here, which simplifies the exposition considerably. The results can be carried over to the more general case of space dependent initial state under certain smoothness assumptions on the initial state as a function of x.

By virtue of the facts quoted in the previous section, we can formulate sufficient conditions for Assumption 4.1 in terms of the shape function h of $\mathcal{P}$, if $\mathcal{P}$ is a Preisach operator:

Assumption 4.2 Let $I \subset \mathbb{R}$ denote some closed interval such that $0 \in \operatorname{int}(I)$. We assume that $\mathcal{P}$ is a Preisach operator with domain I, $\mathcal{P}[0] = 0$, and shape function $h \in C^2(I^2)$, which has the following properties:

(i)

$$\int_0^\infty \sup_{s\in\mathbb{R}} |\partial_1\partial_2 h(s-r, s+r)|\, dr < \infty\,, \tag{4.4}$$

and

$$\int_0^y \int_{-y}^y |\partial_1\partial_2 h(s-r, s+r)|\, ds\, dr \le c_1|y| + c_2 \quad \forall y \ge 0 \tag{4.5}$$

with some constants c_1, c_2 independent of y.

(ii)

$$\underline{\eta} \le \partial_2 h(e_1, e_2) \le \overline{\eta} \quad \forall e_1, e_2 \in I\,.$$

(iii)

$$\partial_2^2 h(e_1, e_2) \begin{cases} \le 0 & \text{for } e_1 \le e_2 \\ \ge 0 & \text{for } e_1 \ge e_2 \end{cases} \qquad \forall e_1, e_2 \in I\,.$$

(iv)

$$\chi_I(x) = \inf\{|h(e_1, e_2)| \mid |e_1 - e_2| = x\,,\ e_1, e_2 \in I\} > 0 \quad \forall x > 0\,.$$

Lemma 4.3 *Assumption 4.2 is sufficient for Assumption 4.1.*

Proof. For any $N \in \mathbb{N}_0$, any $(e_0, \ldots, e_{N-1}) \in S_A$ and any $x \in \mathbb{R}$, by applying deletion rules (3.2), (3.3), (3.4), (3.5), one can find an irreducible substring $s' = (\tilde{e}_0, \ldots, \tilde{e}_{\tilde{N}-1}, x)$ of $s = (e_0, \ldots, e_{N-1}, x)$ such that $\mathcal{P}_f(s') = \mathcal{P}_f(s)$; w.l.o.g. we may omit the tildes and assume that $s = (e_0, \ldots, e_{N-1}, x) \in \overline{S}_0$. Hence, we can make use of (3.8) to conclude that

$$l'_N(x) = \partial_2 h(e_{N-1}, x)\,, \quad l''_N(x) = \partial_2^2 h(e_{N-1}, x)\,.$$

This yields the implication of (ii), (iii) in Assumption 4.1 by the respective parts of Assumption 4.2.

The corresponding assertions for (i) and (iv) follow from Proposition 3.6. □

If we additionally postulate the symmetry (3.12) as well as $\wp$ being compactly supported, with the domain I being (by a proper scaling) just equal to the unit interval $[-1, 1]$, then the assumptions simplify to:

Assumption 4.4 We assume that $\mathcal{P}$ is a Preisach operator with domain $[-1,1]$, and shape function h according to (3.14) where $\overline{h} \in C^2(\bigtriangledown \cap [-1,1]^2)$ – with $\bigtriangledown$ as in (3.13) – has the following properties:

$$\underline{\eta} \leq \partial_2 \overline{h} \leq \overline{\eta}\,, \quad -\underline{\eta} \geq \partial_1 \overline{h} \geq -\overline{\eta} \quad \text{in} \bigtriangledown \cap [-1,1]^2$$

$$\partial_2^2 \overline{h} \leq 0\,, \quad \partial_1^2 \overline{h} \leq 0 \quad \text{in} \bigtriangledown \cap [-1,1]^2$$

$$\chi_{[-1,1]}(x) = \inf\{|\overline{h}(e_1,e_2)| \mid |e_1 - e_2| = x\,,\ e_1, e_2 \in \bigtriangledown \cap [-1,1]\} > 0 \quad \forall x > 0$$

$$\left.\begin{array}{l} \overline{h}(e,e) = 0\,, \quad \partial_1 \overline{h}(e,e) = -\partial_2 \overline{h}(e,e)\,, \quad \partial_1 \overline{h}(e,-e) = -\partial_2 \overline{h}(e,-e)\,, \\ \partial_1^2 \overline{h}(e,e) = -\partial_2^2 \overline{h}(e,e)\,, \quad \partial_1^2 \overline{h}(e,-e) = \partial_2^2 \overline{h}(e,-e)\,, \quad \partial_1 \partial_2 \overline{h}(e,e) = 0 \end{array}\right\}$$
$$\forall e \in [-1,0]\,. \tag{4.6}$$

Note that the estimates (4.4), (4.5) are automatically satisfied for compactly supported h and that the equalities (4.6) ensure continuity of all derivatives of order ≤ 2 for h defined as in (3.14) across the interfaces $e_2 = e_1$ and $e_2 = -e_1$. Actually, since only *piecewise* C^2 smoothness is demanded, we can even relax the conditions on the shape function: It suffices to assume $\overline{h}$ to be piecewise C^2 and the second line in (4.6) can be omitted.

Along the lines of the proof of Theorem 3.5.3 in [3], the following existence result follows.

Proposition 4.5 *Suppose that Assumption 4.1 is satisfied and that* $u_0 \in H^2(0,L)$*, and* $g_0 \in W^{2,1}(0,T) \cap W^{1,\infty}(0,T)$*,* $f \in W^{3,1}(0,T) \cap W^{2,\infty}(0,T)$*. If either* $I = \mathbb{R}$ *or* $\|u_0\|_{H^2}$, $\|u_1\|_{H^1}$, $\|f'''\|_{L^1}$, $\|f''\|_{L^\infty}$, $\|g_0''\|_{L^1}$, $\|g_0'\|_{L^\infty}$ *are sufficiently small, then the initial boundary value problem (4.1), (4.2), (4.3) has a weak solution* u *that satisfies*

$$u_{tt}, u_{xt}, \mathcal{P}[u_x]_t, \mathcal{P}[u_x]_x, \textstyle\int_0^\cdot \mathcal{P}[u_x(\cdot,\tau)]_{xx}\, d\tau \in L^\infty(0,T;L^2(0,L))\,,$$
$$u_x, \mathcal{P}[u_x] \in C((0,L)\times(0,T))\,.$$

Proof. We here provide an informal argument analogous to the one preceding Theorem 3.5.3 in [3]; details can be done following the lines of the proof in [3].

The proof is based on the convexity inequality

$$\frac{d}{dt}\left(\frac{1}{2}\mathcal{P}[e]_t\, e_t\right) \leq \mathcal{P}[e]_t e_{tt} \tag{4.7}$$

that follows under the assumptions made on $\mathcal{P}$ from Proposition 3.4.2 in [3], and on considering the energy functional

$$E(t) = \frac{1}{2}\int_0^L (\rho u_{tt}^2 + \mathcal{P}[u_x]_t u_{xt})dx\,,$$

that due to the piecewise strict monotonicity satisfies

$$E(t) \geq \frac{\rho}{2}\|u_{tt}\|^2_{L^2(0,L)} + \frac{\underline{\eta}}{2}\|u_{xt}\|^2_{L^2(0,L)}\,. \tag{4.8}$$

Adding (4.7) with $e := u_x$ to the identity

$$\frac{d}{dt}\left(\frac{\rho}{2}u_{tt}^2\right) = \frac{1}{\rho}\mathcal{P}[u_x]_x\mathcal{P}[u_x]_{xt}$$

that follows from the PDE, and using the identity $u_{xtt} = \frac{1}{\rho}\mathcal{P}[u_x]_{xx}$ we arrive at

$$\begin{aligned}
\frac{d}{dt}E(t) &\leq \int_0^L \frac{1}{\rho}\mathcal{P}[u_x]_x\mathcal{P}[u_x]_{xt}\,dx + \int_0^L \frac{1}{\rho}\mathcal{P}[u_x]_t\mathcal{P}[u_x]_{xx}\,dx \\
&= \frac{1}{\rho}\mathcal{P}[u_x(L,\cdot)]_t(t)\mathcal{P}[u_x(L,\cdot)]_x(t) - \frac{1}{\rho}\mathcal{P}[u_x(0,\cdot)]_t(t)\mathcal{P}[u_x(0,\cdot)]_x(t) \\
&= (\mathcal{P}[u_x(L,\cdot)]_t)(t)f_L''(t) - g_0'(t)u_{tt}(0,t) \\
&= \left(g_0'(t) + \int_0^L (\mathcal{P}[u_x(x,\cdot)]_{xt})(t)\,dx\right)f_L''(t) \\
&\quad + g_0'(t)\left(-f_L''(t) + \int_0^L u_{xtt}(x,t)\,dx\right),
\end{aligned}$$

where we have done integration by parts in the second line. Therewith, we obtain

$$\begin{aligned}
E(t) &\leq E(0) + \int_0^t \left(\left(\int_0^L \rho u_{ttt}(x,\tau)\,dx\right)f_L''(\tau) + g_0'(\tau)\int_0^L u_{xtt}(x,\tau)\,dx\right)d\tau \\
&= E(0) - \int_0^t\int_0^L \rho u_{tt}(x,\tau)f_L'''(\tau)\,dx\,d\tau - \int_0^t\int_0^L g_0''(\tau)u_{xt}(x,\tau)\,dx\,d\tau \\
&\quad + \int_0^L \rho u_{tt}(x,t)f_L''(t)\,dx - \int_0^L \rho u_{tt}(x,0)f_L''(0)\,dx \\
&\quad + \int_0^L g_0'(t)u_{xt}(x,t)\,dx - \int_0^L g_0'(0)u_{xt}(x,0)\,dx \\
&\leq |E(0)| + \sqrt{L}\rho(\|f_L'''\|_{L^1(0,T)} + 2\|f_L''\|_{L^\infty(0,T)}) \sup_{t\in[0,T]} \|u_{tt}\|_{L^2(0,L)} \\
&\quad + \sqrt{L}(\|g_0''\|_{L^1(0,T)} + 2\|g_0'\|_{L^\infty(0,T)}) \sup_{t\in[0,T]} \|u_{xt}\|_{L^2(0,L)}
\end{aligned}$$

so that by (4.8) the supremum of the energy $\overline{E} := \sup_{t\in[0,T]} E(t)$ satisfies

$$\overline{E} \leq |E(0)| + \underbrace{\sqrt{2L}\max\{\sqrt{\rho}(\|f_L'''\|_{L^1} + 2\|f_L''\|_{L^\infty}), \frac{1}{\sqrt{\eta}}(\|g_0''\|_{L^1} + 2\|g_0'\|_{L^\infty})\}}_{=:C(f_L,g_0)}\sqrt{\overline{E}},$$

from which

$$\sqrt{\overline{E}} \leq \sqrt{|E(0)|} + C(f_L, g_0)$$

follows. In case $I \neq \mathbb{R}$ we have to make sure that the right hand side is sufficiently small in order to guarantee (via (4.8), $u_{tt} = \mathcal{P}[u_x]_x$ and continuity of $\mathcal{P}^{-1}$)

that $u_x(x,t) \in I$ for all $x \in [0,L]$, $t \in [0,T]$. To this end, note that $|E(0)|$ can be estimated by

$$\begin{aligned} |E(0)| &= |\frac{1}{2}\int_0^L (\frac{1}{\rho}\mathcal{P}[u_{0x}]_x^2 + \mathcal{P}[u_x]_t(t=0)u_{1x})dx| \\ &\leq \frac{1}{2}\Big(\frac{1}{\rho}\mathrm{Lip}_{\mathcal{P}}^2 \|u_{0x}\|_{H^1}^2 + \mathrm{Lip}_{\mathcal{P}} \|u_{1x}\|_{L^2}^2\Big), \end{aligned}$$

where $\mathrm{Lip}_{\mathcal{P}}$ is a local Lipschitz constant of $\mathcal{P}$ at zero (note that here $\mathcal{P}$ is applied to strings of length one) and the estimate $\mathcal{P}[u_{0x}]_x(x) \leq \mathrm{Lip}_{\mathcal{P}} \|u_{0x}(x)\|_{H^1}$ can be checked by using difference quotients.

The exact proof is based on a space discretized setting, that leads to a system of ODEs for which global (with respect to time) existence results can be applied. After deriving energy estimates of the kind described above in this semidiscrete context, one can pass to the limits in space discretization to obtain the assertions of the Proposition.□

Note that in practice $I = \mathbb{R}$ will be achievable by an appropriate extension of the hysteresis operator, so that the smallness restriction of the initial and boundary data does not get active.

Numerical solution

For computing a solution of the initial boundary value problem with hysteresis (4.1), (4.2), (4.3), we use the following algorithm, that consists of an outer time stepping loop according to Newmark's scheme in its effective stiffness formulation (cf., e.g., [14]) and an inner loop for resolving the hysteresis nonlinearity (cf., e.g., [2] and Section 4.2 in the context of magnetics):

Algorithm 4.6

$n = 0$

$t_0 = 0$, $u_0 = u_0$, $\dot{u}_0 = u_1$, $\ddot{u}_0 = 0$

Choose integration parameters $\beta_H \in [0, \frac{1}{2}]$, $\gamma_H \in [0,1]$ and time step size $\underline{dt}$

Initialize $c_{d\,0}$ (e.g. $c_{d\,0} = c(u_{0\,x})$ with c as in (5.8))

While $t < T$ do

  $n = n + 1$

  $t_n = t_{n-1} + \underline{dt}$

  $\tilde{u}_n = u_{n-1} + \underline{dt}\,\dot{u}_{n-1} + (1 - 2\beta_H)\frac{(\underline{dt})^2}{2}\ddot{u}_{n-1}$

  $\tilde{\dot{u}}_n = \dot{u}_{n-1} + \underline{dt}(1 - \gamma_H)\ddot{u}_{n-1}$

$$
\left\{
\begin{array}{l}
c^0_{d\,n} = c_{d\,n-1} \\
\text{For } k = 0, 1, 2, 3, \ldots \\
\quad \text{Solve the elliptic boundary value problem} \\
\quad \left\{
\begin{array}{l}
\frac{\rho}{\beta_H (\underline{dt})^2} \underline{du} - (c^k_{d\,n} \underline{du}_x)_x = \frac{\rho}{\beta_H (\underline{dt})^2} (\tilde{u}_n - \tilde{u}_{n-1}) \quad \text{in } (0, L) \\
(c^k_{d\,n} \underline{du}_x)(0) = g_0(t_n) - g_0(t_{n-1}) \,, \quad \underline{du}(L) = f_L(t_n) - f_L(t_{n-1})
\end{array}
\right. \\
\quad \text{for } \underline{du}. \\
\quad u^{k+1}_n = u_{n-1} + \underline{du} \\
\quad p^{k+1}_n = \mathcal{P}(u_{0\,x}, u_{1\,x}, \ldots, u_{n-1\,x}, u^{k+1}_{n\,x}) \quad (*) \\
\quad c^{k+1}_{d\,n} = \dfrac{p^{k+1}_n - p_{n-1}}{u^{k+1}_{n\,x} - u_{n-1\,x}} \quad (**)
\end{array}
\right\}
$$

$$
\begin{aligned}
u_n &= u^{k+1}_n \\
p_n &= p^{k+1}_n \\
\ddot{u}_n &= \tfrac{1}{\beta_H (\underline{dt})^2} (u_n - \tilde{u}_n) \\
\dot{u}_n &= \tilde{\dot{u}} + \gamma_H \underline{dt}\, \ddot{u}_n
\end{aligned}
$$

The inner iteration (within the braces) is supposed to solve

$$
\frac{\rho}{\beta_H (\underline{dt})^2} u_n - \mathcal{P}(u_{0\,x}, u_{1\,x}, \ldots, u_{n-1\,x}, u_{n\,x})_x = \frac{\rho}{\beta_H (\underline{dt})^2} \tilde{u}_n \,,
$$

which – assuming that this equation already holds for n replaced by $n-1$ – is equivalent to

$$
\begin{aligned}
&\frac{\rho}{\beta_H (\underline{dt})^2} \underline{du} - \underbrace{\Big(\mathcal{P}(u_{0\,x}, u_{1\,x}, \ldots, u_{n-1\,x}, u_{n-1\,x} + \underline{du}_x) - \mathcal{P}(u_{0\,x}, u_{1\,x}, \ldots, u_{n-1\,x}) \Big)_x}_{\approx (c_{d\,n} \underline{du}_x)_x} \\
&\quad = \frac{\rho}{\beta_H (\underline{dt})^2} (\tilde{u}_n - \tilde{u}_{n-1}) \,,
\end{aligned}
$$

for $\underline{du} = u_n - u_{n-1}$. Due to the monotonicity of $\mathcal{P}$, the space dependent coefficient $c^{k+1}_{d\,n}$ as defined in $(**)$ is always larger than or equal to $\underline{\eta}$ so that we really deal with elliptic problems in the inner iteration.

Here it is essential to do the evaluation of the Preisach operator in $(*)$ in an efficient way, since this has to be done for each space point or, in a discretized setting, for each finite element (cf. Remark 3.4). In view of the facts summarized in the previous section, we do not apply formula (1.1) containing a double integral directly but for each finite element and every time step

a) store its history in a list containing the corresponding string actualized according to the deletion rules (3.2), (3.3), (3.4), (3.5).

b) compute the output value by means of the sum (3.8),

cf. [26].

4.2 A parabolic PDE with hysteresis

The magnetics problem from Section 2.2 (partially with unit constants) reads as

$$\mu_0 \vec{u}_t + \boldsymbol{\mathcal{P}}[\vec{u}]_t + \text{curl curl}\, \vec{u} = f \quad \text{in } \Omega \times [0,T] , \tag{4.9}$$

with boundary conditions

$$\vec{u} \cdot \vec{n} = 0 \quad \text{on } \partial\Omega \times [0,T] \tag{4.10}$$

and initial condition

$$\vec{u}(\cdot, 0) = \vec{u}_0 \quad \text{in } \Omega \tag{4.11}$$

for given $f : \Omega \times [0,T] \to \mathrm{IR}^3$ – representing (a smoothed version of) the curl of the impressed current density – and $\vec{u}_0 : \Omega \to \mathrm{IR}^3$.

Remark 4.7 Existence and stability (implying uniqueness) for parabolic equations with hysteresis in their principal part have been established by Visintin [29], [30] and Hilpert [5], respectively, see Section 3.3 in [3]. More precisely, they consider problems of the form

$$u_t + \mathcal{P}[u]_t - \Delta u = f \quad \text{in } \Omega \tag{4.12}$$

with homogeneous Dirichlet boundary conditions, as well as initial conditions, and a piecewise increasing continuous hysteresis operator $\mathcal{P}$ satisfying a linear growth estimate. We wish to mention that the existence result is not directly applicable and the proofs are also not immediately transferable to the magnetics problem (4.9), (4.10), (4.11), for the following reasons:

- The proof uses a compactness argument (see (3.8) in Chapter 3 of [3]) which cannot be carried over to the present situation due to the lack of compactness of the embedding $H(\text{curl}) \cap H(\text{div}) \to L^2$ (cf. [1]). By the way, the embedding $H_0(\text{curl}) \cap H(\text{div}^0) = \{\vec{u} \in H_0(\text{curl}) \mid \vec{u} \times \vec{n} = 0 \text{ on } \partial\Omega \text{ and } \text{div}\vec{u} = 0\} \hookrightarrow L^2$ would be compact, but in our context $\text{div}\vec{u} = 0$ is not satisfied, since $\vec{u}$ stands for the magnetic field $\vec{H}$ which is in general not solenoidal (only $\vec{B}$ is, by (2.10)).
- For the latter reason, also the identity $\text{curl}\,\text{curl} = \Delta - \text{grad}\,\text{div}$ that is valid in convex domains, cannot be used for transferring (4.9) to (4.12).

For a well-posedness proof of (4.9) see [10]. Moreover we wish to refer to [31] for a well-posedness proof based on a two scale model.

Numerical solution

Combining a trapeziodal scheme with an iterative resolution of the hysteresis nonlinearity in each time step we arrive at (cf. [2] for a similar method in a magnetic vector potential formulation)

Algorithm 4.8
$n = 0$
$t = 0$, $\vec{u}_0 = \vec{u}_0$, $\dot{\vec{u}}_0 = 0$
Choose integration parameter $\gamma_P \in (0, 1]$ and time step size $\underline{dt}$
Initialize $C_{d\,0}$ (e.g. $C_{d\,0} = \mathrm{diag}((c(\vec{u}_0)_i)_{i=1,2,3})$ with c as in (5.8))
While $t < T$ do
  $n = n + 1$
  $t = t + \underline{dt}$
  $\tilde{\vec{u}}_n = \vec{u}_{n-1} + (1 - \gamma_P)\underline{dt}\dot{\vec{u}}_{n-1}$
  $C^0_{d\,n} = C_{d\,n-1}$
  For $k = 0, 1, 2, 3, \ldots$
    Solve the elliptic boundary value problem

$$\begin{cases} \frac{1}{\gamma_P \underline{dt}}(\mu_0 I_3 + C^k_{d\,n})\underline{d\vec{u}} + \text{curl curl } \underline{d\vec{u}} = f(t_n) - \text{curl curl } \tilde{\vec{u}}_n & \text{in } \Omega \\ \underline{d\vec{u}} \cdot \vec{n} = 0 \text{ on } \partial\Omega \end{cases}$$

    for $\underline{d\vec{u}}$
    $\vec{u}^{k+1}_n = \tilde{\vec{u}}_{n-1} + \underline{d\vec{u}}$
    $\vec{p}^{k+1}_n = \boldsymbol{\mathcal{P}}(\vec{u}_0, \vec{u}_1, \ldots, \vec{u}_{n-1}, \vec{u}^{k+1}_n)$

$$C^{k+1}_{d\,n} = \mathrm{diag}\left(\left(\frac{(\vec{p}^{k+1}_n - \vec{p}_{n-1})_i}{(\vec{u}^{k+1}_n - \vec{u}_{n-1})_i}\right)_{i=1,2,3}\right)$$

  $\vec{u}_n = \vec{u}^{k+1}_n$
  $p_n = p^{k+1}_n$
  $\dot{\vec{u}}_n = \frac{\vec{u}_n - \tilde{\vec{u}}_n}{\gamma_P \underline{dt}}$

Again, monotonicity of the components $\mathcal{P}_i$ in (2.11) guarantee nonnegativity of the coefficient matrices $C^k_{d\,n}$ in the elliptic boundary value problems.

5 Hysteresis identification in PDEs

In this section, we develop some iterative methods for identifying the hysteresis operator $\mathcal{P}$ in the model problems (4.1) and (4.9), respectively. Here we first of all concentrate on the hyperbolic case (4.1) and later on shortly show applicability also to the parabolic problem (4.9).

5.1 The hyperbolic model problem for piezoelectricity

We consider the inverse problem of identifying the operator $\mathcal{P}$ in

$$\rho u_{tt} - (\mathcal{P}[u_x])_x = 0 \quad x \in (0, L), \quad t \in (0, T), \tag{5.1}$$

with boundary conditions

$$\begin{aligned} &\mathcal{P}[u_x(0, \cdot)] = g_0 \text{ or } u(0, \cdot) = f_0 \\ &u(L, t) = f_L(t) \quad t \in (0, T), \end{aligned} \tag{5.2}$$

and initial conditions

$$u(x,0) = u_0(x), \quad u_t(x,0) = u_1(x) \quad x \in (0,L), \tag{5.3}$$

from additional boundary measurements

$$\mathcal{P}[u_x(L,\cdot)](t) = g_L(t) \quad t \in (0,T), \tag{5.4}$$

Identifiability in the static case

Since the hysteresis operator affects the boundary conditions, we could immediately identify its action on the function $u_x(L,\cdot)$ from (5.4) – however, we do not know the function $u_x(L,\cdot)$, (only $u(L,\cdot) = f_L$ is given). It is an approach often found in the engineering literature to assume that for each t, $u_x(\cdot,t)$ is basically constant on $[0,L]$ and therefore equal to $\frac{f_L(t)-f_0(t)}{L}$, where $f_0(t) = u(0,t)$. As a matter of fact, this is really the case if we set ρ to zero and if $\mathcal{P}$ is invertible. If ρ is different from zero, there might be a considerable difference between $u_x(L,\cdot)$ and $\frac{f_L(t)-f_0(t)}{L}$, see Figure 5.1. In this situation, it is crucial to use the PDE model (5.1) for identifying the hysteresis operator rather than the simplifying assumption of space constant u_x to avoid large modelling errors in the identified Preisach operator, see Figure 5.2.

Still, we expect to gain some insight into the question of identifiability in the special case $\rho = 0$ in (4.1), which from a physical point of view is the static situation, where (4.1) contains no time derivative of u.

Note that we have only a function of one variable, namely g_L as additional information. Hence, we cannot expect to be able to recover the full Preisach weight function (or alternatively, the shape function) as a function of two variables. Still the input function f_L can be chosen in such a way that as much information as possible is recovered, see [7], [25], and the references therein. This is usually done by doing excitation along so-called incremental triangles as they can often be found in the engineering literature in this context, see Figure 5.3, and note that the excitation plotted in Figure 5.1 is a smoothed version of these incremental triangles.

Due to the deletion properties of the Preisach operator, these incremental triangles have the advantage that almost the full input history can be forgotten after passing one of the local extrema. This fact guarantees that if $e := u_x(L,\cdot)$ is known (e.g., in case $\rho = 0$), then there is a one-to-one correspondence between the shape function and the measurements: Denote by $\beta_1, \beta_2, \ldots, \alpha_1, \alpha_2 \ldots$ the sequences of local minima and maxima of e and consider, for some $t > 0$ the case that $e(t)$ lies between a preceding local minimum β_j and a subsequent local maximum α_j. Then, due to the fact that $|\beta_1| < |\alpha_1| < |\beta_2| < |\alpha_2| < \ldots$, $\mathcal{P}$ has forgotten all local extrema before β_j, i.e. the input string $s = (\beta_1, \alpha_1, \beta_2, \alpha_2, \ldots, \beta_j, e(t))$ satisfies

$$s \succeq s' = \begin{cases} (\beta_j, e(t)) & \text{if } \beta_j < e(t) < |\beta_j| \\ (e(t)) & \text{if } |\beta_j| < e(t) < \alpha_j\,. \end{cases}$$

Hence, by (3.8), we get

$$\mathcal{P}[e](t) = \begin{cases} \mathcal{P}_f(\beta_j) + h(\beta_j, e(t)) & \text{if } \beta_j < e(t) < |\beta_j| \\ \mathcal{P}_f(e(t)) & \text{if } |\beta_j| < e(t) < \alpha_j \end{cases} \tag{5.5}$$

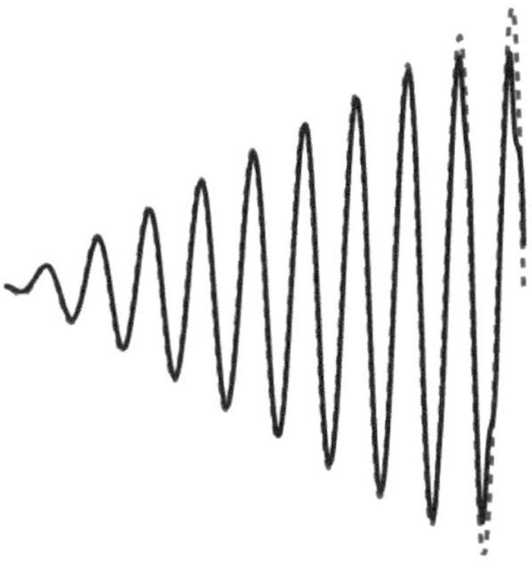

Figure 5.1 Actual value of $u_x(L,\cdot)$ (solid) compared to its space constant approximation $\frac{f_L - f_0(t)}{L}$ (dashed), for $\rho = 0.1$ and $\mathcal{P}$ the hysteresis operator with saturation according to Example 1 in Section 6.

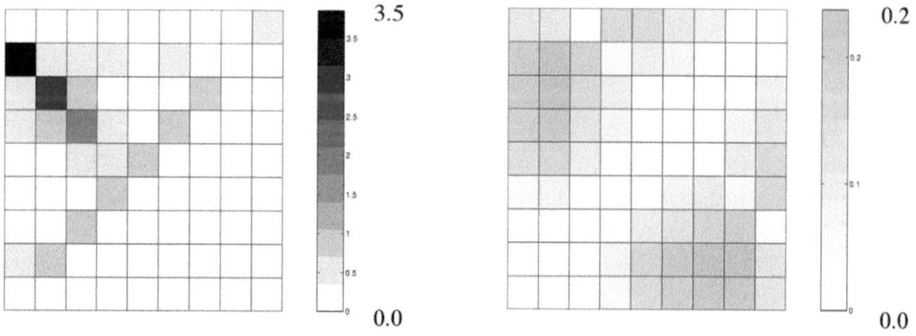

Figure 5.2 Error in recovered Preisach weight function (left, compare to absolute values of $\wp$ in Figure 6.2) as well as relative error in shape function (right) due to the approximation $u_x(L,\cdot) \approx \frac{f_L - f_0(t)}{L}$, for $\rho = 0.1$ and $\mathcal{P}$ the hysteresis operator with saturation according to Example 1 in Section 6.

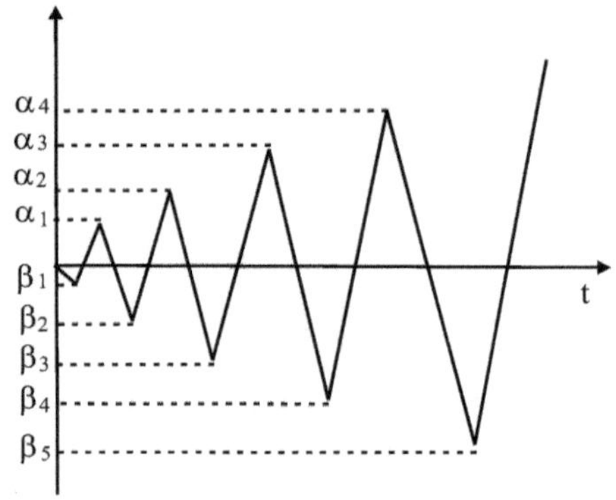

Figure 5.3 Incremental triangles.

which in case of symmetry (3.12) by (3.16) becomes

$$\mathcal{P}[e](t) = \begin{cases} -\frac{1}{2}\overline{h}(\beta_j, -\beta_j) + \overline{h}(\beta_j, e(t)) & \text{if } \beta_j < e(t) < |\beta_j| \\ \frac{1}{2}\overline{h}(-e(t), e(t)) & \text{if } |\beta_j| < e(t) < \alpha_j \,. \end{cases}$$

The case $\alpha_j \geq e(t) \geq \beta_{j+1}$ goes analogously and yields

$$\mathcal{P}[e](t) = \begin{cases} \mathcal{P}_f(\alpha_j) - h(e(t), \alpha_j) & \text{if } \alpha_j > e(t) > -|\alpha_j| \\ \mathcal{P}_f(e(t)) & \text{if } -|\alpha_j| > e(t) > \beta_j \end{cases} \tag{5.6}$$

as well as in the symmetric case

$$\mathcal{P}[e](t) = \begin{cases} \frac{1}{2}\overline{h}(-\alpha_j, \alpha_j) - h(e(t), \alpha_j) & \text{if } \alpha_j > e(t) > -|\alpha_j| \\ -\frac{1}{2}\overline{h}(e(t), -e(t)) & \text{if } -|\alpha_j| > e(t) > \beta_j \,. \end{cases}$$

Obviously, the final value mapping $\mathcal{P}_f$ appearing as a first term in the Everett sum (3.8) can be identified beforehand from two simple additional sets of measurements: One with input increasing from 0 to 1 and one with input decreasing from 0 to -1. Therewith, the identities (5.5), (5.6) provide direct access to values of the shape function.

We conclude that in the purely elliptic case $\rho = 0$ and therewith $u_x(L, \cdot) = \frac{f_L - f_0}{L}$, we can uniquely recover the final value mapping $\mathcal{P}_f$ for strings of length one, as well as the shape function on the subset of the Preisach unit triangle that is defined by a union of lines parallel to the axes as indicated in Figure 5.4. Note that by (3.10) we can immediately deduce the respective values of h at corresponding points below the main diagonal.

These parallel lines induce a natural (semi-)discretization of the shape function on the Preisach unit triangle. More generally, we cannot expect to get more information on the shape function than along those lines parallel to the α and β axes, whose offsets are determined by the extremal values of $u_x(L, \cdot)$. Thus the quality of approximation

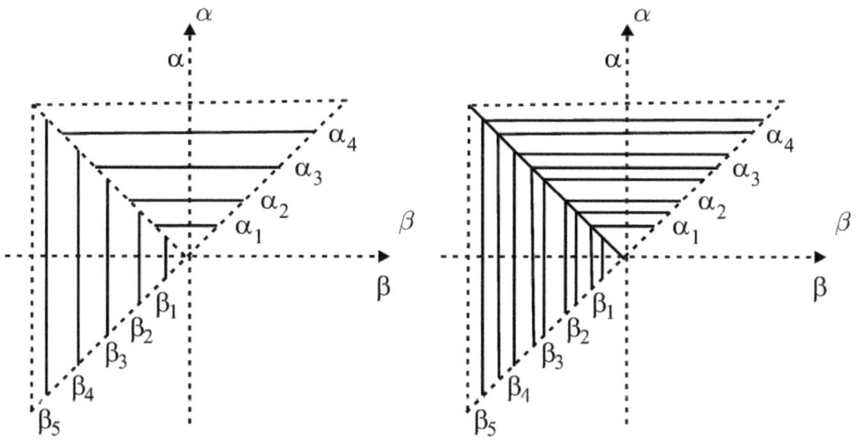

Figure 5.4 Values within the Preisach unit triangle where the shape function can be determined from $\mathcal{P}[\frac{f_L - f_0}{L}]$, with $\frac{f_L - f_0}{L}$ according to Figure 5.3 in the general case (left) as well as the symmetric case according to (3.11) (right).

for the Preisach operator depends on the excitation that indirectly influences $u_x(L, \cdot)$: The denser the local extrema of $u_x(L, \cdot)$ lie in the interval $[-1, 1]$, the better is the approximation of the identified hysteresis operators.

These (semidiscrete) identifiability arguments do not go through in the hyperbolic case $\rho > 0$. Namely, for this purpose, we would have to be able to steer u via the boundary conditions g_0, f_0, f_L in such a way that the input $e = u_x(x, \cdot)$ of the hysteresis operator $\mathcal{P}$ follows the incremental triangles for $x = L$, or at least for some fixed $x \in [0, L]$. Due to the lack of a maximum principle in the hyperbolic case, this does not seem to be straightforward, though. Here methods of (optimal) control might be valuable (cf., e.g., [20], [24] in the context of piezoelectricity).

Some iterative identification methods

I: An alternating iteration

The probably most straightforward approach for recovering $\mathcal{P}$ is the following: Given a preliminary approximation to $\mathcal{P}$, we use the initial-boundary value problem (5.1), (5.2), (5.3) to compute u. Then we insert $u_x(L, \cdot)$ into (5.4) to identify an improved version of $\mathcal{P}$ from the additional boundary measurements.

We discretize the searched for $\mathcal{P}$ by writing it as a linear combination of elementary hysteresis operators $\mathcal{P}_\lambda$

$$\mathcal{P} = \sum_{\lambda \in \Lambda} a_\lambda \mathcal{P}_\lambda \,. \tag{5.7}$$

Here $\mathcal{P}_\lambda$ may be chosen as simple relays, cf. Figure 1.2, which corresponds to a piecewise constant approximation of the weight function. In that case, obviously the set Λ consists of index pairs, corresponding to different up- and down-switching thresholds in case of $\mathcal{P}_\lambda$ being relays. However, also a smoother basis of elementary hysteresis operators may be used. In view of the discussion in the previous section, the discretization (5.7) should reflect the natural semidiscretization induced by the extremal values of $u_x(L,\cdot)$ as discussed above; for simplicity of exposition, we will here use a fixed set $(\mathcal{P}_\lambda)_{\lambda\in\Lambda}$ throughout the identification iterations, though.

Therewith, the alternating iteration can be written as follows:

Algorithm 5.1
Set $u_x^0(L,\cdot)$ to an initial guess for $u_x(L,\cdot)$.
For $n = 1, 2, 3, \ldots$
Compute $\mathbf{a}^n := (a_\lambda)_{\lambda\in\Lambda}$ from the equality
$\sum_{\lambda\in\Lambda} a_\lambda \mathcal{P}_\lambda[u_x^{n-1}(L,\cdot)] = g_L$.
Solve the initial-boundary value problem with hysteresis

$$\begin{aligned} \rho u_{tt} - (\textstyle\sum_{\lambda\in\Lambda} a_\lambda^n \mathcal{P}_\lambda[u_x])_x &= 0 \\ u(\cdot,0) &= u_0 \\ u_t(\cdot,0) &= u_1 \\ \textstyle\sum_{\lambda\in\Lambda} a_\lambda^n \mathcal{P}_\lambda[u_x^{n-1}(0,\cdot)] &= g_0 \\ u(L,\cdot) &= f_L \end{aligned}$$

for $u = u^n$.

Each step of the method consists of

1. Computation of the coefficients $\mathbf{a}^n = (a_\lambda^n)_{\lambda\in\Lambda}$ (and therewith, of the hysteresis operator $\mathcal{P}^n$) from an input-output hysteresis model. This can be treated as a least squares problem for a linear system of equations.
2. Solution of an initial-boundary value problem for a PDE with hysteresis. This can numerically be done by Algorithm 4.6.

For a method for solving task 1 we refer to a paper by Hoffmann and Meyer [6], where a numerical solution method as well as a convergence analysis are provided.

Newton type methods

The alternating iteration from Algorithm 5.1 can be regarded as a simple fixed point iteration and has the advantage of being simple and modular. The drawback that for this method convergence can be guaranteed only for small inhomogeneities u_0, u_1, f_L, g_0, whereas we are interested in the case of large excitations, motivates to also consider Newton type methods. However, here we are faced with the problem that hysteresis operators are not differentiable, which can be seen immediately from the fact that, e.g., at the maximal point of the main hysteresis loop in Figure 1.1, the derivative of the curve is not determined uniquely – it may be tangential to the upgoing or to the downgoing branch or lie in between. Actually this behaviour may occur at each point of the main loop due to possibly emerging minor loops.

Therefore, we use Newton methods that are based on approximating $\mathcal{P}'$ by the derivative of a differentiable approximation $\mathcal{S}$ of $\mathcal{P}$. We choose $\mathcal{S}$ to be the superposition operator

$$\mathcal{S} : e \mapsto c \circ e$$

with c being defined as the centerline of the main hysteresis loop, i.e.,

$$c(e) = \frac{\mathcal{P}_f(-1) + h(-1,e) + \mathcal{P}_f(1) + h(1,e)}{2} \tag{5.8}$$

which in the symmetric case (3.11) becomes

$$c(e) = \frac{\overline{h}(-1,e) - \overline{h}(-1,-e)}{2},$$

see Figure 5.5, left. (An alternative for defining c is the so-called commutation curve, consisting of all turning points occurring during a decremental triangle input; we mention in passing that the latter is just the way B-H curves in magnetics are defined from the general hysteretic B-H relation.)

Therewith, the derivative of $\mathcal{S}$, is the multiplication operator defined by

$$(\mathcal{S}'[e]\underline{de})(t) = c'(e(t))\underline{de}(t)\,.$$

II: An all-at-once Newton type iteration

Considering the whole nonlinear problem of forward PDE solution and inverse hysteresis identification in one equation, we arrive at a formulation of the problem as an operator equation

$$F(u, \mathbf{a}) = 0 \tag{5.9}$$

with

$$F(u,\mathbf{a}) = \begin{pmatrix} \rho u_{tt} - (\sum_{\lambda\in\Lambda} a_\lambda \mathcal{P}_\lambda[u_x])_x \\ u(\cdot,0) - u_0 \\ u_t(\cdot,0) - u_1 \\ \sum_{\lambda\in\Lambda} a_\lambda \mathcal{P}_\lambda[u_x(0,\cdot)] - g_0 \\ u(L,\cdot) - f_L \\ \sum_{\lambda\in\Lambda} a_\lambda \mathcal{P}_\lambda[u_x(L,\cdot)] - g_L \end{pmatrix}. \tag{5.10}$$

We do not specify function spaces in which (5.9) is supposed to hold, as we intend to use this formulation just in order to derive a Newton type algorithm. With F as in (5.10), and with substituting $\mathcal{S}'_\lambda$ for $\mathcal{P}'_\lambda$, a modified Newton step

$$(u^n, \mathbf{a}^n) = -\tilde{F}'(u^{n-1}, \mathbf{a}^{n-1})^{-1} F(u^{n-1}, \mathbf{a}^{n-1})$$

for (5.9) reads as follows

$$u^n = u^{n-1} + w\,, \quad \mathbf{a}^n = \mathbf{a}^{n-1} + \mathbf{b}\,,$$

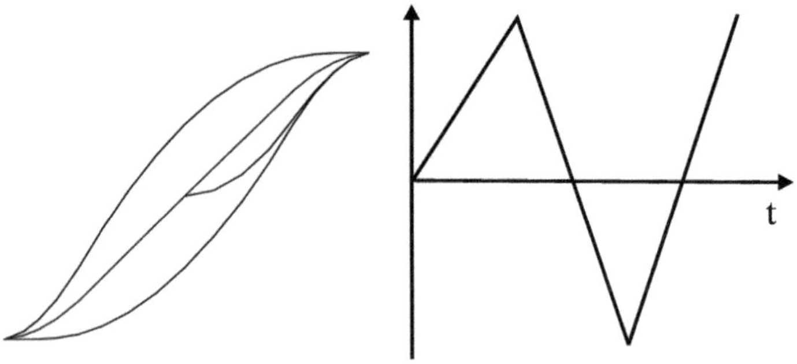

Figure 5.5 Centerline of the main hysteresis loop (left) and corresponding input signal (right).

where $(w, \mathbf{b})$ solve

$$\begin{aligned}
\rho w_{tt} - (\textstyle\sum_{\lambda\in\Lambda} a_\lambda^{n-1} \mathcal{S}'_\lambda[u_x^{n-1}] w_x)_x &= -\rho u_{tt}^{n-1} + (\textstyle\sum_{\lambda\in\Lambda} (a_\lambda^{n-1} + b_\lambda) \mathcal{P}_\lambda[u_x^{n-1}])_x \\
w(x,0) &= u_0 - u^{n-1}(x,0) \\
w_t(x,0) &= u_1 - u_t^{n-1}(x,0) \\
\textstyle\sum_{\lambda\in\Lambda} a_\lambda^{n-1} \mathcal{S}'_\lambda[u_x^{n-1}(0,\cdot)] w_x(0,\cdot) &= g_0 - \textstyle\sum_{\lambda\in\Lambda} (a_\lambda^{n-1} + b_\lambda) \mathcal{P}_\lambda[u_x^{n-1}(0,\cdot)] \\
w(L,\cdot) &= f_L - u^{n-1}(L,\cdot)
\end{aligned} \tag{5.11}$$

$$\sum_{\lambda\in\Lambda} b_\lambda \mathcal{P}_\lambda[u_x^{n-1}(L,\cdot)] = g_L - \sum_{\lambda\in\Lambda} a_\lambda^{n-1} \Big(\mathcal{P}_\lambda[u_x^{n-1}(L,\cdot)] + \mathcal{S}'_\lambda(u_x^{n-1}(L,\cdot) w_x(L,\cdot) \Big) . \tag{5.12}$$

Here w and $\mathbf{b} = (b_\lambda)_{\lambda\in\Lambda}$ are coupled via the right hand sides in the first and fourth line of (5.11) as well as (5.12). This coupling can be realized via an inner iteration for w and $\mathbf{b}$, such that the following algorithm for solving (5.9) results:

Algorithm 5.2
Set u^0 to an initial guess for u, and $\mathbf{a}^0$ to an initial guess for $\mathbf{a}$
For $n = 1, 2, 3, \ldots$
  $w^{(0)} = 0$
  For $k = 1, 2, 3, \ldots$
    Compute $\mathbf{b}^{(k)} := \mathbf{b}$ from the equality (5.12) with $w = w^{(k-1)}$
    Solve the linear initial-boundary value problem (5.11) with $\mathbf{b} = \mathbf{b}^{(k)}$
      to compute $w^{(k)} = w$
  Set $u^n = u^{n-1} + w^{(k)}$, $\mathbf{a}^n = \mathbf{a}^{n-1} + \mathbf{b}^{(k)}$

Each inner iteration of this method consists of

1. Computation of the coefficients $(b_\lambda^{(k)})_{\lambda\in\Lambda}$ from an input-output hysteresis model.
2. Solution of an initial-boundary value problem for a linear PDE.

Note that here we have to repeatedly solve linear PDEs in place of solving a PDE with hysteresis once.

III: A reduced Newton type iteration

An alternative formulation of the identification problem as a nonlinear operator equation

$$F(\mathbf{a}) = 0 \tag{5.13}$$

results from formulation via the operator

$$F(\mathbf{a}) = \sum_{\lambda\in\Lambda} a_\lambda \mathcal{P}_\lambda[u_x^{\mathbf{a}}(L,\cdot)] - g_L$$

where $u^{\mathbf{a}}$ solves

$$\begin{aligned} \rho u_{tt} - (\textstyle\sum_{\lambda\in\Lambda} a_\lambda \mathcal{P}_\lambda[u_x])_x &= 0 \\ u(\cdot,0) &= u_0 \\ u_t(\cdot,0) &= u_1 \\ \textstyle\sum_{\lambda\in\Lambda} a_\lambda \mathcal{P}_\lambda[u_x(0,\cdot)] &= g_0 \\ u(L,\cdot) &= f_L\,. \end{aligned} \tag{5.14}$$

Newton's method for (5.13) with the substitution $\mathcal{S}'$ for $\mathcal{P}'$ results in

Algorithm 5.3
Set $\mathbf{a}^0$ to an initial guess for $\mathbf{a}$
For $n = 0, 1, 2, 3, \ldots$

Solve the initial-boundary value problem with hysteresis (5.14) for $u^{\mathbf{a}^n} = u$

Solve the linear initial-boundary value problems

$$\left.\begin{aligned} \rho w_{\lambda\,tt} - (\textstyle\sum_{\tilde\lambda\in\Lambda} a^n_{\tilde\lambda}\mathcal{S}'_{\tilde\lambda}[u_x^{\mathbf{a}^n}]w_{\lambda\,x})_x &= (\mathcal{P}_\lambda[u_x^{\mathbf{a}^n}])_x \\ w_\lambda(x,0) &= 0 \\ w_{\lambda\,t}(x,0) &= 0 \\ \textstyle\sum_{\tilde\lambda\in\Lambda} a^n_{\tilde\lambda}\mathcal{S}'_{\tilde\lambda}[u_x^{\mathbf{a}^n}(0,\cdot)]w_{\lambda\,x}(0,\cdot) &= -\mathcal{P}_\lambda[u_x^{\mathbf{a}^n}(0,\cdot)] \\ w_\lambda(L,\cdot) &= 0 \end{aligned}\right\}\ \lambda\in\Lambda \tag{5.15}$$

for $(w_\lambda)_{\lambda\in\Lambda}$

Compute $\mathbf{b} := (b_\lambda)_{\lambda\in\Lambda}$ from the equality

$$\begin{aligned} &\sum_{\lambda\in\Lambda} b_\lambda \left(\mathcal{P}_\lambda[u^{\mathbf{a}^n}(L,\cdot)] + \sum_{\tilde\lambda\in\Lambda} a^n_{\tilde\lambda}\mathcal{S}'_{\tilde\lambda}[u_x^{\mathbf{a}^n}(L,\cdot)]w_{\lambda\,x}(L,\cdot)\right) \\ &\qquad = g_L - \sum_{\lambda\in\Lambda} a^n_\lambda \mathcal{P}_\lambda[u_x^{\mathbf{a}^n}(L,\cdot)] \end{aligned} \tag{5.16}$$

and set $\mathbf{a}^{n+1} := \mathbf{a}^n + \mathbf{b}$

Instead of solving card(Λ) linear problems (5.15) and inserting into (5.16), one can use the equivalent formulation

$$\begin{aligned}
\rho w_{tt} - (\textstyle\sum_{\lambda\in\Lambda} a_\lambda^n \mathcal{S}_\lambda'[u_x^{\mathbf{a}^n}]w_x)_x &= (\textstyle\sum_{\lambda\in\Lambda} b_\lambda \mathcal{P}_\lambda[u_x^{\mathbf{a}^n}])_x \\
w(x,0) &= 0 \\
w_t(x,0) &= 0 \\
\textstyle\sum_{\lambda\in\Lambda} a_\lambda^n \mathcal{S}_\lambda'[u_x^{\mathbf{a}^n}(0,\cdot)]w_x(0,\cdot) &= -\textstyle\sum_{\lambda\in\Lambda} b_\lambda \mathcal{P}_\lambda[u_x^{\mathbf{a}^n}(0,\cdot)] \\
w(L,\cdot) &= 0
\end{aligned} \tag{5.17}$$

with $w = \sum_{\lambda\in\Lambda} b_\lambda w_\lambda$,

$$\sum_{\lambda\in\Lambda} b_\lambda \mathcal{P}_\lambda[u_x^{\mathbf{a}^n}(L,\cdot)] = g_L - \sum_{\lambda\in\Lambda} a_\lambda^n \left(\mathcal{P}_\lambda[u_x^{\mathbf{a}^n}(L,\cdot)] + \mathcal{S}_\lambda'[u_x^{\mathbf{a}^n}(L,\cdot)]w_x(L,\cdot)\right) \tag{5.18}$$

and implement the coupling between (5.17) and (5.18) in an iterative manner:

Algorithm 5.4
Set $\mathbf{a}^0$ to an initial guess for $\mathbf{a}$
For $n = 0, 1, 2, 3, \ldots$
  Solve the initial-boundary value problem with hysteresis (5.14) for $u^{\mathbf{a}^n} = u$
  $w^{(0)} = 0$
  For $k = 1, 2, 3, \ldots$
    Compute $\mathbf{b}^{(k)} = \mathbf{b}$ from the equality (5.18) with $w := w^{(k-1)}$
    Solve the linear initial-boundary value problem (5.17) with $\mathbf{b} = \mathbf{b}^{(k)}$
      to compute $w^{(k)} = w$
  Set $\mathbf{a}^{n+1} := \mathbf{a}^n + \mathbf{b}^{(k)}$

Therewith, the algorithm can again be composed from the modules

1. Solution of an initial-boundary value problem for a PDE with hysteresis.
2. Computation of the coefficients $(b_\lambda^{(k)})_{\lambda\in\Lambda}$ from an input-output hysteresis model.
3. Solution of an initial-boundary value problem for a linear PDE.

Note that (5.18) is identical with (5.12), but the two algorithms differ substantially from each other, since Algorithm 5.4 involves solution of a PDE with hysteresis in each step.

Remark 5.5 For the proposed algorithms, convergence to the solution of a discrete identification problem according to (5.7) can be expected. Convergence to the actual infinite dimensional hysteresis operator is closely related to the still open question of identifiability, though. For simple dimensionality reasons (f_L and g_L are univariate functions, while the weight $\wp$ in (1.1) is bivariate) the latter cannot be expected to be answered for single excitation and measurement functions f_L and g_L but only for appropriately chosen sequences of functions f_L, g_L. Note the correspondence to different parameter identification problems in PDEs such as the electrical impedance tomography problem, where a space dependent parameter (thus a function of two or three variables) is determined from the Dirichlet-to-Neumann map (i.e. all Neumann boundary values corresponding to imposed Dirichlet boundary data).

5.2 The parabolic problem of magnetics

We wish to identify the hysteresis operator $\boldsymbol{\mathcal{P}}$ in

$$\mu_0 \vec{u}_t + \boldsymbol{\mathcal{P}}[\vec{u}]_t + \text{curl curl}\, \vec{u} = f \quad \text{in } \Omega \times [0,T], \tag{5.19}$$

with boundary conditions

$$\vec{u} \cdot \vec{n} = 0 \quad \text{on } \partial\Omega \times [0,T] \tag{5.20}$$

and initial condition

$$\vec{u}(\cdot, 0) = \vec{u}_0 \quad \text{in } \Omega, \tag{5.21}$$

from measurements of the magnetic flux

$$\Phi = \int_{A_c} (\mu_0 \vec{u} + \boldsymbol{\mathcal{P}}[\vec{u}]) \cdot \vec{n}\, d\Gamma. \tag{5.22}$$

For this purpose, we just carry over the iterative solution approaches from the previous section to the magnetics problem, with a discretization of the hysteresis operator analogous to (5.7), taking into account that we deal with a vectorial hysteresis operator (2.11)

$$\boldsymbol{\mathcal{P}} = \sum_{\xi \in \Xi} a_\xi \boldsymbol{\mathcal{P}}_\xi. \tag{5.23}$$

Here $\Xi = \Lambda \times \{1,2,3\}$ with Λ as in the scalar case (5.7) and

$$\boldsymbol{\mathcal{P}}_\xi[\vec{u}] = \boldsymbol{\mathcal{P}}_{(\lambda,i)}[\vec{u}] = \mathcal{P}_\lambda[u_i]\vec{e}_i, \quad i \in \{1,2,3\}.$$

I: An alternating iteration

Algorithm 5.6
Set $\vec{u}^0$ to an initial guess for $\vec{u}$
For $n = 1, 2, 3, \ldots$
  Compute $\mathbf{a}^n := (a_\xi)_{\xi \in \Xi}$ from the equality
    $\sum_{\xi \in \Xi} a_\xi \int_{A_c} \boldsymbol{\mathcal{P}}_\xi[\vec{u}^{n-1}] \cdot \vec{n}\, d\Gamma = \Phi - \int_{A_c} \mu_0 \vec{u}^{n-1} \cdot \vec{n}\, d\Gamma$
  Solve the initial-boundary value problem with hysteresis

$$\begin{aligned} \mu_0 \vec{u}_t + \textstyle\sum_{\xi \in \Xi} a_\xi \boldsymbol{\mathcal{P}}_\xi[\vec{u}]_t + \text{curl curl}\, \vec{u} &= f \quad \text{in } \Omega \times [0,T] \\ \vec{u} \cdot \vec{n} &= 0 \quad \text{on } \partial\Omega \times [0,T] \\ \vec{u}(\cdot, 0) &= \vec{u}_0 \quad \text{in } \Omega \end{aligned}$$

  for $\vec{u}^n = \vec{u}$

II: An all-at-once Newton type iteration

We write the parabolic initial boundary value problem with hysteresis (5.19), (5.20), (5.21) and the additional measurements (5.22) in one equation and apply the approximate Newton type method using the centerline (or commutation curve) multiplication operators $\mathcal{S}_\lambda$ for approximating a local linearization of $\mathcal{P}_\lambda$. The coupling between the increments $\vec{w}$ and $\mathbf{b}$ is realized in an inner iteration.

Algorithm 5.7
Set $\vec{u}^0$ to initial guess for $\vec{u}$, and $\mathbf{a}^0$ to an initial guess for $\mathbf{a}$.
For $n = 1, 2, 3, \ldots$
  $\vec{w}^{(0)} = 0$.
  For $k = 1, 2, 3, \ldots$
    Compute $\mathbf{b}^{(k)} := (b_\xi^{(k)})_{\xi\in\Xi}$ from the equality

$$\sum_{\xi\in\Xi} b_\xi \int_{A_c} \boldsymbol{\mathcal{P}}_\xi[\vec{u}^{n-1}] \cdot \vec{n}\, d\Gamma$$
$$= \Phi - \int_{A_c} \Big(\mu_0(\vec{u}^{n-1} + \vec{w}^{(k-1)}) + \sum_{\xi\in\Xi} a_\xi (\boldsymbol{\mathcal{P}}_\xi[\vec{u}^{n-1}] + \mathcal{S}'_\xi[\vec{u}^{n-1}]\vec{w}^{(k-1)})\Big) \cdot \vec{n}\, d\Gamma$$

    Solve the linear initial-boundary value problem

$$\mu_0 \vec{w}_t + \sum_{\xi\in\Xi} a_\xi (\mathcal{S}'_\xi[\vec{u}^{n-1}]\, \vec{w})_t + \text{curl curl}\, \vec{w}$$
$$= f - \mu_0 \vec{u}_t^{n-1} - \sum_{\xi\in\Xi} (a_\xi + b_\xi^{(k)}) \boldsymbol{\mathcal{P}}_\xi[\vec{u}^{n-1}]_t - \text{curl curl}\, \vec{u}^{n-1} \quad \text{in } \Omega \times [0,T]$$
$$\vec{w} \cdot \vec{n} = -\vec{u}^{n-1} \cdot \vec{n} \quad \text{on } \partial\Omega \times [0,T]$$
$$\vec{w}(\cdot, 0) = \vec{u}_0 - \vec{u}^{n-1}(\cdot, 0) \quad \text{in } \Omega$$

    for $\vec{w}^{(k)} = \vec{w}$.
  Set $\vec{u}^n = \vec{u}^{n-1} + \vec{w}^{(k)}$, $\mathbf{a}^n = \mathbf{a}^{n-1} + \mathbf{b}^{(k)}$.

III: A reduced Newton type iteration

Stipulating the initial-boundary value problem (5.19), (5.20), (5.21) to hold for each $\mathbf{a}$ under consideration and iterating for $\mathbf{a}$ on the basis of the measurement equation (5.22), we arrive at

Algorithm 5.8
Set $\mathbf{a}^0$ to an initial guess for $\mathbf{a}$
For $n = 0, 1, 2, 3, \ldots$
  Solve the initial-boundary value problem with hysteresis

$$\begin{aligned} \mu_0 \vec{u}_t + \sum_{\xi\in\Xi} a_\xi^n \boldsymbol{\mathcal{P}}_\xi[\vec{u}]_t + \text{curl curl}\, \vec{u} &= f \quad \text{in } \Omega \times [0,T] \\ \vec{u} \cdot \vec{n} &= 0 \quad \text{on } \partial\Omega \times [0,T] \\ \vec{u}(\cdot, 0) &= \vec{u}_0 \quad \text{in } \Omega \end{aligned} \tag{5.24}$$

  for $\vec{u}^{\mathbf{a}^n} = \vec{u}$

  Solve the linear initial-boundary value problems

$$\left.\begin{aligned} \mu_0 \vec{w}_{\xi\, t} + \sum_{\tilde{\xi}\in\Xi} a_{\tilde{\xi}}^n (\mathcal{S}'_{\tilde{\xi}}[\vec{u}^{\mathbf{a}^n}]\, \vec{w}_\xi)_t + \text{curl curl}\, \vec{w}_\xi &= -\boldsymbol{\mathcal{P}}_\xi[\vec{u}^{\mathbf{a}^n}]_t \quad \text{in } \Omega \times [0,T] \\ \vec{w}_\xi \cdot \vec{n} &= 0 \quad \text{on } \partial\Omega \times [0,T] \\ \vec{w}_\xi(\cdot, 0) &= 0 \quad \text{in } \Omega \end{aligned}\right\}$$

  for $(\vec{w}_\xi)_{\xi\in\Xi}$

Compute $\mathbf{b} := (b_\xi)_{\xi\in\Xi}$ from the equality
$\sum_{\xi\in\Xi} b_\xi \int_{A_c} \left(\mu_0 \vec{w}_\xi + \boldsymbol{\mathcal{P}}_\xi[\vec{u}^{\mathbf{a}^n}] + \sum_{\tilde\xi\in\Xi} a^n_{\tilde\xi} \mathcal{S}'_{\tilde\xi}[\vec{u}^{\mathbf{a}^n}]\vec{w}_\xi\right) \cdot \vec{n}\, d\Gamma$
$= \Phi - \int_{A_c} \left(\mu_0 \vec{u}^{\mathbf{a}^n} + \sum_{\xi\in\Xi} a^n_\xi \boldsymbol{\mathcal{P}}_\xi[\vec{u}^{\mathbf{a}^n}]\right) \cdot \vec{n}\, d\Gamma$
and set $\mathbf{a}^{n+1} := \mathbf{a}^n + \mathbf{b}$.

An inner iterative coupling analogous to Algorithm 5.4 results in

Algorithm 5.9
Set $\mathbf{a}^0$ to an initial guess for $\mathbf{a}$
For $n = 0, 1, 2, 3, \dots$
Solve the initial-boundary value problem with hysteresis (5.24) for $\vec{u}^{\mathbf{a}^n} = \vec{u}$
$\vec{w}^{(0)} = 0$, $\mathbf{b}^{(0)} = 0$
For $k = 1, 2, 3, \dots$
Solve the linear initial-boundary value problem

$$\begin{aligned} \mu_0 \vec{w}_t + \sum_{\xi\in\Xi} a^n_\xi (\mathcal{S}'_\xi[\vec{u}^{\mathbf{a}^n}]\, \vec{w})_t + \operatorname{curl}\operatorname{curl} \vec{w} &= -\sum_{\xi\in\Xi} b^{(k-1)}_\xi \boldsymbol{\mathcal{P}}_\xi[\vec{u}^{\mathbf{a}^n}]_t \quad \text{in } \Omega\times[0,T] \\ \vec{w}\cdot\vec{n} &= 0 \quad \text{on } \partial\Omega\times[0,T] \\ \vec{w}(\cdot,0) &= 0 \quad \text{in } \Omega \end{aligned}$$

for $\vec{w}^{(k)} = \vec{w}$
Compute $\mathbf{b}^{(k)} = \mathbf{b}$ from the equality
$\sum_{\xi\in\Xi} b_\xi \int_{A_c} \boldsymbol{\mathcal{P}}_\xi[\vec{u}^{\mathbf{a}^n}] \cdot \vec{n}\, d\Gamma$
$= \Phi - \int_{A_c} \left(\mu_0(\vec{u}^{\mathbf{a}^n} + \vec{w}^{(k)}) + \sum_{\xi\in\Xi} a_\xi (\boldsymbol{\mathcal{P}}_\xi[\vec{u}^{\mathbf{a}^n}] + \mathcal{S}'_\xi[\vec{u}^{\mathbf{a}^n}]\vec{w}^{(k)})\right) \cdot \vec{n}\, d\Gamma$
Set $\mathbf{a}^{n+1} := \mathbf{a}^n + \mathbf{b}^{(k)}$

6 Numerical results

We tested the proposed methods for the hyperbolic case (for results in the parabolic case we refer to [11]) with three different hysteresis operators to be identified. All of these examples are symmetric (cf. (3.14)) with normalized domain $I = [-1, 1]$:

1. Hysteresis with saturation (see Figure 6.1 top left)

$$\overline{h}_{\text{sat}}(e_1, e_2) = \frac{2}{(1+e_{\text{sat}})^2}(\varphi(e_2) - \varphi(e_1)),$$

   where

$$\varphi(e) = \begin{cases} e + \frac{(e+e_{\text{sat}})^2}{2(1-e_{\text{sat}})} & \text{if } -1 \le e \le -e_{\text{sat}} \\ e & \text{if } -e_{\text{sat}} \le e \le e_{\text{sat}} \\ e - \frac{(e-e_{\text{sat}})^2}{2(1-e_{\text{sat}})} & \text{if } e_{\text{sat}} \le e \le 1 \end{cases}$$

   and $e_{\text{sat}} \in [0, 1]$ determines the value of saturation for the input, i.e., the main loop becomes flatter as soon as the modulus of the input value exceeds e_{sat}. Note that

Figure 6.1 Main hysteresis loops of four different prototype hysteresis operators.

saturation is an important phenomenon in the context of both piezoelectricity and magnetics.

2. Narrow hysteresis (see Figure 6.1 top right)

$$\overline{h}_{\text{narrow}}(e_1, e_2) = \frac{1}{2 - \varepsilon/3} \begin{cases} \frac{(e_2-e_1)^2}{\varepsilon} - \frac{(e_2-e_1)^3}{3\varepsilon^2} & \text{if } (e_2 - e_1) \leq \varepsilon \\ e_2 - e_1 - \frac{\varepsilon}{3} & \text{if } (e_2 - e_1) \leq \varepsilon\,. \end{cases}$$

Here $\varepsilon > 0$ is a small parameter determining the width of the main hysteresis loop.

3. Broad hysteresis (see Figure 6.1 bottom left)

$$\overline{h}_{\text{broad}}(e_1, e_2) = e_2 - e_1 + (2 - (e_2 - e_1)) \ln(2 - (e_2 - e_1))\,.$$

As an initial guess in the iterative identification algorithms, we chose the most simple hysteresis operator with constant weight function $\wp \equiv \frac{1}{2}$ (see Figure 6.1 bottom right), which results in the following shape function

$$\overline{h}_{\text{std}}(e_1, e_2) = \frac{1}{2}(e_2 - e_1)^2\,.$$

Figure 6.2 shows the first few weight function iterates of the alternating iteration Algorithm 5.1 applied to the saturation hysteresis example, while in Figure 6.3, we display the respective errors in the Everett function. The same is done with randomly perturbed data (one per cent noise level) in Figures 6.4, 6.5. Here we continued the iteration beyond the optimal stopping index $n = 2$ in order to demonstrate the noise effect. To obtain reasonable results in the noisy case, we used a discrepancy principle type a posteriori rule, that stopped the iteration as soon as the residual fell below the noise level. This criterion was very well able to detect the optimal stopping index in all our computational experiments with noisy data.

It becomes evident that reconstruction of the shape function h is much more stable than recovery of the weight function $\wp$. This reflects the fact that identification of $\wp$ involves twice differentiation according to (3.9). Instability of weight function identification can be seen from the singular values of the sensitivity matrix $\frac{\partial g_L}{\partial \mathbf{a}}$, that decay like $\frac{1}{(j+3)^2}$ if we consider $\mathbf{a}$ as the coefficients of $\mathcal{P}$ in a discretization (5.7) with elementary relays, so that $\mathbf{a}$ corresponds to a pixel wise discretization of the weight function, see Figure 6.6.

For the other hysteresis examples (narrow and broad), results obtained with the alternating iteration Algorithm 5.1 are displayed in Figures 6.7, 6.8.

An efficiency comparison between alternating iteration Algorithm 5.1 and reduced Newton Algorithm 5.4 is summarized in Table 6.1 and shows similar computing times for both algorithms. Algorithm 5.2 showed some convergence difficulties, probably due to the fact that only linearized versions of the PDE with hysteresis are solved, there.

		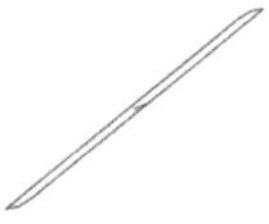		
initial residual	0.5123	0.2307	0.8077	
alternating iteration	5 39.63 1.6e-4	5 39.29 1.8e-4	5 40.73 0.9e-4	no. steps CPU sec. residual
Newton	3 (1.3) 35.09 2.8e-4	4 (1.5) 49.58 0.9e-4	4 (1.25) 45.63 1.9e-4	no. steps (av. inner its.) CPU sec. residual

Table 6.1 Comparison of alternating iteration with reduced Newton.

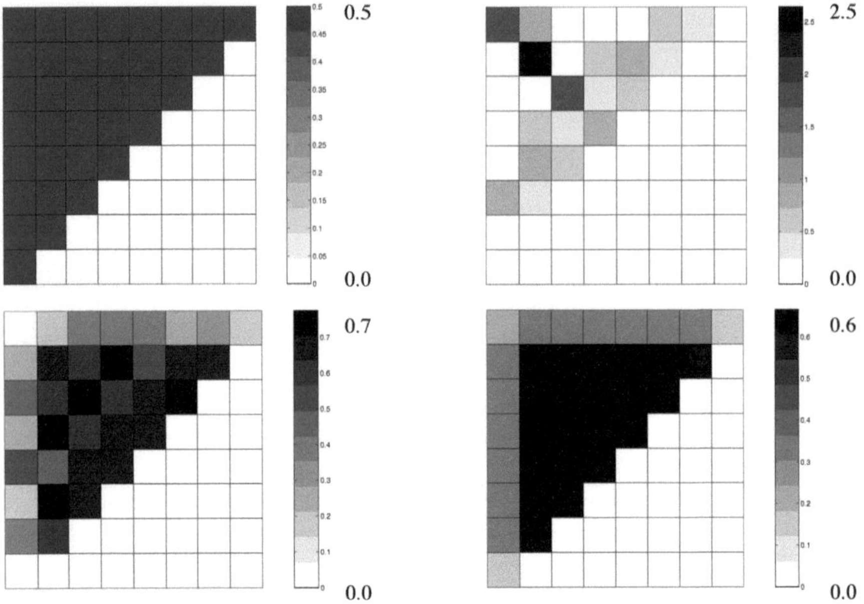

Figure 6.2 Weight functions $\wp_n$ obtained by the alternating iteration for the hysteresis with saturation: $n = 0$ (top left), $n = 1$ (top right), $n = 2$ (bottom left), $n = 3$ (bottom right) with exact data.

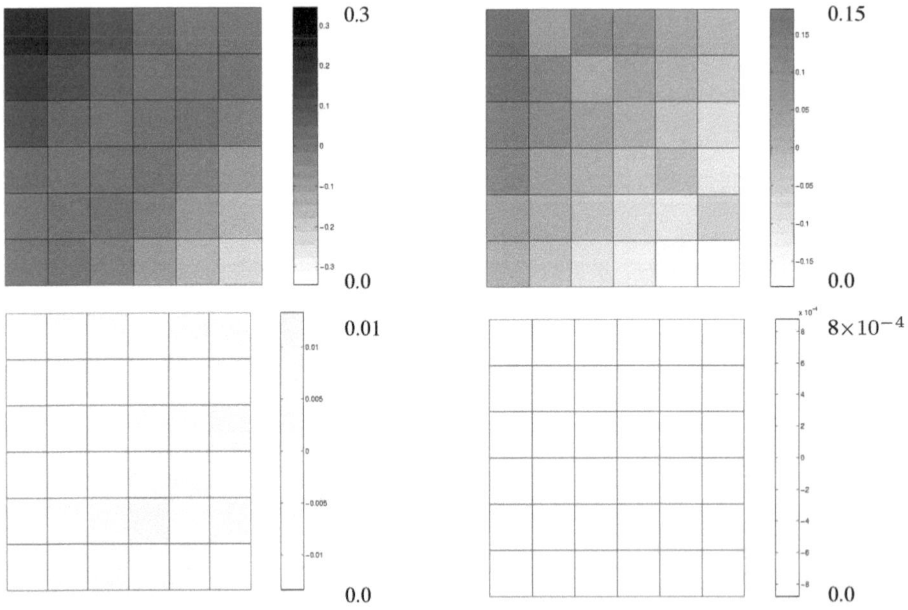

Figure 6.3 Relative error in Everett functions $h_n - h$ obtained by the alternating iteration for the hysteresis with saturation: $n = 0$ (top left), $n = 1$ (top right), $n = 2$ (bottom left), $n = 3$ (bottom right) with exact data.

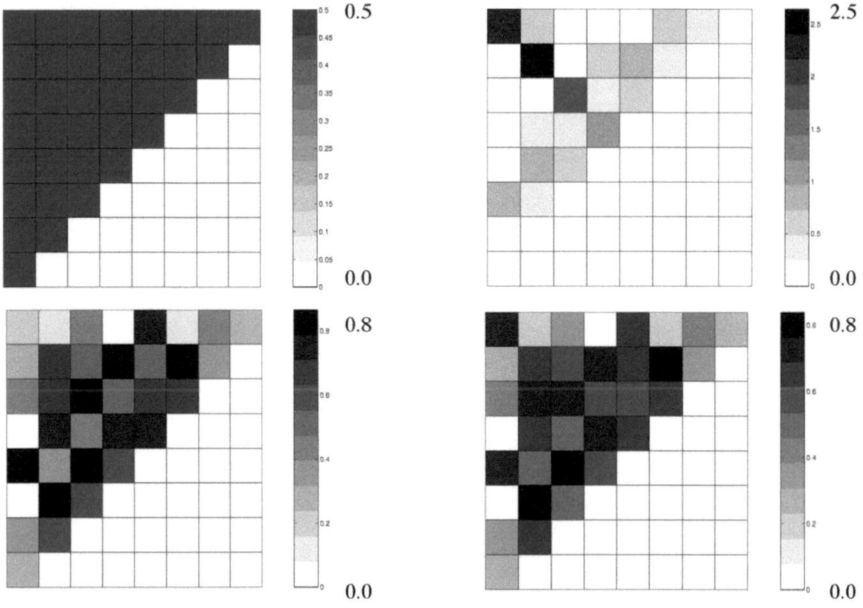

Figure 6.4 Weight functions $\wp_n$ obtained by the alternating iteration for the hysteresis with saturation: $n = 0$ (top left), $n = 1$ (top right), $n = 2$ (bottom left), $n = 3$ (bottom right) with noisy data.

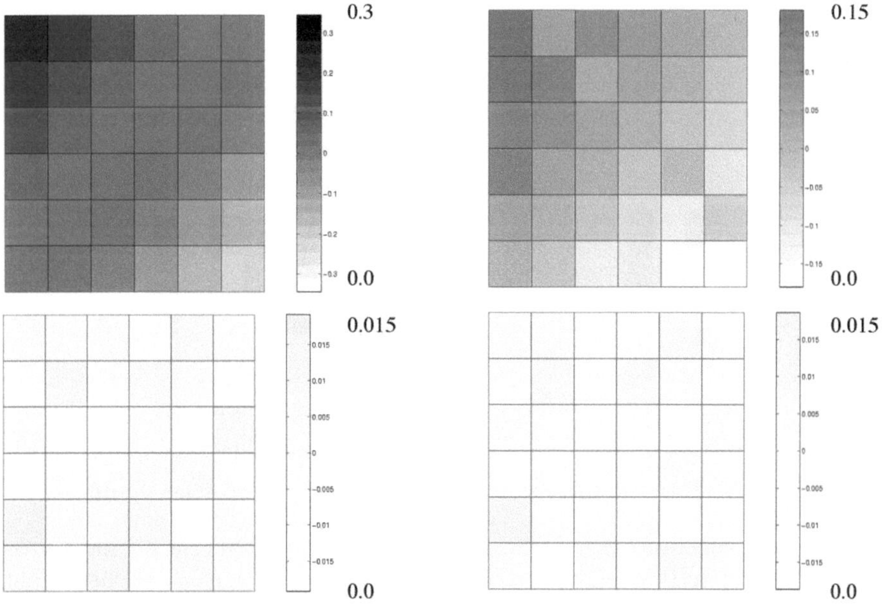

Figure 6.5 Relative error in Everett functions $h_n - h$ obtained by the alternating iteration for the hysteresis with saturation: $n = 0$ (top left), $n = 1$ (top right), $n = 2$ (bottom left), $n = 3$ (bottom right) with noisy data.

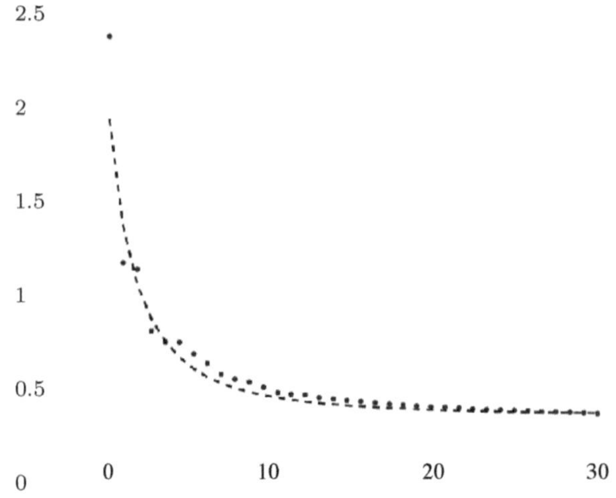

Figure 6.6 Singular values σ_j of the sensitivity matrix in weight function identification (dots) compared to $(j+3)^{-2}$ (dashed).

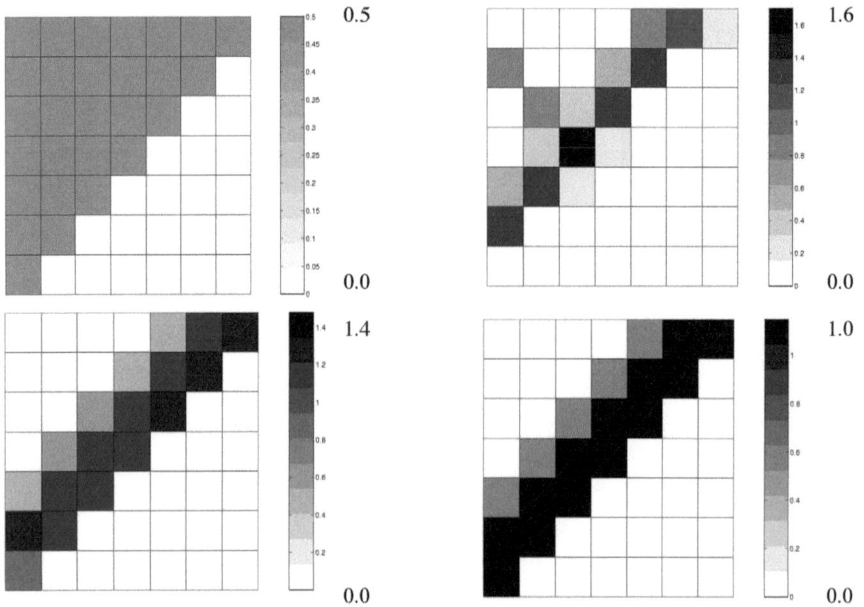

Figure 6.7 Weight functions $\wp_n$ obtained by the alternating iteration for the narrow hysteresis: $n = 0$ (top left), $n = 1$ (top right), $n = 2$ (bottom left), $n = 3$ (bottom right) with exact data.

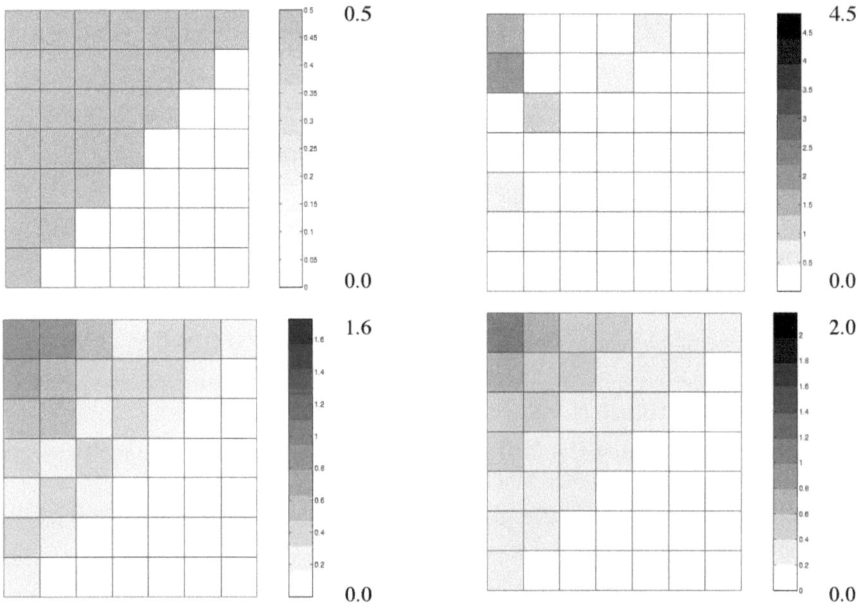

Figure 6.8 Weight functions $\wp_n$ obtained by the alternating iteration for the broad hysteresis: $n = 0$ (top left), $n = 1$ (top right), $n = 2$ (bottom left), $n = 3$ (bottom right) with exact data.

7 Conclusions and remarks

In this paper, we have investigated hysteresis in PDEs by means of two model problems, and proposed and tested some iterative methods for hysteresis identification in PDEs. In our exposition, we concentrated on a hyperbolic PDE motivated from piezoelectricty. For a more detailed treatment of the direct and inverse problem in magnetics, leading to a parabolic PDE, we refer to [10] and [11], respectively.

Future research will be concerned with a mathematically rigorous justification of the approaches shown here. Especially, the question of identifiability will have to be investigated in more detail.

Moreover, adaptive discretization of the hysteresis operators according to the local maxima and minima of the boundary values $u_x(L, \cdot)$ could be of interest. On the other hand, optimal steering of the boundary values to enhance identifiability is a task, that may be treated by means of control techniques.

Bibliography

[1] C. Amrouche, C. Bernardi, M. Dauge, and V. Girault, *Vector potentials in three-dimensional non-smooth domains*, Math. Methods Applied Sci. 21 (1998), pp. 823–864.

[2] J.P.A. Bastos and N. Sadowski, *Electromagnetic Modeling by Finite Elements*, Marcel Dekker, New York, 2003.

[3] M. Brokate and J. Sprekels, *Hysteresis and Phase Transitions*, Springer, New York, 1996.

[4] D.H. Everett, *A general approach to hysteresis*, Trans. Faraday Soc. 51 (1955), pp. 1551–1557.

[5] M. Hilpert, *On uniqueness of evolution problems with hysteresis*, Mathematical Models for Phase Change Problems (Basel) (J.F. Rodrigues, ed.), Birkhäuser, 1989.

[6] K.-H. Hoffmann and G.H. Meyer, *A least squares method for finding the Preisach Hysteresis operator from measurements*, Numer. Math. 55 (1989), pp. 695–710.

[7] K.-H. Hoffmann, J. Sprekels, and A. Visintin, *Identification of hysteresis loops*, J. Comput. Phys. 78 (1988), pp. 215–230.

[8] IEEE-UFFC, *IEEE Standard on Piezoelectricity*, 1985.

[9] M. Jais, *Inverse Probleme parabolischer Differentialgleichungen mit Hysterese*, Ph.D. thesis, Technical University of Munich, 2005.

[10] B. Kaltenbacher, *Hysteresis modelling in magnetics via Preisach operators: a regularity result and some numerical solution strategies*, Tech. report, University of Erlangen, 2006, in preparation.

[11] ______, *Iterative Identification of Hysteresis in Maxwell's equations*, COMPEL (2007), to appear.

[12] B. Kaltenbacher, M. Kaltenbacher, and S. Reitzinger, *Identification of nonlinear $B - H$ curves based on magnetic field computations and using multigrid preconditioners for ill-posed problems*, Eur. J. Appl. Math. 14 (2003), pp. 15–38.

[13] B. Kaltenbacher, T. Lahmer, M. Mohr, and M. Kaltenbacher, *PDE based determination of piezoelectric material tensors*, Eur. J. Appl. Math. 17 (2006), pp. 383–416.

[14] M. Kaltenbacher, *Numerical Simulation of Mechatronic Sensors and Actuators*, Springer, Berlin–Heidelberg–New York, 2004.

[15] M. Kamlah, *Erroelectric and ferroelastic piezoceramics – modeling of electromechanical hysteresis phenomena*, Continuum Mech. Thermodyn. 13 (2001), pp. 219–268.

[16] M. Kamlah and Ch. Tsakmakis, *Phenomenological modeling of the nonlinear electromechanical coupling in ferroelectrics*, Int. J. Solids Structures 36 (1999), pp. 669–695.

[17] M.A. Krasnoselskii and A.V. Pokrovskii, *Systems with Hysteresis*, Springer, Heidelberg, 1989.

[18] P. Krejčí, *Global Behaviour of solutions to the wave equation with hysteresis*, Adv. Math. Sci. Appl. 2 (1993), pp. 1–23.

[19] ______, *Hysteresis, Convexity, and Dissipation in Hyperbolic Equations*, Gakkotosho, 1996, Tokyo.

[20] K. Kuhnen, *Modeling, identification and compensation of complex hysteretic nonlinearities – A modified Prandtl-Ishlinskii approach*, European Journal of Control 9 (2003), pp. 407–418.

[21] Ch.M. Landis, *Fully coupled, multi-axial, symmetric constitutive laws for polycrystalline ferroelectric ceramics*, J. Mech. Phys. Solids 50 (2002), pp. 127–152.

[22] I.D. Mayrgoyz, *Mathematical Models of Hysteresis*, Springer, New York, 1991.

[23] V. Mehling, Ch. Tsakmakis, and D. Gross, *A simple, fully coupled three-dimensional phenomenological model for ferroelectric ceramics*, ECCOMAS Coupled Problems 2005 (Barcelona) (M. Papadrakis, E. Onate, and B. Schrefler, eds.), CIMNE, 2005.

[24] B. Miara, *Controlabilité d'un corps piézoelectrique*, C.R. Acad. Sci. Paris 333 (2001), pp. 267–270.

[25] U. Reiländer, *Das Großsignalverhalten Piezoelektrischer Aktoren*, Ph.D. thesis, Technical University of Munich, 2003.

[26] R. Simkovics, *Nichtlineares piezoelektrisches Finite-Elemente-Verfahren zur Modellierung piezokeramischer Aktoren*, Ph.D. thesis, University of Erlangen, 2002.

[27] R.C. Smith, S. Seelecke, Z. Ounaies, and J. Smith, *A free energy model for hysteresis in ferroelectric materials*, J. Intell. Mat. Sys. Struct. 14 (2003), pp. 719–739.

[28] E. Della Torre, *Magnetic Hysteresis*, IEEE Press, 1999, New York.

[29] A. Visintin, *On the Preisach model for hysteresis*, Nonlinear Analysis T.M.A. 9 (1984), pp. 977–996.

[30] ———, *Differential Models of Hysteresis*, Springer, Berlin, 1994.

[31] ———, *Vector Preisach model and Maxwell's equations*, Physica B 306 (2001), pp. 21–25.

Author information

Barbara Kaltenbacher, Institute for Stochastics and Applications, University of Stuttgart, Germany.
Email: `barbara.kaltenbacher@mathematik.uni-stuttgart.de`

Manfred Kaltenbacher, Department of Sensor Technology, University of Erlangen-Nuremberg, Germany.
Email: `manfred.kaltenbacher@lse.eei.uni-erlangen.de`

Radon Series Comp. Appl. Math **1**, 47–87

Multilevel methods for anisotropic elliptic problems

Johannes Kraus and Svetozar Margenov

Key words. Elliptic problems, multilevel preconditioning, hierarchical splitting, CBS constant, anisotropy, nonconforming finite element

AMS classification. 65F10, 65N20, 65N30

1 Abstract

Various preconditioners based on multilevel extensions of two-level finite element methods are known to lead to iterative methods which have an optimal order computational complexity with respect to the size (or discretization parameter) of the system.

First author: The support by the Austrian Academy of Sciences is gratefully acknowledged.

The methods can be on block matrix factorized form, recursively extended via certain matrix polynomial approximations of the arising Schur complement matrices or on additive, i.e., block diagonal form, using stabilizations of the condition number at certain levels.

This survey is devoted to the class of algebraic multilevel methods which were introduced and studied by Axelsson and Vassilevski for the case of isotropic elliptic problems in [14, 15], and are referred now as AMLI. Only two parameters, the constant in the strengthened Cauchy–Bunyakowski–Schwarz inequality (CBS constant) and the degree of anisotropy of the differential operator (or the shape of elements of the initial, coarsest mesh, which is analyzed equivalently) determine the optimality of the methods and their robustness. The coefficient jumps are shaped automatically, while as we shall see, the robustness with respect to anisotropy needs special constructions.

The focus of the paper is on some recent achievements in the AMLI methods for strongly anisotropic FEM elliptic systems. The cases of coefficient and mesh anisotropy for conforming (Courant) and nonconforming (Crouzeix–Raviart) linear finite elements are considered.

An important part of the AMLI algorithms is the treatment of the systems with the pivot block matrix, which arises on each finer level in a recursive refinement method and corresponds to the added degrees of freedom on that level. This block is well-conditioned for model type problems but becomes increasingly ill-conditioned when the coefficient matrix becomes more anisotropic or, equivalently, when the mesh aspect ratio increases. It is also important that in the nonconforming case the finite element spaces corresponding to two successive levels of mesh refinements are not nested. To handle this, a proper two-level basis is considered, which enables to fit the general framework for the construction of two-level preconditioners for conforming finite elements and to generalize the method to the multilevel case.

All presented methods lead to preconditioners with spectral equivalence bounds which hold uniformly with respect to both the problem and discretization parameters. As a result, the total computational complexity of the related PCG solvers is optimal. Such robust results have not been achieved by other methods.

2 Introduction

In this paper we consider the elliptic boundary value problem

$$\begin{aligned} Lu \equiv -\nabla \cdot (\mathbf{a}(\mathbf{x})\nabla u(\mathbf{x})) &= f(\mathbf{x}) \quad \text{in } \Omega, \\ u &= 0 \quad \text{on } \Gamma_D, \\ (\mathbf{a}(\mathbf{x})\nabla u(\mathbf{x}))\cdot \mathbf{n} &= 0 \quad \text{on } \Gamma_N, \end{aligned} \tag{2.1}$$

where Ω is a polygonal domain in $\mathbb{R}^2$, $f(\mathbf{x})$ is a given function in $L^2(\Omega)$, the coefficient matrix $\mathbf{a}(\mathbf{x})$ is symmetric positive definite and uniformly bounded in Ω, $\mathbf{n}$ is the outward unit vector normal to the boundary $\Gamma = \partial\Omega$, and $\Gamma = \bar{\Gamma}_D \cup \bar{\Gamma}_N$. We assume that the elements of the diffusion coefficient matrix $\mathbf{a}(\mathbf{x})$ are piecewise smooth functions on $\bar{\Omega}$.

The weak formulation of the above problem reads as follows: given $f \in L^2(\Omega)$ find $u \in \mathcal{V} \equiv H^1_D(\Omega) = \{v \in H^1(\Omega) : v = 0 \text{ on } \Gamma_D\}$, satisfying

$$\begin{aligned} \mathcal{A}(u,v) &= (f,v) \quad \forall v \in H^1_D(\Omega), \quad \text{where} \\ \mathcal{A}(u,v) &= \int_\Omega \mathbf{a}(\mathbf{x})\nabla u(\mathbf{x}) \cdot \nabla v(\mathbf{x}) d\mathbf{x}. \end{aligned} \tag{2.2}$$

We assume that the domain Ω is discretized by the triangulation $\mathcal{T}_h$ which is obtained by a proper number of refinement steps of a given coarser triangulation $\mathcal{T}_H$. We assume also that $\mathcal{T}_H$ is aligned with the discontinuities of the coefficient matrix $\mathbf{a}(\mathbf{x})$ so that over each element $T \in \mathcal{T}_H$ the entries of $\mathbf{a}(\mathbf{x})$ are smooth functions.

The variational problem (2.2) is then discretized using the finite element method, i.e., the continuous space $\mathcal{V}$ is replaced by a finite dimensional subspace $\mathcal{V}_h$. Then the finite element formulation is: find $u_h \in \mathcal{V}_h$, satisfying

$$\begin{aligned} \mathcal{A}_h(u_h,v_h) &= (f,v_h) \quad \forall v_h \in \mathcal{V}_h, \quad \text{where} \\ \mathcal{A}_h(u_h,v_h) &= \sum_{e\in\mathcal{T}_h} \int_e \mathbf{a}(e)\nabla u_h \cdot \nabla v_h d\mathbf{x}. \end{aligned} \tag{2.3}$$

Here $\mathbf{a}(e)$ is a piecewise constant symmetric positive definite matrix, defined by the integral averaged values of $\mathbf{a}(\mathbf{x})$ over each element from the coarsest triangulation $\mathcal{T}_H$, i.e.,

$$\mathbf{a}(e) = \frac{1}{|T|}\int_T \mathbf{a}(\mathbf{x}) d\mathbf{x}, \qquad \forall e \subset T \in \mathcal{T}_H.$$

We note that in this way strong coefficient jumps across the boundaries between adjacent finite elements from $\mathcal{T}_H$ are allowed.

The resulting discrete problem to be solved is then a linear system of equations

$$A_h \mathbf{u}_h = \mathbf{f}_h, \tag{2.4}$$

with A_h and $\mathbf{f}_h$ being the corresponding global stiffness matrix and global right hand side, and h being the discretization (meshsize) parameter for the underlying triangulation $\mathcal{T}_h$ of Ω.

The stiffness matrix is symmetric, positive definite and sparse. In the case of advanced real life applications A_h is presumably very large. For such problems, the advantages of the iterative solution methods increase with the size of the problem. In addition, the spectral condition number of the stiffness matrix behaves as $\kappa(A_h) = O(h^{-2})$ which deteriorates the convergence rate of the simpler iterative methods with the rise of the number of degrees of freedom $N = O(h^{-2})$.

The aim of the preconditioning is to overcome (or to relax) the meshsize dependency of the iterations' count. Let us consider the linear system

$$A\mathbf{u} = \mathbf{f}$$

with symmetric and positive definite matrix A. The following estimate characterizes the Preconditioned Conjugate Gradient (PCG) method (see, e.g., [4]):

$$n_{it} \leq \frac{1}{2}\sqrt{\kappa\left(C^{-1}A\right)}\ln\left(\frac{2}{\epsilon}\right) + 1, \tag{2.5}$$

where C is a symmetric and positive definite preconditioning matrix (also called preconditioner) and n_{it} is the related number of PCG iterations sufficient to get a prescribed relative accuracy of $\epsilon > 0$. The general strategy for efficient preconditioning simply follows from the estimate (2.5). It reads as follows:

- The condition number of the preconditioned matrix is much less than the original one, i.e., $\kappa\left(C^{-1}A\right) \ll \kappa(A)$.
- The computational complexity to solve the preconditioned system is much smaller than the complexity to solve the original problem, i.e., $\mathcal{N}\left(C^{-1}\mathbf{v}\right) \ll \mathcal{N}\left(A^{-1}\mathbf{v}\right)$.

One could say that the above conditions are contradictory. It is really true that when $\kappa\left(C^{-1}A\right)$ tends to its minimal value, the preconditioner should tend to A, and $\mathcal{N}\left(C^{-1}\mathbf{v}\right) \rightarrow \mathcal{N}\left(A^{-1}\mathbf{v}\right)$. Fortunately, such kind of reasonings are too pessimistic according to the recent state-of-the-art of the preconditioning algorithms.

Definition 2.1 The preconditioner is called optimal if the related PCG algorithm has optimal order of computational complexity $O(N)$, that is, if $\kappa\left(C^{-1}A\right) = O(1)$ and $\mathcal{N}\left(C^{-1}\mathbf{v}\right) = O(N)$.

The existence of optimal iterative solution methods has been an open question before the early 1960s. Now, the optimal order multigrid and multilevel methods are well-known in the community of researchers and practitioners dealing with large-scale scientific computations and their advanced applications.

The paper is devoted to some recent achievements in the development of robust PCG solvers for FEM elliptic systems (2.4). The methods to be discussed belong to the class of block matrix approximate factorization preconditioners of the algebraic multilevel iteration, AMLI, type.

The AMLI methods are based on multilevel finite element meshes and were originally introduced by Axelsson and Vassilevski in [14, 15] for the case of isotropic elliptic problems discretized by conforming linear finite elements. They are optimal with respect to the meshparameter (problem size) and can handle straightforwardly arbitrary coefficient jumps on the coarsest mesh used. During the years after, a lot of work was done to improve the computational efficiency using such techniques as varying degree of the stabilization matrix polynomial, or balancing the size of the coarsest level problem (so called short AMLI), etc., see, e.g., [5, 12, 16, 17, 50, 51].

The survey has a particular focus to the robustness for systems which are severely ill-conditioned due to some problem parameters taking near certain limit values. Examples of such parameters are the anisotropy, aspect ratio of the mesh and domain geometry, Poisson ratio for nearly incompressible materials etc. Furthermore, the condition number may increase rapidly when the discretization mesh is refined (due to both a smaller mesh parameter and possible irregularity of the mesh elements). In finding a good solution method one should preferably search for efficient preconditioners for the parameter free conjugate gradient iterative solution method.

The paper is mostly addressed to the robustness with respect to coefficient and/or mesh anisotropy. However, some related results for other ill-conditioned problems, e.g., the case of almost incompressible elasticity, will be also briefly commented.

Given the finite element matrix in a recursively defined hierarchical basis, the construction of the AMLI method is purely algebraic and the conditions for its optimality are also algebraically formulated. Only two parameters, the constant in the Cauchy–Bunyakowski–Schwarz inequality (the CBS constant) and the degree of anisotropy determine the robust optimality of the method. As was already mentioned, the robustness with respect to the coefficient jumps is shaped automatically. This is not the case with respect to the anisotropy.

The opportunity locally (at the current macroelement level) to estimate the CBS constant $\gamma \in (0, 1)$ is crucial for the theory of the AMLI methods, see, e.g., [10, 14, 19], as well as the review papers [6, 27, 51]. First robust estimates of the CBS constant are published by Maitre and Musy in 1982, [39]. They consider a piecewise Laplacian operator in an arbitrarily shaped coarser triangulation. The next important observation is that, for linear finite elements, the mesh and the coefficient anisotropies are equivalent with respect to the related CBS constant estimates. Some recent results on the CBS constant for 2D and 3D elliptic problems discretized by conforming finite elements can be found in [8, 21, 45, 48]. An important generalization of these studies concerns the coupled problem of linear elasticity where the robustness of the CBS constant is in the term of almost incompressibility, i.e., the estimates are uniform with respect to the Poisson ratio, see, e.g., [1, 2, 21, 40, 41].

The uniform estimate of the CBS constant is not enough to construct a good AMLI method. To guarantee the robustness we need also an optimal order solver (preconditioner) for the systems with the pivot block of the hierarchical two level stiffness matrix. This block corresponds to the added degrees of freedom on that level, and becomes increasingly ill-conditioned with the degree of anisotropy. First optimal results for problems with varying direction of dominating anisotropy were published by Margenov and Vassilevski in [43] where the study is restricted to the case of orthotropy. Later on, the problem was solved in a very general setting of mesh and coefficient anisotropy, see, e.g., [11, 13].

The above short survey was limited to the case of AMLI methods based on the hierarchical two-level representation of the stiffness matrices where the current Schur complements are approximated by (a matrix polynomial of) the coarser level stiffness matrix. An alternative approach was recently proposed by Kraus in [33] where the approximation of the global Schur complement is based on the assembly of the macroelement ones. The related theory is still not completed. However, the already available robust estimates (see also [9]), and the collected numerical tests, well indicate some attractive advantages of this approach.

The Galerkin finite element method is the most traditional tool for discretizing the elliptic problem (2.1). However, for applications related to highly heterogeneous porous media the mixed finite element methods have proven to be more accurate and locally mass conservative. While applying the mixed FEM, the continuity of the velocity normal to the boundary between two adjacent finite elements could be enforced by Lagrange multipliers. In [3] Arnold and Brezzi have demonstrated that after the elimination of the unknowns representing the pressure and the velocity from the algebraic system, the resulting Schur system for the Lagrange multipliers is equivalent to a discretization of (2.1) by Galerkin method using Crouzeix–Raviart nonconforming linear finite elements. First studies on AMLI preconditioners for Crouzeix–Raviart FEM systems were published by Margenov and Vassilevski in [44]. The problem of robustness was completely solved in [22, 23] (see also [35]), where the obtained uniform estimates fully correspond to the related results for conforming linear FEM. An important point to make is that in the case of nonconforming elements, the finite element spaces corresponding to two successive levels of mesh refinement are not nested in general, which requires the construction of a proper hierarchical two-level basis.

The expanding interest in nonconforming finite elements is additionally motivated by their stable approximation of problems with small parameters, where the standard conforming elements suffer from so-called locking effects. For such problems, the final goal is to get a locking-free solver for systems obtained by locking-free discretization. Such totally locking-free AMLI preconditioners for Crouzeix–Raviart FEM systems of pure displacement almost incompressible elasticity problems are studied in [24, 31].

Novel optimal order AMLI preconditioners for Rannacher–Turek nonconforming FEM systems [28, 29] and discontinuous Galerkin FEM systems [36, 37] were recently published. These studies are addressed to the further development of robust AMLI solvers for ill-conditioned parameter dependent problems.

The paper is organized as follows. The theoretical background of the AMLI methods is briefly presented in Section 3. Next Sections 4 and 5 contain a survey on the robust AMLI algorithms for conforming (Courant) and nonconforming (Crouzeix–Raviart) linear finite elements respectively. All reported results are uniform with respect to both, coefficient and mesh anisotropy. Section 6 is devoted to local Schur complement based multilevel preconditioners. Short concluding remarks are given at the end.

3 Algebraic multilevel iteration methods

The material selected in this section follows the spirit of the AMLI methods as presented originally in [14, 15]. At the same time, some needed generalizations are made to support the cases of nonconforming FEM and/or discontinuous Galerkin discretizations.

3.1 Additive and multiplicative two-level methods: Condition number estimates

Consider now a general matrix A, which is assumed to be symmetric positive definite and partitioned in a two-by-two block form

$$A = \begin{bmatrix} A_{11} & A_{12} \\ A_{21} & A_{22} \end{bmatrix}. \tag{3.1}$$

The coupling of this partitioning is characterized by the corresponding constant $\gamma \in [0,1)$ in the Cauchy–Bunyakowski–Schwarz (CBS) inequality. It measures the strength of the off-diagonal blocks of $A_{12} = A_{21}^T$ in relation to the diagonal blocks and can be simply defined as

$$\gamma^2 = \rho\left(A_{22}^{-1} A_{21} A_{11}^{-1} A_{12}\right). \tag{3.2}$$

The basic two-level preconditioners to A will be defined under the assumptions

$$A_{11} \le C_{11} \le (1+\delta_1)A_{11} \text{ and } A_{22} \le C_{22} \le (1+\delta_2)A_{22}. \tag{3.3}$$

The inequalities (3.3) are in a positive semidefinite sense where C_{ii} are symmetric and positive definite matrices for some positive constants δ_i, $i = 1, 2$.

(a) The symmetric preconditioner of block-diagonal (additive) form is introduced as

$$M_A = \begin{bmatrix} C_{11} & 0 \\ 0 & C_{22} \end{bmatrix}.$$

The following general estimate of the spectral condition number of $M_A^{-1}A$ holds:

$$\varkappa(M_A^{-1}A) \le \frac{1+\gamma}{\Delta_0(1-\gamma)}, \tag{3.4}$$

where

$$\Delta_0 = \frac{4(1+\gamma)}{(1+\delta_2)\left[1+\Delta+\sqrt{(\Delta-1)^2+4\Delta\gamma^2}\right]}, \quad \Delta = \frac{1+\delta_1}{1+\delta_2}.$$

When $C_{ii} = A_{ii}$, $i = 1, 2$, the estimate (3.4) reduces to

$$\varkappa(M_A^{-1}A) \le \frac{1+\gamma}{1-\gamma}. \tag{3.5}$$

(b) The full block-matrix factorization preconditioner (multiplicative, or of block Gauss–Seidel form) is based on the exact block-matrix factorization of A,

$$A = \begin{bmatrix} A_{11} & 0 \\ A_{21} & S \end{bmatrix} \begin{bmatrix} I_1 & A_{11}^{-1}A_{12} \\ 0 & I_2 \end{bmatrix},$$

where $S = A_{22} - A_{21}A_{11}^{-1}A_{12}$ is the exact Schur complement of A. The multiplicative preconditioner is then of the form

$$M_F = \begin{bmatrix} C_{11} & 0 \\ A_{21} & C_{22} \end{bmatrix} \begin{bmatrix} I_1 & C_{11}^{-1}A_{12} \\ 0 & I_2 \end{bmatrix}. \tag{3.6}$$

Then

$$\varkappa(M_F^{-1}A) \le \frac{1}{1-\gamma^2}\left\{1+\frac{1}{2}\left[\delta_1+\delta_2+\sqrt{(\delta_1-\delta_2)^2+4\delta_1\delta_2\gamma^2}\right]\right\}. \tag{3.7}$$

If $C_{ii} = A_{ii}$, $i = 1, 2$, then estimate (3.7) reduces to

$$\varkappa(M_F^{-1}A) \le \frac{1}{1-\gamma^2}. \tag{3.8}$$

Detailed proofs of (3.4), (3.5) and (3.7), and an analysis of other versions of constructing M_A and M_F are found, for instance, in [4].

The classical theory of the optimal order two-level preconditioners for FEM systems was first developed in [10, 19], see also [5]. The crucial question is how to construct a two-by-two splitting of the stiffness matrix such that the related CBS constant is far away from the upper limit of one.

The general framework requires to define two nested finite element spaces $\mathcal{V}_H \subset \mathcal{V}_h$, that correspond to two consecutive (regular) mesh refinements. The well studied case of conforming linear finite elements is the starting point in the theory of two-level and multilevel methods. Let $\mathcal{T}_H$ and $\mathcal{T}_h$ be two successive mesh refinements of the domain Ω, which correspond to $\mathcal{V}_H$ and $\mathcal{V}_h$. Let $\{\phi_H^{(k)}, k = 1, 2, \ldots, N_H\}$ and $\{\phi_h^{(k)}, k = 1, 2, \ldots, N_h\}$ be the standard finite element nodal basis functions. We split the mesh-points $\mathbf{N}_h$ from $\mathcal{T}_h$ into two groups: the first group contains the nodes $\mathbf{N}_H$ from $\mathcal{T}_H$ and the second one consists of the rest, where the latter are the newly added node-points $\mathbf{N}_{h\setminus H}$ from $\mathcal{T}_h \setminus \mathcal{T}_H$. Next we define the so-called hierarchical basis functions

$$\{\widetilde{\phi}_h^{(k)}, k = 1, 2, \ldots, N_h\} = \{\phi_H^{(l)} \text{ on } \mathcal{T}_H\} \cup \{\phi_h^{(m)} \text{ on } \mathcal{T}_h \setminus \mathcal{T}_H\}. \tag{3.9}$$

Let then $\widetilde{A}_h$ be the corresponding hierarchical stiffness matrix. Under the splitting (3.9) both matrices A_h and $\widetilde{A}_h$ admit in a natural way a two-by-two block structure

$$A_h = \begin{bmatrix} A_{11} & A_{12} \\ A_{21} & A_{22} \end{bmatrix} \begin{matrix} \}N_{h\setminus H} \\ \}N_H \end{matrix} \tag{3.10}$$

and

$$\widetilde{A}_h = \begin{bmatrix} \widetilde{A}_{11} & \widetilde{A}_{12} \\ \widetilde{A}_{21} & \widetilde{A}_{22} \end{bmatrix} = \begin{bmatrix} A_{11} & \widetilde{A}_{12} \\ \widetilde{A}_{21} & A_H \end{bmatrix} \begin{matrix} \}N_{h\setminus H} \\ \}N_H \end{matrix} . \tag{3.11}$$

It is well known, that for the considered linear conforming finite elements, the transformation matrix which relates the nodal point vectors for the standard and the hierarchical basis functions has the form $J = \begin{bmatrix} I_1 & J_{12} \\ 0 & I_2 \end{bmatrix}$, i.e.,

$$\mathbf{v} = \begin{bmatrix} \mathbf{v}_1 \\ \mathbf{v}_2 \end{bmatrix} = J \begin{bmatrix} \widetilde{\mathbf{v}}_1 \\ \widetilde{\mathbf{v}}_2 \end{bmatrix}, \qquad \begin{matrix} \mathbf{v}_1 = \widetilde{\mathbf{v}}_1 + J_{12}\widetilde{\mathbf{v}}_2 \\ \mathbf{v}_2 = \widetilde{\mathbf{v}}_2 \end{matrix} .$$

Clearly, the hierarchical stiffness matrix $\widetilde{A}_h$ is more dense than A_h and therefore its action on a vector is computationally more expensive. The transformation matrix J, however, enables us in practical implementations to work with A_h, since $\widetilde{A}_h = J^T A_h J$. It is also important to note that the transformation of the nodal basis stiffness matrix A_h to the hierarchical one $\widetilde{A}_h$ does not change the Schur complements, i.e.,

$$S = A_{22} - A_{21}A_{11}^{-1}A_{12} = \widetilde{A}_{22} - \widetilde{A}_{21}\widetilde{A}_{11}^{-1}\widetilde{A}_{12} = \widetilde{S}. \tag{3.12}$$

As it will be shown in the next section, the introduced hierarchical decomposition allows for uniform estimates of the CBS constant using a simplified local analysis.

3.2 Estimates of the CBS constant: Local analysis

Let $W = V_1 \times V_2$ be a vector splitting, which is consistent with the partitioning (3.1). Let us assume that A is symmetric positive semidefinite and A_{11} is positive definite, and let $W_1 = \{\mathbf{v} = [\mathbf{v}_1^T, \mathbf{0}^T]^T\}$, $W_2 = \{\mathbf{v} = [\mathbf{0}^T, \mathbf{v}_2^T,]^T\}$. The following three lemmas (see, e.g., [4, 27]) form the theoretical background of the CBS constant estimates.

Lemma 3.1

(*a*) $\gamma \le 1$.

(*b*) $\gamma = 1$ *if there exists* $\mathbf{w} = \begin{bmatrix} \mathbf{v}_1 \\ \mathbf{v}_2 \end{bmatrix} \in \ker(A)$ *for which* $\mathbf{v}_2 \notin \ker(A_{22})$.

(*c*) $\gamma < 1$ *if for any* $\mathbf{w} = \begin{bmatrix} \mathbf{v}_1 \\ \mathbf{v}_2 \end{bmatrix} \in \ker(A)$ *it holds that* $\mathbf{v}_2 \in \ker(A_{22})$.

(*d*) *Under the assumption of (c),*

$$\gamma = \sup_{\mathbf{v}_i \in V_i \setminus \ker(A_{ii}),\, i=1,2} \frac{\mathbf{v}_1{}^t A_{12} \mathbf{v}_2}{\{\mathbf{v}_1{}^T A_{11} \mathbf{v}_1 \;\; \mathbf{v}_2{}^T A_{22} \mathbf{v}_2\}^{1/2}}.$$

Lemma 3.2 *Let A be symmetric positive semidefinite matrix satisfying condition (c) from Lemma 3.1. Then for any* $\mathbf{v}_2 \in V_2 \setminus \ker(A_{22})$

$$1 - \gamma^2 \le \frac{\mathbf{v}_2{}^T S \mathbf{v}_2}{\mathbf{v}_2^T A_{22} \mathbf{v}_2} \le 1,$$

where the left-hand side inequality is sharp and the right-hand side inequality is sharp if $\ker(A_{12}) \ne \{0\}$.

Now, we are ready to show how the CBS constant can be estimated locally (see also [37]). Let us assume that

$$A = \sum_{E \in \mathcal{E}} A_E, \qquad \mathbf{v} = \sum_{E \in \mathcal{E}} \mathbf{v}_E, \tag{3.13}$$

where A_E are symmetric positive semidefinite local matrices, $\mathcal{E}$ is some index set, and the summation is understood as assembling. The global splitting naturally induces the block two-by-two representation of the local matrix A_E and the related vector $\mathbf{v}_E$, namely,

$$A_E = \begin{bmatrix} A_{E:11} & A_{E:12} \\ A_{E:21} & A_{E:22} \end{bmatrix}, \quad \mathbf{v}_E = \begin{bmatrix} \mathbf{v}_{E:1} \\ \mathbf{v}_{E:2} \end{bmatrix}. \tag{3.14}$$

Lemma 3.3 *Let the local matrix A_E, $E \in \mathcal{E}$, satisfy the condition (c) from Lemma 3.1. Then*

$$\gamma \le \max_{E \in \mathcal{E}} \gamma_E \tag{3.15}$$

where γ_E stands for the local CBS constant corresponding to A_E.

In the case of conforming linear finite elements, the hierarchical two-level splitting (3.11) fully satisfies the conditions of Lemma 3.3. The index set $\mathcal{E}$ consists of the standard macroelements $E \in \mathcal{T}_h$, and A_E are the macroelement stiffness matrices. Then, Lemma 3.2 implies the following simple rule to compute the local CBS constant:

$$\gamma_E^2 = 1 - \mu_1, \tag{3.16}$$

where μ_1 is the minimal eigenvalue of the generalized eigenproblem

$$\widetilde{S}_E \mathbf{v}_{E:2} = \mu A_e \mathbf{v}_{E:2}, \qquad \mathbf{v}_{E:2} \neq \text{const}. \tag{3.17}$$

Here, the macroelement $E \in \mathcal{T}_h$ is a refinement of some coarser grid element $e \in \mathcal{T}_H$.

In the case of nonconforming finite elements, the local analysis follows the same scheme, where the definition of a two-level hierarchical basis is a key question (see, e.g., [22, 28, 42]). The techniques in the recently studied cases of discontinuous Galerkin methods (see [36, 37]) are more specific. The local matrices there can be associated with the macroelement edges (faces). How to split the contributions from the macroelement matrices to the edge ones is a new important question. These last examples well demonstrate the potential for novel applications provided by the generalized setting of Lemma 3.3. What is important to note, is the leading general principle of local analysis.

3.3 From two-level to multilevel: Introduction to AMLI methods

Let $A_h \mathbf{u}_h = \mathbf{f}_h$ be the algebraic formulation of a given FEM elliptic problem to be solved by the PCG method. Consider a sequence of nested meshes (triangulations) $\mathcal{T}_0 \subset \mathcal{T}_1 \subset \cdots \subset \mathcal{T}_\ell$ of the domain of definition Ω, their corresponding sets of nodes $\mathbf{N}_0 \subset \mathbf{N}_1 \subset \cdots \subset \mathbf{N}_\ell$ with cardinalities $N_0 < N_1 < \cdots < N_\ell$ and the finite element stiffness matrices associated with each triangulation, $A^{(0)}, A^{(1)}, \ldots, A^{(\ell)}$, where it is assumed that $A_h = A^{(\ell)}$.

The multilevel methods have evolved from two-level methods. The straightforward recursive extension leads to the class of hierarchical basis (HB) methods which are almost optimal (logarithmic growth of the number of iterations) for certain 2D problems only [20, 38]. In the general case, the condition number of the HB methods grows exponentially with the number of levels ℓ. Therefore, in order to obtain multilevel preconditioners of both additive or multiplicative type, which result in an optimal convergence rate, i.e.,

$$\varkappa({M^{(\ell)}}^{-1} A^{(\ell)}) = O(1),$$

and optimal computational complexity (linearly proportional to the number of degrees of freedom at the finest discretization level), HB preconditioners are combined with various types of stabilization techniques.

One particular purely algebraic stabilization technique is the so-called Algebraic Multilevel Iteration (AMLI) method, where a specially constructed matrix polynomial P_{β_k} of degree β_k is used on some (all) levels $k = 1, \ldots, \ell$. The AMLI methods are originally introduced and studied in a multiplicative form, see [14, 15].

Starting from the coarsest mesh (level 0) with $M_F^{(0)} = A^{(0)}$, the basic idea is to apply the two-level preconditioner (3.6) recursively at all levels $k = 1, 2, \ldots, \ell$ of mesh refinement, i.e.,

$$M_F^{(k)} = \begin{bmatrix} C_{11}^{(k)} & 0 \\ \widetilde{A}_{21}^{(k)} & C_{22}^{(k)} \end{bmatrix} \begin{bmatrix} I & C_{11}^{(k)^{-1}} \widetilde{A}_{12}^{(k)} \\ 0 & I \end{bmatrix}. \tag{3.18}$$

Here $C_{11}^{(k)}$ is some preconditioner for the pivot block of the (hierarchical) stiffness matrix

$$\widetilde{A}^{(k)} = \begin{bmatrix} \widetilde{A}_{11}^{(k)} & \widetilde{A}_{12}^{(k)} \\ \widetilde{A}_{21}^{(k)} & \widetilde{A}_{22}^{(k)} \end{bmatrix}$$

at level k and the matrix $C_{22}^{(k)}$ is implicitly defined by the equation

$$C_{22}^{(k)^{-1}} = \left[I - P_{\beta_k} \left(M_F^{(k-1)^{-1}} \widetilde{A}^{(k-1)} \right) \right] \widetilde{A}^{(k-1)^{-1}}, \tag{3.19}$$

where $M_F^{(k-1)}$ and $\widetilde{A}^{(k-1)}$ denote the multiplicative preconditioner and the stiffness matrix at level $(k-1)$, respectively, and $\widetilde{A}^{(0)} = A^{(0)}$ by definition. Then, a properly shifted Chebyshev polynomial P_{β_k} of degree $\beta_k \in \{2, 3\}$, satisfying the conditions

$$0 \le P_{\beta_k}(t) < 1, \quad 0 < t \le 1, \quad P_{\beta_k}(0) = 1,$$

can be used in order to stabilize the condition number of the linear AMLI preconditioner [14, 15].

The main result from this analysis is that the AMLI preconditioner has optimal computational complexity, if $\beta_k = \beta$ and

$$\frac{1}{\sqrt{1-\gamma^2}} < \beta < \rho, \tag{3.20}$$

where $\rho \approx \dfrac{N_{k+1}}{N_k}$ is the reduction factor of the number of degrees of freedom.

The additive AMLI is straightforwardly obtained if

$$M_A^{(k)} = \begin{bmatrix} C_{11}^{(k)} & 0 \\ 0 & C_{22}^{(k)} \end{bmatrix} \tag{3.21}$$

is used instead of (3.18). Then, the optimality condition (3.20) is modified to

$$\sqrt{\frac{1+\gamma}{1-\gamma}} < \beta < \rho. \tag{3.22}$$

As one can see, the condition (3.22) is rather restrictive which is among the motivations for studies of AMLI methods with varying polynomial degrees β_k, see, e.g., [5, 16, 17, 50, 51].

It is important to note, that the CBS constant is not only used to analyze the related two-level and multilevel preconditioners. It is also involved in the construction of the acceleration matrix polynomial P_{β_k}. In other words, the smaller γ means the faster the PCG method convergence in a very general setting.

Now we are ready to formulate the problem of robust AMLI preconditioning of anisotropic elliptic problems. What we need, in fact, is: a) to have a proper estimate of the CBS constant which is uniform with respect to the anisotropy and b) to construct a robust preconditioner $C_{11}^{(k)}$ to the pivot block $\widetilde{A}_{11}^{(k)}$.

Remark 3.4 In many cases, including the classical formulation of the AMLI method for conforming linear elements, $\widetilde{A}_{11}^{(k)} = A_{11}^{(k)}$ and $\widetilde{A}_{22}^{(k)} = A^{(k-1)}$. However, these equalities can be violated in general.

4 Robust AMLI algorithms: Conforming linear finite elements

4.1 Some basic relations

A general important technique for finite element methods is to transform the arising integral over an arbitrary element to a standard reference element. We will use here the transformation method for a planar domain using triangular finite elements.

First we show that the analysis for an arbitrary triangle (e) with coordinates (x_i, y_i), $i = 1, 2, 3$, can be done on the reference triangle $(\widetilde{e})$, with coordinates $(0,0)$, $(1,0)$ and $(0,1)$. Transforming the finite element functions between these triangles, the element bilinear form $\mathcal{A}_e(\cdot\,,\cdot)$ becomes (see, e.g., [5, 11]):

$$\begin{aligned}
\mathcal{A}_e(u,v) &= \mathcal{A}_{\widetilde{e}}(\widetilde{u},\widetilde{v}) \\
&= \int_{\widetilde{e}} \left[\frac{\partial \widetilde{u}}{\partial \widetilde{x}}, \frac{\partial \widetilde{u}}{\partial \widetilde{y}}\right] \left[\begin{array}{cc} (x_2-x_1) & (y_2-y_1) \\ (x_3-x_1) & (y_3-y_1) \end{array}\right]^{-1} \left[\begin{array}{cc} a_{11} & a_{12} \\ a_{21} & a_{22} \end{array}\right] \\
&\quad \times \left[\begin{array}{cc} (x_2-x_1) & (x_3-x_1) \\ (y_2-y_1) & (y_3-y_1) \end{array}\right]^{-1} \left[\frac{\partial \widetilde{v}}{\partial \widetilde{x}}, \frac{\partial \widetilde{v}}{\partial \widetilde{y}}\right]^T \left|\frac{\partial(x,y)}{\partial(\widetilde{x},\widetilde{y})}\right| d\widetilde{e}
\end{aligned}$$

where $0 < \widetilde{x}, \widetilde{y} < 1$, i.e., it takes the form

$$\mathcal{A}_{\widetilde{e}}(\widetilde{u},\widetilde{v}) = \int_{\widetilde{e}} \sum_{i,j} \widetilde{a}_{ij} \frac{\partial \widetilde{u}}{\partial \widetilde{x}_i} \frac{\partial \widetilde{v}}{\partial \widetilde{x}_j}\, d\widetilde{e}, \tag{4.1}$$

where the coefficients $\widetilde{a}_{ij}$ depend on both the coordinates in e (or, equivalently, the angles in e) and the coefficients a_{ij} in the differential operator. A similar form holds in case of 3D problems.

We conclude that it suffices for the analysis of local uniform bounds for the finite element method to consider the (macro)element stiffness matrices for the reference triangle and arbitrary coefficients $[a_{ij}]$, or alternatively, for the operator $-\Delta$ (where $[a_{ij}] = \left[\begin{array}{cc} 1 & 0 \\ 0 & 1 \end{array}\right]$) and an arbitrary triangle e.

Following the standard FEM assembling procedure we can write the global stiffness matrix A in the form

$$A = \sum_{e \in \mathcal{T}_k} L_e{}^T A_e L_e, \tag{4.2}$$

where A_e is the element stiffness matrix, L_e stands for the restriction mapping of the global vector of unknowns to the local one corresponding to element $e \in \mathcal{T}_k$.

Consider now the case of Laplace operator and an arbitrary shaped linear triangular finite element. Then, the element stiffness matrix A_e can be written in the form [7]

$$A_e = \frac{1}{2}\begin{bmatrix} b+c & -c & -b \\ -c & a+c & -a \\ -b & -a & a+b \end{bmatrix}, \tag{4.3}$$

where a, b, and c equal the cotans of the angles in $e \in \mathcal{T}_h$.

In the local analysis, without loss of generality we will assume that $|a| \leq b \leq c$. This can be concluded from the following lemma, see [11].

Lemma 4.1 *Let $\theta_1, \theta_2, \theta_3$ be the angles in an arbitrary triangle. Then with $a = \cot\theta_1$, $b = \cot\theta_2$, $c = \cot\theta_3$ it holds*

(i) $a = (1 - bc)/(b + c)$.

(ii) If $\theta_1 \geq \theta_2 \geq \theta_3$, then $|a| \leq b \leq c$.

(iii) $a + b > 0$.

The next representation of the element stiffness matrix simply follows from Lemma 4.1.

$$A_e = \frac{c}{2}\begin{bmatrix} \beta+1 & -1 & -\beta \\ -1 & \alpha+1 & -\alpha \\ -\beta & -\alpha & \alpha+\beta \end{bmatrix}, \tag{4.4}$$

$\alpha = a/c$, $\beta = b/c$, and $(\alpha, \beta) \in D$, where

$$D = \{(\alpha,\beta) \in \mathbb{R}^2 : -\frac{1}{2} < \alpha \leq 1, \max\{-\frac{\alpha}{\alpha+1}, |\alpha|\} \leq \beta \leq 1\}. \tag{4.5}$$

The next purely algebraic inequality plays an important role in the following considerations of robust AMLI algorithms for linear finite elements.

Lemma 4.2 *For all $(\alpha, \beta) \in D$ there holds the inequality*

$$\frac{\alpha\beta + \alpha + \beta + 1}{(\alpha+\beta+1)(\alpha+\beta+2)} > \frac{4}{15}. \tag{4.6}$$

For more details, including the proofs, see [11].

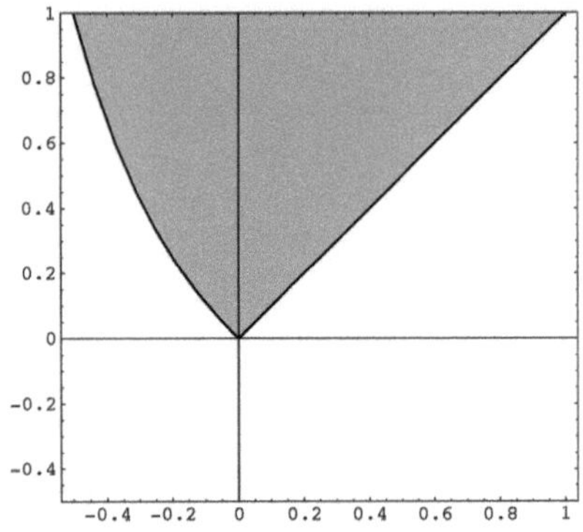

Figure 4.1 Domain of the parameters (α, β).

4.2 Uniform estimates of the constant in the strengthened CBS inequality

Consider two consecutive meshes $\mathcal{T}_k \subset \mathcal{T}_{k+1}$. The following uniform refinement procedure will be set as a default assumption. The current coarse triangle $e \in \mathcal{T}_k$ is subdivided in four congruent triangles by joining the mid-edge nodes to get the macroelement $E \in \mathcal{T}_{k+1}$. The related macroelement stiffness matrix (3.14) consists of blocks which are 3×3 matrices and the local eigenproblem (3.17) to compute γ_E has a reduced dimension of 2×2.

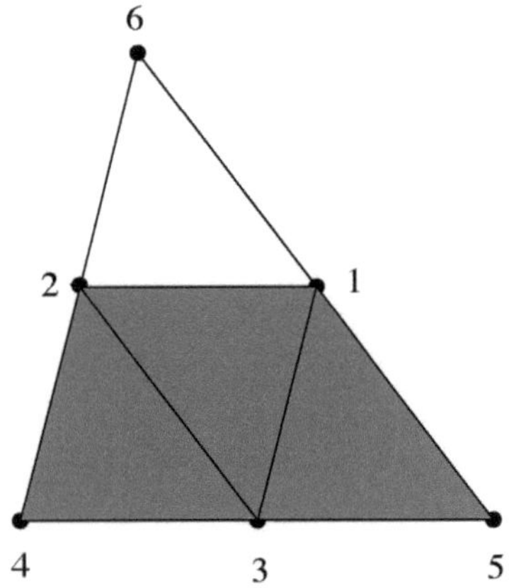

Figure 4.2 Support for the piecewise linear basis function ϕ_3.

In the so arising six node-points of the macroelement we can also use hierarchical basis functions, where we keep the linear basis functions in the vertex nodes and add piecewise quadratic basis functions in the mid-edge nodes with support on the whole triangle. This alternative approach is called the p-version ($p = 2$). Let us denote by $\widehat{\gamma}_E$ the corresponding CBS constant.

The following relation between γ_E and $\widehat{\gamma}_E$ holds.

Theorem 4.3 ([39, 5]) *Let us consider a piecewise Laplacian elliptic problem on an arbitrary finite element triangular mesh $\mathcal{T}_k$, and let each element from $\mathcal{T}_k$ be refined into four congruent elements to get $\mathcal{T}_{k+1}$. Then*

$$\widehat{\gamma}_E^2 = \frac{4}{3}\gamma_E^2, \tag{4.7}$$

where $\widehat{\gamma}_E$, γ_E are the local CBS constants for the hierarchical piecewise quadratic and the piecewise linear finite elements, respectively.

Corollary 4.4 *The local estimate*

$$\gamma_E^2 < \frac{3}{4} \tag{4.8}$$

holds uniformly with respect to the mesh anisotropy.

The next fundamental result follows directly from the local estimate (4.8), the equivalence relation (4.1) and Lemma 3.3.

Theorem 4.5 *Consider the problem (2.3) discretized by conforming linear finite elements, where the coarsest grid $\mathcal{T}_0$ is aligned with the discontinuities of the coefficient $\mathrm{a}(e)$, $e \in \mathcal{T}_0$. Let us assume also that $\mathcal{T}_k \subset \mathcal{T}_{k+1}$ are two consecutive meshes where each element from $\mathcal{T}_k$ is refined into four congruent elements to get $\mathcal{T}_{k+1}$. Then, the estimate*

$$\gamma^2 < \frac{3}{4} \tag{4.9}$$

of the CBS constant holds uniformly with respect to the coefficient jumps, mesh or/and coefficient anisotropy, and the refinement level k.

We present here some further achievements in the robust estimates of the CBS constant. The estimate (4.9) is generalized in [8] to the case when each element from the current coarser mesh is subdivided into m^2, $m \geq 2$, congruent elements using a uniform refinement (with a multiplicity of m). Then, the following universal estimate holds

$$\gamma^2 < \frac{m^2-1}{m^2}, \qquad m = 2, 3, \ldots . \tag{4.10}$$

A similar result is obtained for the case of 3D linear finite elements. Let us assume that the current element from the coarser mesh is divided into m^3 congruent tetrahedrons. Then, the uniform estimate for $m = 2$ reads as

$$\gamma^2 < \frac{9}{10},$$

which is extended to the universal estimate

$$\gamma^2 < \frac{(m^2-1)(m^2+2)}{m^2(m^2+1)}, \qquad m = 2, 3, \ldots, \tag{4.11}$$

see [8, 21] and the references therein.

An important generalization of the AMLI theory concerns the Lamé system of linear elasticity. Here, the coefficient anisotropy is introduced by the Poisson ratio. The general scheme of analyzing the CBS constant is applied. Let us note, that the obtained estimates are the same as for the scalar elliptic problems, being robust with respect to the mesh anisotropy and the (almost incompressible) Poisson ratio, see, e.g., [1, 2, 40, 41].

4.3 Additive preconditioning of the pivot block

When applicable, we will skip the superscripts of the pivot block and its approximation. Here, we will write A_{11}, C_{11}, instead of $A_{11}^{(k+1)}$, $C_{11}^{(k+1)}$. The construction and the analysis of the preconditioners C_{11} are based on a macroelement-by-macroelement assembling procedure. Following (4.2), we write A_{11} in the form

$$A_{11} = \sum_{E \in \mathcal{T}_{k+1}} L_E{}^T A_{E:11} L_E. \tag{4.12}$$

Now, we use the representation (4.3) of the element stiffness matrix, and get $A_{E:11}$ in the form

$$A_{E:11} = r_T \begin{bmatrix} a_T + b_T + c_T & -c_T & -b_T \\ -c_T & a_T + b_T + c_T & -a_T \\ -b_T & -a_T & a_T + b_T + c_T \end{bmatrix},$$

where: a) the factor r_T depends on the shape of $T \in \mathcal{T}_0$ and on the related coefficients matrix $\mathbf{a}(\mathbf{e})$ which is one and the same for all elements $e \in \mathcal{T}_k \cap T$, and b) $|a_T| \le b_T \le c_T$ equal the cotans of the angles in some triangle $\widetilde{T}$ associated with T. Following the scaled representation (4.4) we get

$$A_{E:11} = r_T c_T \begin{bmatrix} \alpha + \beta + 1 & -1 & -\beta \\ -1 & \alpha + \beta + 1 & -\alpha \\ -\beta & -\alpha & \alpha + \beta + 1 \end{bmatrix}, \tag{4.13}$$

where $\alpha = a_T/c_T$, $\beta = b_T/c_T$, and $(\alpha, \beta) \in D$ as defined in (4.5). Then, the additive preconditioner of A_{11} is defined as follows

$$C_{11} = \sum_{E \in \mathcal{T}_{k+1}} L_E{}^T C_{E:11} L_E, \tag{4.14}$$

where

$$C_{E:11} = 2 r_T c_T \begin{bmatrix} \alpha + \beta + 1 & -1 & 0 \\ -1 & \alpha + \beta + 1 & 0 \\ 0 & 0 & \alpha + \beta + 1 \end{bmatrix}. \tag{4.15}$$

As one can see, the local matrix $C_{E:11}$ is obtained by preserving only the *strongest* off-diagonal entries.

Theorem 4.6 ([11, 13]) *The additive preconditioner of A_{11} has a relative condition number uniformly bounded by*

$$\kappa\left({C_{11}}^{-1}A_{11}\right) < \frac{1}{4}(11+\sqrt{105}) \approx 5.31. \tag{4.16}$$

This condition number estimate holds independently on shape and size of each element and on the coefficient matrix $\mathrm{a}(e)$ *of the FEM problem.*

This additive preconditioner was first introduced and analyzed in [13]. A new proof, based on the algebraic inequality from Lemma 4.2, has been presented in [11].

4.4 Multiplicative preconditioning of the pivot block

Let us partition the nodes corresponding to the block A_{11} into two groups where the first one contains the centers of parallelogram superelements Q (see Figure 4.3) which are weakly connected in the sense of the relations between the coefficients $|a_T| \le b_T \le c_T$. It is important to note that the parallelograms $Q \subset T \in \mathcal{T}_0$, i.e., it is not allowed to be composed of triangles of neighbor elements from the coarsest triangulation $\mathcal{T}_0$. With respect to this partitioning, A_{11} admits the following two-by-two block-factored form

$$A_{11} = \begin{bmatrix} D_{11} & F_{11} \\ {F_{11}}^T & E_{11} \end{bmatrix} = \begin{bmatrix} D_{11} & 0 \\ {F_{11}}^T & S_{11} \end{bmatrix} \begin{bmatrix} I & {D_{11}}^{-1}F_{11} \\ 0 & I \end{bmatrix}, \tag{4.17}$$

where S_{11} stands for the related Schur complement. We define now C_{11} as the symmetric block Gauss–Seidel preconditioner of A_{11}, i.e.,

$$C_{11} = \begin{bmatrix} D_{11} & 0 \\ {F_{11}}^T & E_{11} \end{bmatrix} \begin{bmatrix} I & {D_{11}}^{-1}F_{11} \\ 0 & I \end{bmatrix}. \tag{4.18}$$

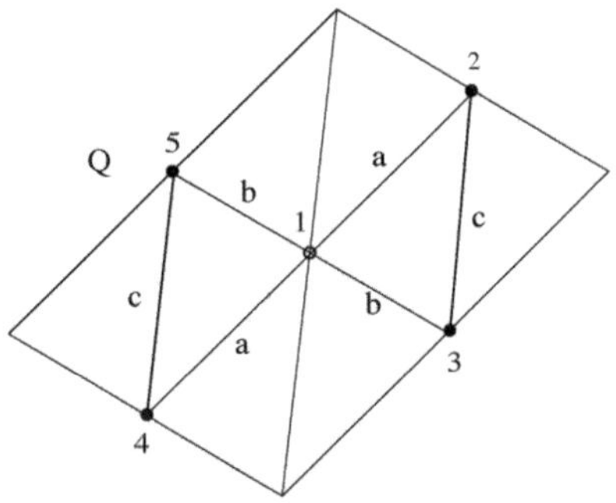

Figure 4.3 Multiplicative preconditioner. Block partitioning of the nodes of the superelement Q.

Since D_{11} is a diagonal matrix it follows that the Schur complement S_{11} can be assembled from the corresponding superelement Schur complements $S_{Q:11} = E_{Q:11} - {F_{Q:11}}^T {D_{Q:11}}^{-1} F_{Q:11}$.

A similar construction was first introduced in [43] for the particular case of triangulations $\mathcal{T}_0$ consisting of right triangles with legs parallel to the coordinate axes and a diagonal coefficient matrix $\mathbf{a}(e)$. The general case is studied in [11]. As for the additive algorithm, the local analysis is based on the algebraic inequality from Lemma 4.2.

Theorem 4.7 ([11]) *The multiplicative preconditioner of A_{11} has a relative condition number uniformly bounded by*

$$\kappa \left({C_{11}}^{-1} A_{11} \right) < \frac{15}{8} = 1.875. \tag{4.19}$$

As one can observe, this estimate considerably improves the result (4.16) for the additive preconditioner.

4.5 Optimal complexity solution algorithms for systems with C_{11}

The ability of solving systems with the considered preconditioning matrices C_{11} efficiently is determined by their connectivity pattern, assuming that rapid solution methods are used at this step of the algorithm.

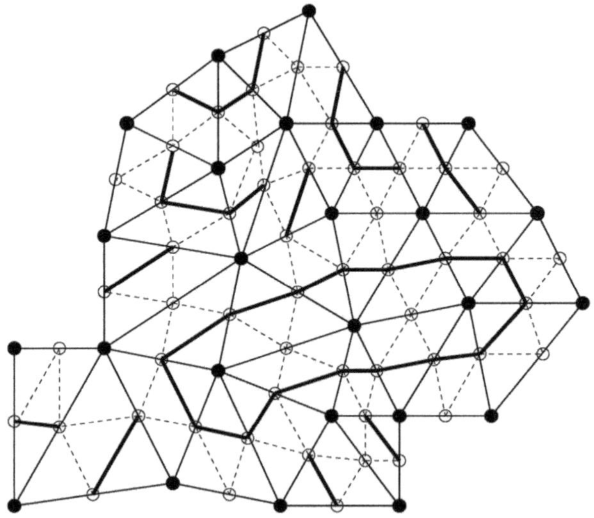

Figure 4.4 An example of the connectivity pattern for the additive preconditioner.

Consider first the additive preconditioner. Due to the form of the corresponding element matrices $C_{E:11}$, the connectivity pattern of the preconditioner C_{11} is such as shown in Figure 4.4. In some more details, the coupled nodes form either a single point, a polyline or a polygon. Therefore, there are no cross-points. This means, that the matrix C_{11} has a generalized tridiagonal structure (see [13] and also [47]), that is, the solution of linear systems with C_{11} has a computational cost which is proportional to the related problem size.

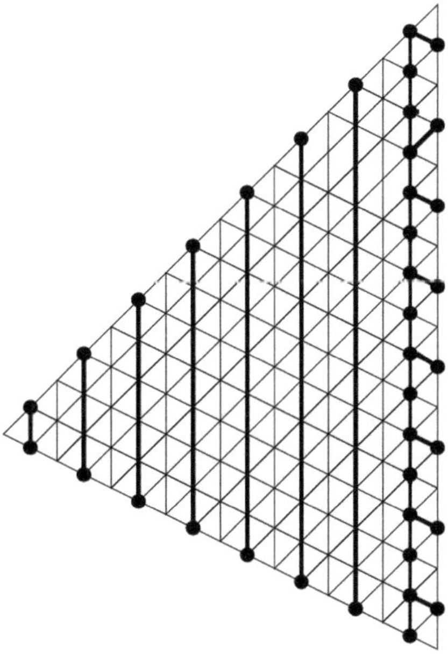

Figure 4.5 Connectivity pattern of the reduced problem corresponding to $T \in \mathcal{T}_0$.

The construction of an efficient solver for C_{11} in the multiplicative case is also almost trivial, see [11]. For this purpose, one uses the *nested dissection* (ND) algorithm which is known as a rapid solution method for sparse linear systems. If the graph representing the connectivity of the matrix is planar, then the computational complexity of the ND algorithm is $\mathcal{N}_{\mathrm{ND}} = O(n^{3/2})$, where n stands for the size of the problem.

Now, let us denote by $\overline{N}$ the size of C_{11}. At the first step of the algorithm we eliminate the unknowns corresponding to the nodes from the interior of the triangles from the coarsest mesh $\mathcal{T}_0$. This is performed by solving tridiagonal systems (see Figure 4.5) and therefore requires a number of arithmetic operations proportional to $\overline{N}$, i.e., $\mathcal{N}_1^M = O(\overline{N})$. The obtained reduced problem has a planar graph of connectivity. Its size is $O(\overline{N}^{1/2})$, and therefore, the solution of the reduced problem by the ND algorithm has a computational cost $\mathcal{N}_2^M = O(\overline{N}^{3/4})$, and the total complexity is $\mathcal{N}^M = \mathcal{N}_1^M + \mathcal{N}_2^M = O(\overline{N})$. Therefore, the multiplicative preconditioner has an optimal computational complexity with respect to both problem and discretization parameters.

Let us summarize the outcome of the results presented in this section. The estimate $\gamma^2 < 3/4$, in a combination with the considered preconditioners of the pivot block $A_{11}^{(k+1)}$, ensures the optimal complexity of the multiplicative AMLI algorithm (3.18) with polynomial degree $\beta \in \{2, 3\}$. All related estimates are robust with respect to both, mesh and/or coefficient anisotropy. The universal robust estimates (4.10) and (4.11) of the CBS constant for uniform refinement with multiplicity $m \geq 2$ provide new opportunities for balanced algorithms with varying polynomial degree β_k. In this respect, it is still an open question, how to construct robust preconditioners of the pivot block $A_{11}^{(k+1)}$, when $m > 2$.

5 Robust AMLI algorithms: Nonconforming linear finite elements

For the nonconforming Crouzeix–Raviart finite element, where the nodal basis functions are defined at the midpoints along the edges of the triangle rather than at its vertices (cf. Figure 5.1), the natural vector spaces $\mathcal{V}_H(E) := \text{span}\,\{\phi_I, \phi_{II}, \phi_{III}\}$ and $\mathcal{V}_h(E) := \text{span}\,\{\phi_i\}_{i=1}^{9}$ (cf. the macro-element in Figure 5.1(b)) are no longer nested, i.e., $\mathcal{V}_H(E) \nsubseteq \mathcal{V}_h(E)$. This makes the direct construction with $\mathcal{V}_2(E) := \mathcal{V}_H(E)$, as used for conforming elements, impossible. Consequently, the hierarchical basis functions have to be chosen in a way that for the resulting subspaces $\mathcal{V}_1(E)$ and $\mathcal{V}_2(E)$, and hence, for the whole finite element space $\mathcal{V}$, the direct sum condition

$$\mathcal{V} = \mathcal{V}_1 \oplus \mathcal{V}_2 \tag{5.1}$$

is satisfied again.

5.1 Crouzeix–Raviart finite elements

A simple computation shows that the standard nodal basis element stiffness matrix for a nonconforming Crouzeix–Raviart (CR) linear finite element A_e^{CR} coincides with that of the corresponding conforming linear element up to a factor 4, i.e.,

$$A_e^{\text{CR}} = 2 \begin{bmatrix} b+c & -c & -b \\ -c & a+c & -a \\ -b & -a & a+b \end{bmatrix}, \tag{5.2}$$

cf. (4.3). The construction of the hierarchical stiffness matrix at macro-element level starts with the assembly of four such matrices according to the numbering of the nodal points, as shown in Figure 5.1(b). It further utilizes a transformation, which is based on a proper decomposition of the vector space $\mathcal{V}(E) = \mathcal{V}_h(E)$, which is associated with the fine-grid basis functions related to this macro-element E. This will be addressed in the following.

We consider three different splittings, each of which makes use of half-difference and half-sum basis functions in order to guarantee the condition

$$\mathcal{V}(E) := \text{span}\,\{\Phi_E\} = \mathcal{V}_1(E) \oplus \mathcal{V}_2(E). \tag{5.3}$$

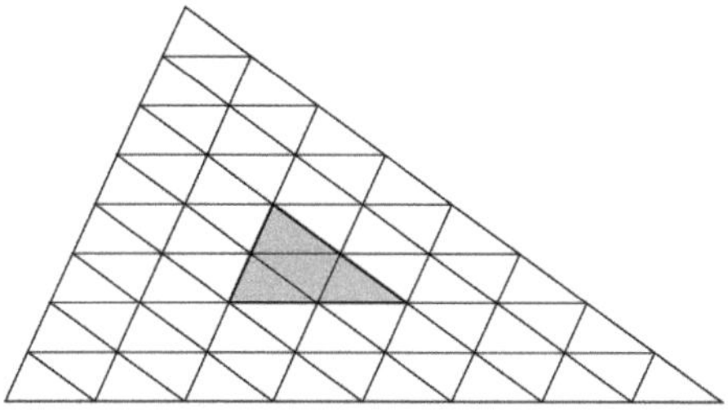

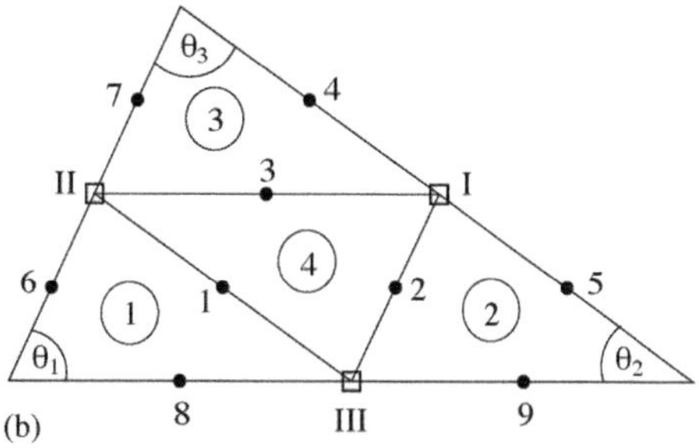

Figure 5.1 Crouzeix–Raviart finite element. (a) Discretization. (b) Macro-element.

Here $\Phi_E := \{\phi^{(i)}\}_{i=1}^{9}$ denotes the set of the "midpoint" basis functions of the four congruent elements in the macro-element E, as depicted in Figure 5.1(b). The splitting of $\mathcal{V}(E)$ can be defined in the general form

$$\begin{aligned}\mathcal{V}_1(E) &:= \operatorname{span}\{\phi_1,\, \phi_2,\, \phi_3,\, \phi_1^D + \phi_4 - \phi_5,\, \phi_2^D + \phi_6 - \phi_7,\, \phi_3^D + \phi_8 - \phi_9\}\,,\\ \mathcal{V}_2(E) &:= \operatorname{span}\{\phi_1^C + \phi_4 + \phi_5,\, \phi_2^C + \phi_6 + \phi_7,\, \phi_3^C + \phi_8 + \phi_9\}\,,\end{aligned} \tag{5.4}$$

where $\phi_i^D := \sum_k d_{ik}\phi_k$ and $\phi_i^C := \sum_k c_{ik}\phi_k$ with $i,k \in \{1,2,3\}$. The transformation matrix corresponding to this general splitting is given by

$$J_E = J_E(C,D) = \begin{bmatrix} I_3 & D & C \\ 0 & J_- & J_+ \end{bmatrix} \quad (\in \mathbb{R}^{9\times 9}), \tag{5.5}$$

where I_3 denotes the 3×3 identity matrix, and C and D are 3×3 matrices whose entries c_{ij} respectively d_{ij} will be specified later. The 3×6 matrices

$$J_- := \frac{1}{2}\begin{bmatrix} 1 & -1 & & & & \\ & & 1 & -1 & & \\ & & & & 1 & -1 \end{bmatrix}^T \quad \text{and} \quad J_+ := \frac{1}{2}\begin{bmatrix} 1 & 1 & & & & \\ & & 1 & 1 & & \\ & & & & 1 & 1 \end{bmatrix}^T \tag{5.6}$$

introduce the so-called half-difference and half-sum basis functions associated with the sides of the macro-element triangle. The matrix J_E transforms the vector of the macro-element basis functions $\phi_E := (\phi^{(i)})_{i=1}^{9}$ to the hierarchical basis vector $\tilde{\phi}_E := (\tilde{\phi}^{(i)})_{i=1}^{9} = J_E^T \phi_E$ and the hierarchical stiffness matrix at macro-element level is obtained as

$$\tilde{A}_E = J_E^T A_E J_E = \begin{bmatrix} \tilde{A}_{E:11} & \tilde{A}_{E:12} \\ \tilde{A}_{E:12}^T & \tilde{A}_{E:22} \end{bmatrix} \begin{matrix} \}\in \mathcal{V}_1(E) \\ \}\in \mathcal{V}_2(E) \end{matrix}\;. \tag{5.7}$$

The related global stiffness matrix is obtained as $\tilde{A}_h := \sum_{E\in\mathcal{T}_H} \tilde{A}_E$.

The transformation matrix $J = J(C, D)$ such that $\tilde{\phi} = J^T \phi$ is then used for the transformation of the global matrix A_h to its hierarchical form $\tilde{A}_h = J^T A_h J$, and (by a proper permutation of rows and columns) the latter admits the 3×3-block representation

$$\tilde{A}_h = \begin{bmatrix} \tilde{A}_{11} & \tilde{A}_{12} & \tilde{A}_{13} \\ \tilde{A}_{12}^T & \tilde{A}_{22} & \tilde{A}_{23} \\ \tilde{A}_{13}^T & \tilde{A}_{23}^T & \tilde{A}_{33} \end{bmatrix} \begin{matrix} \left.\vphantom{\begin{matrix} a \\ b \end{matrix}}\right\} \in \mathcal{V}_1 \\ \} \in \mathcal{V}_2 \end{matrix} \tag{5.8}$$

according to the interior, half-difference and half-sum basis functions, which are associated with the locally introduced splitting (5.4). The upper-left 2×2 block is thus related to the vector space $\mathcal{V}_1$, while the lower-right block $\tilde{A}_{33}$ relates to $\mathcal{V}_2$. Note that due to the structure of J_E, the relation $\tilde{A}_{11} = A_{11}$ still holds.

5.2 Two-level splittings: First reduce (FR) and decomposition with differences and aggregates (DA)

The so-called "First Reduce" (FR) splitting [22, 35, 44] can be described as a combination of a basis transformation involving the matrix

$$J_{\pm} := \begin{bmatrix} I & 0 & 0 \\ 0 & J_- & J_+ \end{bmatrix}, \tag{5.9}$$

where J_- and J_+ are given by (5.6), and a reduction step. In the latter the degrees of freedom corresponding to the so-called interior basis functions are eliminated. If

$$\bar{A}_h := J_{\pm}^T A_h J_{\pm} = \begin{bmatrix} \bar{A}_{11} & \bar{A}_{12} & \bar{A}_{13} \\ \bar{A}_{12}^T & \bar{A}_{22} & \bar{A}_{23} \\ \bar{A}_{13}^T & \bar{A}_{23}^T & \bar{A}_{33} \end{bmatrix} \begin{matrix} \left.\vphantom{\begin{matrix} a \\ b \end{matrix}}\right\} \in \mathcal{V}_1 \\ \} \in \mathcal{V}_2 \end{matrix} \tag{5.10}$$

denotes the matrix after the transformation step, then the unknowns related to block $\bar{A}_{11}$ $(= A_{11})$ are first eliminated and the system with $\bar{A}_h$ is reduced to a system with its Schur complement

$$B = \begin{bmatrix} \bar{A}_{22} & \bar{A}_{23} \\ \bar{A}_{23}^T & \bar{A}_{33} \end{bmatrix} - \begin{bmatrix} \bar{A}_{12}^T \\ \bar{A}_{13}^T \end{bmatrix} A_{11}^{-1} \begin{bmatrix} \bar{A}_{12} & \bar{A}_{13} \end{bmatrix}. \tag{5.11}$$

Note that the exact computation of B is cheap since $\bar{A}_{11} = A_{11}$ is block-diagonal with blocks of size 3×3. We consider then the partitioning of the matrix B into a 2×2 block form

$$B = \begin{bmatrix} B_{22} & B_{23} \\ B_{23}^T & B_{33} \end{bmatrix}, \tag{5.12}$$

where B_{22} corresponds to the half-difference basis functions, while the block B_{33} is related to the subspace $\mathcal{V}_2$ and is thus associated with the coarse grid. This partitioning of B is now used for the construction of two-level preconditioners. As shown in the following remark, the FR splitting also fits into the framework of our general splitting.

Remark 5.1 (FR splitting) The reduction step in the FR approach can (due to the symmetry of $\bar{A}_h$ and the fact that $\bar{A}_{11} = A_{11}$) be written in multiplicative form

$$J_B^T \bar{A}_h J_B = \begin{bmatrix} A_{11} & 0 \\ 0 & B \end{bmatrix} \quad \text{with} \quad J_B := \begin{bmatrix} I & -A_{11}^{-1}\bar{A}_{12} & -A_{11}^{-1}\bar{A}_{13} \\ 0 & I & 0 \\ 0 & 0 & I \end{bmatrix}, \tag{5.13}$$

where J_B denotes the matrix corresponding to the reduction step. Combining the transformation (5.10) with this reduction step yields the FR transformation matrix

$$J_{\mathrm{FR}} := J_{\pm} J_B = \begin{bmatrix} I & -A_{11}^{-1}\bar{A}_{12} & -A_{11}^{-1}\bar{A}_{13} \\ 0 & J_- & J_+ \end{bmatrix}. \tag{5.14}$$

Note that J_{FR}, as defined in (5.14), exhibits the structure of the transformation matrix of the general splitting (5.5), which is obtained when choosing $D = -A_{11}^{-1}\bar{A}_{12}$ and $C = -A_{11}^{-1}\bar{A}_{13}$. The FR hierarchical basis matrix thus has the form

$$\tilde{A}_h = J_{\mathrm{FR}}^T A_h J_{\mathrm{FR}} = \begin{bmatrix} A_{11} & 0 \\ 0 & B \end{bmatrix}, \tag{5.15}$$

where B is given by (5.12). That means, in particular, we have $\tilde{A}_{33} = B_{33}$.

On the other hand, any splitting based on differences and aggregates can be written in the product representation (5.14) with a properly defined matrix J_B (actually obtained by replacing the first row of J_B in (5.13) by $[I \;\; D \;\; C]$ and using appropriate matrices C and D).

Since the local constant in the strengthened CBS inequality does not depend on the choice of the matrix D, cf. Section 5.3, we make use of this fact in the following definition of the transformation related to the FR splitting, which is formulated in the framework of differences and aggregates.

Definition 5.2 (First Reduce (FR) splitting) The splitting based on differences and aggregates incorporating the "first reduce" step (in short FR splitting), cf. Margenov and Vassilevski [44] and Blaheta et al. [22], is characterized by using $D = 0$ and $C = -A_{11}^{-1}\bar{A}_{13}$ in the general transformation matrix (5.5). The matrices $\bar{A}_{11} = A_{11}$ and $\bar{A}_{13}$ are to be taken from (5.10).

Definition 5.3 (Differences and Aggregates (DA)) The transformation for the standard splitting based on differences and aggregates (DA splitting), cf. Blaheta et al. [22], follows from the general transformation (5.5) for $D = 0$ and $C = \frac{1}{2}\operatorname{diag}(1, 1, 1)$.

Definition 5.4 (Generalized Differences and Aggregates (GDA)) A so-called generalized DA (GDA) splitting, as considered by Margenov and Synka [42], is retrieved by substituting in (5.5) the matrices $D = 0$ and

$$C = \frac{1}{2}I + \mu(\mathbb{1} - 3I) \tag{5.16}$$

whereby a proper choice of $\mu \in [0, \frac{1}{4}]$ depends on the minimum angle condition used in the triangular mesh.[1]

5.3 Uniform estimates of the CBS constant in case of non-nested spaces

Similar to the conforming finite element spaces the constant γ in the strengthened Cauchy–Bunyakowski–Schwarz (CBS) inequality, which measures the cosine of the abstract angle between the two subspaces $\mathcal{V}_1$ and $\mathcal{V}_2$, i.e.,

$$\gamma = \cos(\mathcal{V}_1, \mathcal{V}_2),$$

is an important quality criterion for the splitting. In this section we will therefore deal again with upper bounds for γ but this time for discretizations using nonconforming Crouzeix–Raviart type finite elements.

Let us consider the hierarchical basis matrix $\tilde{A}_h = J^T A_h J$, partitioned as in (3.11), where J is now defined according to (5.5). Then the CBS constant is given by

$$\gamma^2 = \max_{\mathbf{v}_3 \neq \mathbf{0}} \left[1 - \frac{\mathbf{v}_3^T \tilde{S} \mathbf{v}_3}{\mathbf{v}_3^T \tilde{A}_{33} \mathbf{v}_3} \right] = 1 - \min_{\mathbf{v}_3 \neq \mathbf{0}} \frac{\mathbf{v}_3^T \tilde{S} \mathbf{v}_3}{\mathbf{v}_3^T \tilde{A}_{33} \mathbf{v}_3} \tag{5.17}$$

where $\mathbf{0}$ denotes the vector of all zeros and $\tilde{S}$ is the Schur complement obtained from $\tilde{A}_h$ by eliminating the degrees of freedom corresponding to basis functions in $\mathcal{V}_1$, i.e., by reducing the system to the lower-right block. Let S be the corresponding Schur complement obtained in the same way from A_h. Then the following lemma provides the starting point for our discussion.

Lemma 5.5 *Under the above assumptions we have $\tilde{S} = S$ and thus $\tilde{S}$ is invariant with respect to the matrices C and D, which appear in the transformation matrix J.*

Proof. Follows from direct calculation. □

The matrix $\tilde{A}_{33}$ does not depend on D. As a consequence, the CBS constant does not depend on the choice of the matrix D.

Lemma 5.6 *The CBS constant γ, given by equation (5.17), for the splitting (5.3) associated with the hierarchical basis transformation matrix that is induced by the local transformation (5.5) is invariant to matrix D.*

[1] Here $\mathbb{1}$ denotes the 3×3 matrix of all ones.

We end the discussion of the general splitting with an important observation that is relevant in the context of the multilevel hierarchical splitting, i.e., for the multilevel extension of the two-level preconditioners. There, for both variants, the multiplicative and the additive AMLI method, the construction is such that the matrix block $\tilde{A}_{33}^{(k)}$ for $k = \ell, \ldots, 1$ defines the coarse-level matrix $A^{(k-1)}$. Then, as is well known, the condition

$$\ker\left(\tilde{A}_{E:33}^{(k)}\right) = \ker\left(A_e\right) = \operatorname{span}\left(\mathbf{1}\right) \quad \forall k \in \{\ell, \ldots, 1\} \tag{5.18}$$

is necessary for any splitting to result in a CBS constant γ_E that is strictly less than one. As it was proven in [35] the local estimates of γ, as presented in the rest of this section, hold true for all coarse levels. This is due to the fact that starting with a matrix which has been obtained by means of a global assembly of nonconforming Crouzeix–Raviart (CR) linear finite elements A_e^{CR}, cf. (5.2), the matrices $\tilde{A}_{E:33}^{(k)}$ are of the same type on all levels $k = \ell, \ell - 1, \ldots, 1$.

In the following we shall provide a summary of the main results from References [22, 23, 35] related to a local estimation of the CBS constant and the resulting AMLI methods.

DA splitting

The analysis of the related two-level method is performed locally by considering the corresponding problems at macro-element level. The CBS constant γ for the standard two-level DA-splitting with $C := \frac{1}{2}I$ (cf. Definition 5.3) is uniformly bounded w.r.t. both coefficient and mesh anisotropy by

$$\gamma_{\mathrm{DA}}^2 \leq \gamma_{\mathrm{DA,E}}^2 \leq \frac{3}{4}. \tag{5.19}$$

This estimate is independent of the discretization parameter h and possible jumps aligned with the finite element partitioning $\mathcal{T}_H$.

The extension of the two-level algorithm to the multilevel case is based on successive regular refinement of a given triangulation $\mathcal{T}^{(k)}$ at level k. It requires the recursive application of the two-level splitting of the finite element spaces $\mathcal{V}^{(k)}$ into the spaces $\mathcal{V}_2^{(k)} =: \mathcal{V}^{(k-1)}$ and $\mathcal{V}_1^{(k)} := \mathcal{V}^{(k)} \setminus \mathcal{V}_2^{(k)}$, where $k = \ell, \ldots, 1$, and $\mathcal{V}^{(0)}$ denotes the vector space corresponding to the coarsest mesh.

A repeated application of the estimate (5.19) is possible because of the (very simple) proportionality relations obtained for the DA splitting, i.e., at every level k there holds

$$\tilde{A}_{E:33}^{(k)} = 4\, S_E^{(k)} = A_e^{(k)}. \tag{5.20}$$

Note that equation (5.20) immediately implies that the two nonzero eigenvalues (corresponding to the eigenspace orthogonal to the constant vector) of the generalized eigenvalue problem

$$S_E \mathbf{v}_{E:3} = \lambda_E \tilde{A}_{E:33} \mathbf{v}_{E:3}, \quad \mathbf{v}_{E:3} \neq \text{const.}, \tag{5.21}$$

are identical and given by

$$\lambda_{\min,E} = \lambda_E^{(1)} = \lambda_E^{(2)} = \frac{1}{4}. \tag{5.22}$$

The bound (5.19) then is an immediate consequence of the equation

$$\gamma^2 \leq \max_{E \in \mathcal{T}} \gamma_E^2 = \max_{E \in \mathcal{T}} (1 - \lambda_{\min,E}). \tag{5.23}$$

GDA splitting

This generalization of the standard DA splitting is based on the assumption of a minimum angle condition, which is commonly used in commercial mesh generators, and on symmetry assumptions for the matrix C (in the transformation J) which lead to a dependence of a single parameter $\mu \in [0, 1/4]$, cf. [42].

The CBS constant, as defined in (5.17), clearly depends on C. We now seek a matrix C that provides the minimal (local) $\gamma_E = \gamma_{\min,E}$ where

$$\begin{aligned} \gamma_{\min,E}^2 &:= \min_C [\gamma_{\min,E}^2] = 1 - \max_C \left[\min_{\mathbf{v}_{E:3} \neq \text{const.}} \frac{\mathbf{v}_{E:3}^T S_E \mathbf{v}_{E:3}}{\mathbf{v}_{E:3}^T \tilde{A}_{E:33}(C) \mathbf{v}_{E:3}} \right] \\ &= 1 - \min_{\mathbf{v}_{E:3} \neq \text{const.}} \left[\max_C \frac{\mathbf{v}_{E:3}^T S_E \mathbf{v}_{E:3}}{\mathbf{v}_{E:3}^T \tilde{A}_{E:33}(C) \mathbf{v}_{E:3}} \right] \end{aligned} \tag{5.24}$$

subject to the constraint that C is of the form (5.16) for which we use the short notation $C \in \mathcal{C}_{\mathrm{GDA}}$. A proof of the following theorem can be found in [35].

Theorem 5.7 *Consider a macro element matrix A_E, assembled from four element matrices of the form (5.2) corresponding to the four similar triangles obtained by a regular refinement step, cf. Figure 5.1. Then the local CBS constant $\gamma_E = \gamma_{\mathrm{GDA},E}$ for the GDA splitting, cf. Definition 5.4, is minimal for*

$$\mu = \mu_{\mathrm{opt}} = \frac{s - r}{5s - 4r} \tag{5.25}$$

where $s = a + b + c$ and $r = \sqrt{s^2 - 3}$. The minimal local CBS constant then is given by

$$\gamma_{\mathrm{GDA},\min,E}^2 = \frac{s\,(2s + r)}{4\,(1 + s^2)}. \tag{5.26}$$

FR splitting

The GDA splitting with the optimal choice of the parameter μ as given by (5.25), however, in general is still not the best splitting, which is obtained for the FR basis transformation. A proof of the following theorem can be found in [35].

Theorem 5.8 *Let $\tilde{A}_{E:33}(C)$ be the macro-element matrix related to the lower-right block of (5.8). Then, the minimum value for γ_E, as defined in (5.24), is attained for $C = C_{\text{opt}} := -A_{E:11}^{-1}\bar{A}_{E:13}$ where $\bar{A}_{E:13}$ is the macro-element matrix of the upper-right block of (5.10).*

Remark 5.9 Note that the optimal choice C_{opt}, in Theorem 5.8, provides the harmonic extension obtained via the first reduce approach. Equivalently, C_{opt} can be shown to be the solution of the optimization problem

$$C_{\text{opt}} = \min_{C} \|I - S_E^{-1}\tilde{A}_{E:33}(C)\|_2.$$

Remark 5.10 An explicit expression for the local CBS constant corresponding to the FR splitting was derived in Reference [22]. Using the substitutions $s := a + b + c$ and $t := abc$ this formula reads

$$\gamma_{\text{FR},E}^2 = \frac{3}{8} + \frac{1}{8}\sqrt{\frac{s - 9t}{s - t}}. \tag{5.27}$$

As it can be seen from Figure 5.2 the estimates $\gamma_{\text{GDA,min},E}^2$ as given by (5.26) for the GDA splitting and $\gamma_{\text{FR},E}^2$ as given by (5.27) for the FR splitting coincide for any isosceles triangle.

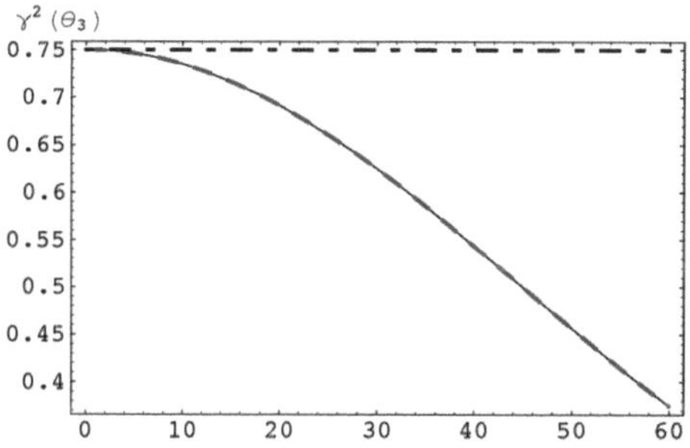

Figure 5.2 Local estimates of γ_E^2 for isosceles triangle depending on minimal angle: FR (solid), GDA (dashed), and DA (dot-dashed) splitting.

For rectangular triangles formula (5.27) yields $\gamma_{\text{FR},E}^2 = 1/2$ independently of the other angles and the value of $\gamma_{\text{GDA,min},E}^2$ approaches the local γ^2-value of the DA splitting (which is always $3/4$) when the minimal angle tends to zero, cf. Figure 5.3.

Thus $\gamma_{\text{GDA,min},E}^2$ for an arbitrary but fixed macro element E is always in between $\gamma_{\text{FR},E}^2$ and $\gamma_{\text{DA},E}^2$ (and since those estimates are sharp) we arrive at the following conclusion, cf. [35]:

Theorem 5.11 *Consider the discrete problem derived from the weak formulation (2.3) of the boundary-value problem (2.2) using linear Crouzeix–Raviart finite elements for discretization, and assume that the coefficient matrix* $\mathrm{a}(e)$ *in (2.3) is piecewise constant*

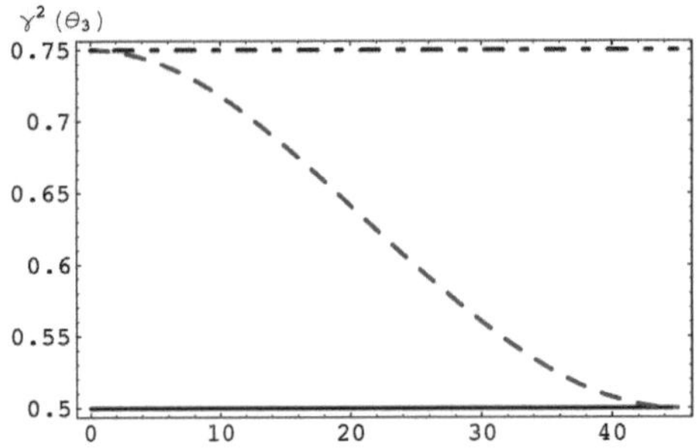

Figure 5.3 Local estimates of γ_E^2 for rectangular triangle depending on minimal angle: FR (solid), GDA (dashed), and DA (dot-dashed) splitting.

and SPD on the coarsest mesh partitioning. Then, comparing the different decompositions of the finite element space according to the standard DA splitting (cf. Definition 5.3), the GDA splitting (cf. Definition 5.4), using the optimal parameter $\mu = \mu_{\text{opt}}$ from (5.25), and the FR splitting (cf. Definition 5.2), we have the following relations between their respective CBS constants:

$$\gamma_{\text{FR}}^2 \le \gamma_{\text{GDA}}^2 \le \gamma_{\text{DA}}^2 \le 3/4. \tag{5.28}$$

5.4 Optimal order preconditioning of the pivot block in the two-level splittings

The splittings FR and DA are considered in this section. The presentation follows the results recently published in [23]. Let us start with FR splitting. After the exact local elimination of the block A_{11} we get the Schur complement B. The problem of optimal order preconditioning of the pivot block in the two-level splitting is now related to the block B_{22} associated with the half-differences of the hierarchical basis. Let us note, that the structure of B_{22}, see (5.12), is the same as of the pivot block A_{11} as appears in the case of conforming elements.

The macroelement block $B_{E:22}$ is found explicitly, namely,

$$B_{E,22} = \frac{2p}{q}\begin{bmatrix} 3\,q+2(1+\beta+\beta^2) & q+2 & -q-2\,\beta^2 \\ q+2 & 3\,q+2(1+\alpha+\,\alpha^2) & -z-2\,\alpha^2 \\ -q-2\,\beta^2 & -q-2\,\alpha^2 & 3\,q+2\,(\alpha^2+\alpha\beta+\beta^2) \end{bmatrix},$$

$q = \alpha + \alpha\beta + \beta$ and $p = 3(\alpha + \alpha\beta_\beta) + 3(\alpha^2 + \alpha\beta + \beta^2) + \alpha\beta(3\alpha + 3\beta + 1)$. The above expressions are rather more complicated than the related representation (4.13) of $A_{E:11}$. Nevertheless, the additive and multiplicative preconditioners discussed in Sections 4.3–4.4 are directly applicable. Moreover, the condition number estimates are completely the same as for the Courant conforming finite elements and the following theorem holds.

Theorem 5.12 ([23]) *For any element size and shape and any problem anisotropy it holds that*

- $\kappa\left(C_{22}^{-1}B_{22}\right) < \dfrac{1}{4}(11+\sqrt{105})$ *if* C_{22} *is the additive preconditioner to* B_{22},
- $\kappa\left(C_{22}^{-1}B_{22}\right) < \dfrac{15}{8}$ *if* C_{22} *is the multiplicative preconditioner to* B_{22},
- *the cost of the application of preconditioners in both cases is proportional to the number of unknowns.*

As for the conforming elements, the estimates of $\kappa\left(C_{22}^{-1}B_{22}\right)$ are derived using the inequality from Lemma 4.2, where the local condition numbers are derived in terms of $(\alpha,\beta)\in D$.

The later results are easily converted to the case of DA splitting. The related pivot block to be considered has the form

$$\begin{bmatrix} \widetilde{A}_{11} & \widetilde{A}_{12} \\ \widetilde{A}_{12} & \widetilde{A}_{22} \end{bmatrix}.$$

Here, the (locally performed) elimination of the block $\widetilde{A}_{11}$ gives rise to a Schur complement equal to

$$\widetilde{A}_{11} - \widetilde{A}_{12}\widetilde{A}_{11}^{-1}\widetilde{A}_{12} = B_{22}$$

which preconditioning was already discussed.

6 Schur complement based multilevel preconditioners

6.1 Hierarchical versus standard nodal-basis: pros and cons

Hierarchical basis matrices, in general, are less sparse than standard nodal basis matrices, and, though they may be generated in the course of a mesh refinement procedure in a quite natural way, one is faced with additional computational costs when (re)constructing them from the nodal basis representation for the purpose of an efficient solution of the linear system only. On the other hand, as we conclude from the previous sections, the theory of hierarchical basis block factorization methods, including various stabilization techniques, is fairly matured whereas alternative approaches, which avoid the hierarchical basis construction, have been far less frequently studied. Their rigorous analysis is only established up to a certain extent. For that reason, some numerical results are included in this section, which should demonstrate the potential of the latter approach.

As a matter of fact, avoiding the hierarchical basis representation, we have to meet certain supplementary requirements in order to develop robust two- and multilevel schemes. A general guideline is to achieve robustness, also with respect to perturbations of the M-matrix property, which is a crucial factor in this case.

In the following we will briefly discuss a technique, originally proposed by Kraus [33], which is based on a particular approximation of the Schur complement. Assuming access to the individual (fine-grid) element stiffness matrices, any element agglomeration procedure, see, e.g., [30, 34], can be used in order to assemble macro-element matrices that provide a basis for the computation of appropriate coarse-level matrices, which makes the hierarchical basis representation (3.11) unnecessary in the construction. However, building a bridge to the methods discussed in the previous sections of this survey subserves the analysis. As an important example, a (locally computable) condition number bound for the related sparse approximation of the Schur complement will be presented in Section 6.4.

6.2 A general two-level preconditioner

We start the description of the nodal-basis approach with the consideration of a general two-level preconditioner. Let the degrees of freedom (DOF) be partitioned into two groups, usually denoted as fine and coarse DOF. We replace then the exact block factorization

$$A = \begin{bmatrix} A_{11} & A_{12} \\ A_{21} & A_{22} \end{bmatrix} = \begin{bmatrix} I & \\ A_{21}A_{11}^{-1} & I \end{bmatrix} \cdot \begin{bmatrix} A_{11} & A_{12} \\ & S \end{bmatrix} \tag{6.1}$$

with an approximate factorization

$$B = \begin{bmatrix} I & \\ A_{21}B_{11}^{-1} & I \end{bmatrix} \cdot \begin{bmatrix} B_{11} & A_{12} \\ & Q \end{bmatrix} = \begin{bmatrix} B_{11} & A_{12} \\ A_{21} & Q + A_{21}B_{11}^{-1}A_{12} \end{bmatrix}. \tag{6.2}$$

Herein, the matrix B_{11} is a preconditioner for the pivot block A_{11} and Q is an approximation of the exact Schur complement $S = A_{22} - A_{21}A_{11}^{-1}A_{12}$. Note that a change to the hierarchical (two-level) basis would only affect the off-diagonal blocks A_{12} and A_{21} in (6.1); the pivot block A_{11} as well as the Schur complement S are the same for both bases [49]. However, as already mentioned, we want to do without this stabilizing modification of the off-diagonal blocks here.

In the following, $\underline{\alpha}, \underline{\beta}, \underline{\gamma}, \overline{\alpha}, \overline{\beta}, \overline{\gamma}$ denote positive constants satisfying

$$\begin{aligned} 0 < \ & \underline{\alpha}, \underline{\beta}, \underline{\gamma} \ \leq 1, \\ 1 \leq \ & \overline{\alpha}, \overline{\beta}, \overline{\gamma} \ < \infty. \end{aligned}$$

It is known that the spectral condition number

$$\kappa(B^{-1}A) = \frac{\lambda_{\max}(B^{-1}A)}{\lambda_{\min}(B^{-1}A)}, \tag{6.3}$$

measuring the quality of the two-level preconditioner B defined via (6.2), depends on the extremal eigenvalues of $B_{11}^{-1}A_{11}$ and $Q^{-1}S$, that is, it involves the bounds

$$\underline{\alpha}\mathbf{v}_1^T A_{11}\mathbf{v}_1 \leq \mathbf{v}_1^T B_{11}\mathbf{v}_1 \leq \overline{\alpha}\mathbf{v}_1^T A_{11}\mathbf{v}_1 \quad \forall \mathbf{v}_1, \tag{6.4}$$

and

$$\underline{\beta}\mathbf{v}_2^T S\mathbf{v}_2 \leq \mathbf{v}_2^T Q\mathbf{v}_2 \leq \overline{\beta}\mathbf{v}_2^T S\mathbf{v}_2 \quad \forall \mathbf{v}_2, \tag{6.5}$$

respectively. However, and this has been shown in [46], avoiding the hierarchical basis representation, a bound

$$\underline{\gamma}\mathbf{v}^T A\mathbf{v} \le \mathbf{v}^T B\mathbf{v} \le \overline{\gamma}\mathbf{v}^T A\mathbf{v} \quad \forall \mathbf{v} \tag{6.6}$$

that is independent of the mesh size can only be obtained if B_{11}^{-1} acts nearly as an exact inverse on all vectors $A_{12}\mathbf{v}_2$ for which $\mathbf{v}_2$ is *smooth* on the coarse grid, i.e., a low energy mode of the Schur complement S. For instance, this requirement is met if the condition

$$\overline{\alpha}\mathbf{v}_2^T A_{21}B_{11}^{-1}A_{12}\mathbf{v}_2 \le (1-\xi)\mathbf{v}_2^T A_{22}\mathbf{v}_2 + \xi\mathbf{v}_2^T A_{21}A_{11}^{-1}A_{12}\mathbf{v}_2 \quad \forall \mathbf{v}_2 \tag{6.7}$$

is fulfilled for some $\xi \le 1$. Note that the assumption (6.7) is loosened if we let ξ be negative.

Now, inequalities (6.4) and (6.7) imply

$$\begin{aligned} \mathbf{v}_2^T(A_{22}-S)\mathbf{v}_2 &= \mathbf{v}_2^T A_{21}A_{11}^{-1}A_{12}\mathbf{v}_2 \\ &\le \overline{\alpha}\mathbf{v}_2^T A_{21}B_{11}^{-1}A_{12}\mathbf{v}_2 \le \mathbf{v}_2^T(A_{22}-\xi S)\mathbf{v}_2 \quad \forall \mathbf{v}_2 \end{aligned} \tag{6.8}$$

and (6.6) can be based on the bounds (6.4), (6.5) and (6.7). For details see [46] where the analysis is carried out for the case $\overline{\alpha} = 1$. We summarize the main result (without this restriction) in the theorem below, cf. [33].

Theorem 6.1 *Let A and B, defined via (6.1) and (6.2), be symmetric nonnegative definite matrices such that A_{11} and B_{11} are invertible. Moreover, assume that (6.4) and (6.5) hold. If, in addition, condition (6.7) is satisfied for some $\xi \le 1$, then the bound (6.6) holds, i.e.,*

$$\kappa(B^{-1}A) \le \frac{\overline{\gamma}}{\underline{\gamma}}, \tag{6.9}$$

where $\underline{\gamma}$ is the smallest root of

$$\gamma^2 - \gamma(\underline{\beta} + \overline{\alpha} - \xi(\overline{\alpha} - \underline{\alpha})) + \underline{\alpha}\underline{\beta} \tag{6.10}$$

and $\overline{\gamma}$ is the largest root of

$$\gamma^2 - \gamma(\overline{\beta} + \overline{\alpha} - \xi(\overline{\alpha} - \underline{\alpha})) + \underline{\alpha}\overline{\beta}. \tag{6.11}$$

For M-matrices condition (6.7) has been analyzed by Notay [46]. It can be shown that if $B_{11} = B_{11,\mathrm{MILU}}$, i.e., B_{11} results from a modified incomplete factorization of A_{11} (where the modification vector is assumed to be $(1, 1, \ldots, 1)^T$), then (6.7) is always satisfied for $\xi = 0$. Moreover, if A arises from a linear finite element discretization of a two-dimensional second order elliptic PDE (on a triangular mesh with no interior angle larger than $\frac{\pi}{2}$) one can actually prove that (6.7) holds for $\xi = \frac{1}{2}$. In general, this assumption can be checked (and a value for ξ can be determined) based on the following considerations (which also apply to SPD non-M matrices): First, we note that $(A_{22} - \xi S - \overline{\alpha} A_{21} B_{11}^{-1} A_{12})$ is the Schur complement of the matrix

$$M = \begin{bmatrix} \frac{1}{\overline{\alpha}} B_{11} & A_{12} \\ A_{21} & A_{22} - \xi S \end{bmatrix}.$$

Thus, since B_{11} is SPD, condition (6.7) holds if and only if M is symmetric positive semidefinite (SPSD). Further, if

$$\mathbf{v}_1^T B_{11} \mathbf{v}_1 \geq \mathbf{v}_1^T \tilde{B}_{11} \mathbf{v}_1 \quad \forall \mathbf{v}_1 \tag{6.12}$$

and

$$\mathbf{v}_2^T S \mathbf{v}_2 \leq \mathbf{v}_2^T \tilde{S} \mathbf{v}_2 \quad \forall \mathbf{v}_2 \tag{6.13}$$

for any SPSD matrices $\tilde{B}_{11}$ and $\tilde{S}$ it is clear that (6.7) follows if the matrix

$$\tilde{M} = \begin{bmatrix} \frac{1}{\overline{\alpha}} \tilde{B}_{11} & A_{12} \\ A_{21} & A_{22} - \xi \tilde{S} \end{bmatrix} \tag{6.14}$$

is SPSD. This motivates the study of the class of two-level preconditioners defined via (6.2) where B_{11} satisfies a row-sum criterion. However, for the Schur complement approximation Q we suggest its computation by assembling local (exact) Schur complements (cf. Section 6.4).

6.3 A variant of incomplete factorization based on exact local factorization

As already mentioned in Section 6.2, a preconditioner B_{11} for the pivot block A_{11} in particular has to be accurate on a certain subspace (cf. (6.7) and (6.8)). We will compare the efficiency of two types of preconditioners $B_{11} = LU$ which satisfy the row-sum criterion

$$A_{11} \cdot (1, 1, \ldots, 1)^T = B_{11} \cdot (1, 1, \ldots, 1)^T. \tag{6.15}$$

The first preconditioner, denoted by $B_{11,\mathrm{MILU}}$, is obtained from classical MILU factorization [4, 25] of A_{11} where off-diagonal fill-in (for L and U) is only allowed in positions in which A_{11} is nonzero. It is well known that even for SPD matrices incomplete factorization methods can suffer from breakdown since they may yield zero (or negative) pivots when an exact factorization would give only positive pivots [26]. However, in the present context the partitioning of the matrix A (cf. (6.1) in Section 6.2) usually causes pivot blocks which prevent possible breakdowns.

For the purpose of comparison, we construct a second preconditioner, now based on local complete factorizations. Here we take advantage of the knowledge of the individual element matrices. A straightforward approach is to regard the small-sized macro element matrices (resulting from assembling element matrices). For that reason we transfer the global ordering of nodes (and DOF) to a local ordering, i.e., we label the interior fine-grid nodes (DOF) first, followed by those on the boundary of the considered macro element, and label the coarse-grid nodes (DOF) last, as exemplified in Figure 6.1. With respect to this local ordering the macro element matrices take a 2×2 block form

$$A_E = \begin{bmatrix} A_{E:11} & A_{E:12} \\ A_{E:21} & A_{E:22} \end{bmatrix}, \tag{6.16}$$

where $A_{E:11}$ and $A_{E:22}$ are the blocks corresponding to fine and coarse-grid DOF, respectively.

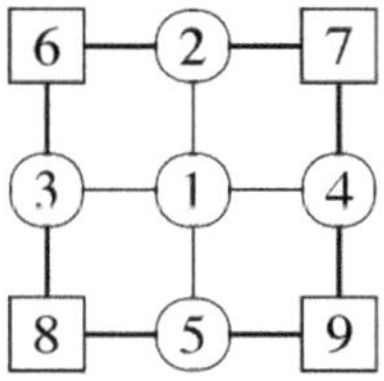

Figure 6.1 Local labeling within macro element for a mesh of quadrilateral elements.

Let us assume that $A_{E:11}$ is non-singular for all macro elements E. Then the factorization

$$A_{E:11} = L_E U_E \quad \forall E \tag{6.17}$$

is well defined where L_E and U_E can be scaled such that the diagonal of L_E is the identity, i.e., $\mathrm{diag}(L_E) = I_E$. Now, since

$$A_{11} = \sum_E A_{E:11}, \tag{6.18}$$

the sum of upper triangular matrices U_E yields an approximate upper triangular factor of the global pivot block A_{11}, i.e.,

$$U := \sum_E U_E. \tag{6.19}$$

Consequently, a preconditioner $B_{11} = B_{11,\mathrm{ILUE}}$ is defined in virtue of

$$B_{11,\mathrm{ILUE}} := LU, \quad U := \sum_E U_E, \quad L := U^T(\mathrm{diag}(U))^{-1}. \tag{6.20}$$

However, in order to satisfy the row-sum criterion (6.15) the diagonal of U in (6.19) needs a recalculation. For any two $N \times N$ matrices $C = (c_{ij})$ and $B_{11} = (b_{ij})$, the latter being a product of the form $B_{11} = U^T(\mathrm{diag}(U))^{-1}U$, where $U = (u_{ij})$ is an upper triangular matrix, the criterion $C \cdot (1, 1, \ldots, 1)^T = B_{11} \cdot (1, 1, \ldots, 1)^T$ is equivalent to

$$u_{kk} = \sum_{j=1}^{N} c_{kj} - \sum_{j=k+1}^{N} u_{kj} - \sum_{i=1}^{k-1} \frac{u_{ik}}{u_{ii}} \sum_{j=i}^{N} u_{ij} \quad \forall k = 1, 2, \ldots, N. \tag{6.21}$$

This a posteriori modification of the diagonal of U can be done efficiently using the algorithm below.

Algorithm 6.2 [A posteriori diagonal modification of U]

for $k = 1$ **to** N

$$s_C(k) := \sum_{j=1}^{N} c_{kj}$$

$$s_U(k) := \sum_{j=k+1}^{N} u_{kj}$$

$$u_{kk} := s_C(k) - s_U(k) - \sum_{i=1}^{k-1} \frac{u_{ik} s_U(i)}{u_{ii}}$$

$$s_U(k) := s_U(k) + u_{kk}$$

end

Modifying the diagonal of U defined by (6.19) according to the aforementioned algorithm yields a preconditioner, henceforth denoted as $B_{11,\mathrm{MILUE}}$, where L is again defined as $L := U^T(\mathrm{diag}(U))^{-1}$.

6.4 Approximation by local Schur complements

In this section we will define a particular approximation Q to the exact Schur complement $S = A_{22} - A_{21}A_{11}^{-1}A_{12}$, and analyze its relative condition number.

The proposed approximation technique is simple. From the neighborhood (macro element) matrices (6.16) one computes the local exact Schur complements

$$S_E = A_{E:22} - A_{E:21}(A_{E:11})^{-1}A_{E:12} \tag{6.22}$$

for all macro elements E, which then serve as element matrices on the next coarser level. The procedure is repeated on all levels, i.e.,

$$A_e^{(k-1)} := S_E^{(k)} \quad \forall E \tag{6.23}$$

at levels $k = \ell, \ell-1, \ldots, 1$. In other words, we assemble the small-sized local Schur complement matrices to a global Schur complement approximation, i.e.,

$$Q^{(k)} := \sum_e A_e^{(k-1)} = \sum_E S_E^{(k)}. \tag{6.24}$$

In the following, whenever it is clear that we refer to some fixed level k, we drop the index k, e.g., writing $Q = \sum_E S_E$, for instance.

Consider now the approximation property (6.5). In the following, we will derive a locally computable eigenvalue bound $\underline{\beta}$, and, we will show that (6.5) holds with $\overline{\beta} = 1$.

First, we recall the following energy minimizing property of the Schur complement,

$$\begin{aligned} \mathbf{v}_2^T S \mathbf{v}_2 &= \min_{\mathbf{v}_1} \begin{bmatrix} \mathbf{v}_1 \\ \mathbf{v}_2 \end{bmatrix}^T A \begin{bmatrix} \mathbf{v}_1 \\ \mathbf{v}_2 \end{bmatrix} \\ &= \min_{\mathbf{v}_1} \begin{bmatrix} \mathbf{v}_1 \\ \mathbf{v}_2 \end{bmatrix}^T \left(\sum_E A_E \right) \begin{bmatrix} \mathbf{v}_1 \\ \mathbf{v}_2 \end{bmatrix} \\ &= \min_{\mathbf{v}_1} \sum_E \left(\begin{bmatrix} \mathbf{v}_{E:1} \\ \mathbf{v}_{E:2} \end{bmatrix}^T A_E \begin{bmatrix} \mathbf{v}_{E:1} \\ \mathbf{v}_{E:2} \end{bmatrix} \right) \quad \forall \mathbf{v}_2, \end{aligned} \tag{6.25}$$

where $\mathbf{v}_E$ denotes the restriction of an arbitrary vector $\mathbf{v}$ to any given macro element E. Now, using (6.24) and (6.25) it is readily seen that (6.5) holds with $\overline{\beta} = 1$:

$$\begin{aligned} \mathbf{v}_2^T Q \mathbf{v}_2 &= \mathbf{v}_2^T \left(\sum_E S_E \right) \mathbf{v}_2 \\ &= \sum_E \mathbf{v}_{E:2}^T S_E \mathbf{v}_{E:2} \\ &= \sum_E \left(\min_{\mathbf{v}_{E:1}} \begin{bmatrix} \mathbf{v}_{E:1} \\ \mathbf{v}_{E:2} \end{bmatrix}^T A_E \begin{bmatrix} \mathbf{v}_{E:1} \\ \mathbf{v}_{E:2} \end{bmatrix} \right) \\ &\leq \min_{\mathbf{v}_1} \sum_E \left(\begin{bmatrix} \mathbf{v}_{E:1} \\ \mathbf{v}_{E:2} \end{bmatrix}^T A_E \begin{bmatrix} \mathbf{v}_{E:1} \\ \mathbf{v}_{E:2} \end{bmatrix} \right) = \mathbf{v}_2^T S \mathbf{v}_2 \quad \forall \mathbf{v}_2. \end{aligned} \tag{6.26}$$

Regarding a lower bound $\underline{\beta}$ the following basic relation, cf. Lemma 3.2, is an important observation: The (hierarchical basis) coarse-level matrix A_H in (3.11) and the Schur complement approximation Q in (6.24) satisfy the inequalities

$$(1 - \gamma^2) \mathbf{v}_2^T A_H \mathbf{v}_2 \leq \mathbf{v}_2^T Q \mathbf{v}_2 \leq \mathbf{v}_2^T A_H \mathbf{v}_2 \quad \forall \mathbf{v}_2, \tag{6.27}$$

which follows by applying Lemma 3.2 at macro-element level and using the fact that both matrices A_H and Q are assembled from their corresponding local counterparts. Using then the right hand side inequality of Lemma 3.2 (again with $A_{22} = A_H$) we deduce from (6.27) the lower bound $\underline{\beta} = 1 - \gamma^2$, where γ is the CBS constant related to the two-level hierarchical splitting (3.9). This simple derivation of a lower bound in terms of γ was presented in [9]. It shows that the approximation (6.24) is reasonable if the corresponding hierarchical basis transformation defines a proper splitting. In particular, we conclude that the condition number bound

$$\kappa(Q^{-1} S) \leq \frac{1}{1 - \gamma^2}$$

holds for arbitrary coefficients in the differential operator (if they are piecewise constant) and even for degenerated elements. We summarize the facts in the following theorem.

Theorem 6.3 *Let Q be the assembly of the local (macro-element matrix) Schur complements S_E, i.e., $Q = \sum_E S_E$, and let S be the global (true) Schur complement matrix in (3.12). Then*

$$(1-\gamma^2)\mathbf{v}_2^T S \mathbf{v}_2 \le \mathbf{v}_2^T Q \mathbf{v}_2 \le \mathbf{v}_2^T S \mathbf{v}_2 \quad \forall \mathbf{v}_2. \tag{6.28}$$

Finally it should be noted that this method can also be applied to non-selfadjoint problems. Efficient Schur complement based multilevel preconditioners for linear isostasy saddle point problems have been studied in [18].

6.5 Numerical results

In this section we will present some numerical results for the nonlinear algebraic multilevel iteration (NLAMLI) algorithm, for details see [32, 33]. This is a slightly modified version of the algorithm originally proposed in [17]. However, we will consider this method employing the components defined in Sections 6.3–6.4.

The first test problem describes anisotropic diffusion. The domain Ω is the unit square. The weak formulation is given by:

Problem 1

$$\begin{aligned} \mathcal{A}(u,v) &:= \int_\Omega (\nabla v)^T D(x,y) \nabla u \, dx dy = (f,v) \quad \forall \quad v \in H_0^1(\Omega) \\ D(x,y) &:= \begin{pmatrix} \epsilon & 0 \\ 0 & 1 \end{pmatrix}, \quad 0 < \epsilon \le 1 \end{aligned} \tag{6.29}$$

In the second test problem, we impose a jump on the direction of strong diffusion along the line $x = 1/2$, i.e., we modify Problem 1 by changing (6.29) accordingly (cf. [13]):

Problem 2 *Replace (6.29) in Problem 1 with*

$$D(x,y) := \begin{cases} \begin{pmatrix} \epsilon & 0 \\ 0 & 1 \end{pmatrix} & \text{for all } (x,y) \in (0,1) \times (0, \tfrac{1}{2}), \\ \begin{pmatrix} 1 & 0 \\ 0 & \epsilon \end{pmatrix} & \text{for all } (x,y) \in (0,1) \times (\tfrac{1}{2}, 1). \end{cases}$$

The two-level preconditioner at level k is given by

$$B^{(k)} = \begin{bmatrix} I^{(k)} & \\ A_{21}^{(k)} (B_{11}^{(k)})^{-1} & I \end{bmatrix} \cdot \begin{bmatrix} B_{11}^{(k)} & A_{12}^{(k)} \\ & Q^{(k)} \end{bmatrix}, \tag{6.30}$$

where the Schur complement approximation from Section 6.4 defines the coarse-grid matrix, i.e.,

$$A^{(k-1)} := Q^{(k)}, \tag{6.31}$$

and the preconditioner $B_{11}^{(k)} = L^{(k)} U^{(k)}$ is obtained from (modified) incomplete factorization of $A_{11}^{(k)}$ (cf. Section 6.3).

In Tables 6.1 and 6.2 we list the number of outer generalized conjugate gradient (GCG) iterations that suffice to reduce the Euclidean norm of the initial residual (corresponding to a random initial guess) by a factor 10^6.[1] We compare two iterative methods, (I) and (II). In the NLAMLI (I) procedure we use a global MILU factorization for preconditioning the A_{11} block at all levels, i.e., $B_{11} = B_{11,\text{MILU}}$. In the second procedure (II) we use the preconditioner $B_{11,\text{MILUE}}$, defined via (6.20) and (6.21), instead. Starting with the first coarse level, two inner GCG iterations are (recursively) performed on every other level, which results in a cheap(er) W-cycle variant.

	$1/h$	8	16	32	64	128	256	512
$\epsilon = 1.0$	(I)	5	6	6	7	7	7	7
	(II)	5	6	7	7	7	7	7
$\epsilon = 0.5$	(I)	5	7	7	8	8	8	8
	(II)	5	7	8	8	8	8	8
$\epsilon = 0.25$	(I)	7	9	9	9	9	9	9
	(II)	5	7	9	9	9	9	9
$\epsilon = 0.1$	(I)	10	16	18	18	18	18	18
	(II)	5	8	10	10	11	11	11
$\epsilon = 0.01$	(I)	18	*	*	*	*	*	*
	(II)	5	7	9	10	11	11	11

Table 6.1 Number of outer GCG iterations for Problem 1.

	$1/h$	8	16	32	64	128	256	512
$\epsilon = 0.5$	(I)	6	7	7	8	8	8	8
	(II)	5	7	8	8	8	8	8
$\epsilon = 0.25$	(I)	8	9	9	9	9	10	10
	(II)	5	8	9	9	10	9	10
$\epsilon = 0.1$	(I)	12	19	20	21	21	21	21
	(II)	6	8	10	11	11	11	11
$\epsilon = 0.05$	(I)	16	47	56	59	60	60	61
	(II)	6	8	10	11	11	11	11
$\epsilon = 0.01$	(I)	16	*	*	*	*	*	*
	(II)	6	8	10	11	11	11	11

Table 6.2 Number of outer GCG iterations for Problem 2.

[1] An asterisk indicates that 250 outer iterations are not sufficient to reach the convergence criterion.

The numerical tests clearly indicate the potential of the considered preconditioning technique. We observe that even with a global MILU factorization (procedure (I)) of the pivot block the corresponding multilevel iteration becomes an optimal-order method if the ratio of anisotropy is not too large. However, for increasing anisotropy the convergence of method (I) deteriorates whereas method (II) yields the desired robustness. In particular, the results for Problem 2, where the parameter ϵ additionally introduces a discontinuity in the PDE coefficients (changing the direction of dominating anisotropy), demonstrate the importance of an adequate preconditioner B_{11} for the pivot block. In the above examples, this requirement obviously can be met using exact local factorization, i.e., by employing $B_{11,\mathrm{MILUE}}$.

This leads to the conclusion that, though its analysis is not complete yet, the Schur complement based algebraic multilevel preconditioning technique, presented in this section, offers an attractive alternative to the hierarchical basis approach.

7 Concluding remarks

Let us summarize some of the major contributions of the considered robust multilevel methods. A complete theory for anisotropic elliptic problems discretized by linear conforming and nonconforming finite elements was presented. The optimal order preconditioning of the pivot blocks in the hierarchical two-by-two block splittings plays a key role in the construction of the AMLI methods for the cases of conforming Courant and nonconforming Crouzeix–Raviart elements. The additive approximation of the matrix $A_{11}^{(k)}$ has a block diagonal structure which makes the implementation of the algorithms more efficient and easy. The multiplicative preconditioner of $A_{11}^{(k)}$ gives a better condition number. However, its optimal complexity depends heavily on using a nested dissection algorithm which makes the implementation more complicated. The last section concerns the robustness of multilevel preconditioners based on local Schur complements' assembly. Althouh the related theory is still not completed, the presented results clearly illustrate some convincing advantages of such an alternative approach. The nonlinear variant of the AMLI method provides a general setting for future applications of the Schur complement based AMLI preconditioners including the case of strongly coupled systems of elliptic partial differential equations.

We end the discussion with concluding remarks related to some hot research topics in the field of multilevel preconditioning methods and algorithms.

- **Theory of the multilevel methods:** a) Novel approaches for hierarchical decomposition of non-nested finite element spaces; b) Generalized techniques for estimates of the CBS constant; c) Theory of the nonlinear multilevel methods.

- **Multilevel methods for non-standard FEM systems:** a) Non-conforming finite elements; b) Mixed FEMs; c) Discontinuous Galerkin approximations.

- **Specific target research problems:** a) Robust methods for 3D elliptic problems; b) Multilevel preconditioning of coupled problems; c) Novel multilevel algorithms for graph-Laplacians.

Bibliography

[1] B. Achchab, O. Axelsson, L. Laayouni, and A. Souissi, *Strengthened Cauchy-Bunyakowski-Schwarz inequality for a three dimensional elasticity system*, Num. Lin. Alg. Appl. 8 (2001), pp. 191–205.

[2] B. Achchab and J.F. Maitre, *Estimate of the constant in two strengthened CBS inequalities for FEM systems of 2D elasticity: Application to Multilevel methods and a posteriori error estimators*, Num. Lin. Alg. Appl. 3 (1996), pp. 147–160.

[3] D.N. Arnold and F. Brezzi, *Mixed and nonconforming finite element methods: Implementation, postprocessing and error estimates*, RAIRO, Model. Math. Anal. Numer. 19 (1985), pp. 7–32.

[4] O. Axelsson, *Iterative Solution Methods*, Cambridge University Press, 1994.

[5] ______, *Stabilization of algebraic multilevel iteration methods; additive methods*, Numerical Algorithms 21 (1999), pp. 23–47.

[6] ______, *A survey of algebraic multilevel iteration (AMLI) methods*, BIT Numerical Mathematics 43 (2003), pp. 863–879.

[7] O. Axelsson and V.A. Barker, *Finite Element Solution of Boundary Value Problems: Theory and Computations*, Academic Press, Orlando, 1984.

[8] O. Axelsson and R. Blaheta, *Two simple derivations of universal bounds for the C.B.S. inequality constant*, Appl. Math. 49 (2004), pp. 57–72.

[9] O. Axelsson, R. Blaheta, and M. Neytcheva, *Preconditioning of boundary value problems using elementwise Schur complements*, Tech. report, Uppsala University, Information Technology, No. 2006-048.

[10] O. Axelsson and I. Gustafsson, *Preconditioning and two-level multigrid methods of arbitrary degree of approximations*, Math. Comp. 40 (1983), pp. 219–242.

[11] O. Axelsson and S. Margenov, *On multilevel preconditioners which are optimal with respect to both problem and discretization parameters*, Computational Methods in Applied Mathematics 3 (2003), pp. 6–22.

[12] O. Axelsson and M. Neytcheva, *Algebraic multilevel iterations for Stieltjes matrices*, Num. Lin. Alg. Appl. 1 (1994), pp. 213–236.

[13] O. Axelsson and A. Padiy, *On the additive version of the algebraic multilevel iteration method for anisotropic elliptic problems*, SIAM J. Sci. Comput. 20 (1999), pp. 1807–1830.

[14] O. Axelsson and P.S. Vassilevski, *Algebraic multilevel preconditioning methods I*, Numer. Math. 56 (1989), pp. 157–177.

[15] ______, *Algebraic multilevel preconditioning methods II*, SIAM J. Numer. Anal. 27 (1990), pp. 1569–1590.

[16] ______, *A black box generalized conjugate gradient solver with inner iterations and variable-step preconditioning*, SIAM J. Matrix Anal. Appl. 12 (1991), pp. 625–644.

[17] ______, *Variable-step multilevel preconditioning methods, I: self-adjoint and positive definite elliptic problems*, Num. Lin. Alg. Appl. 1 (1994), pp. 75–101.

[18] E. Bängtsson and M. Neytcheva, *An agglomerate multilevel preconditioner for linear isostasy saddle point problems*, LNCS, vol. 3743, Springer, 2006, pp. 113–120.

[19] R. Bank and T. Dupont, *An optimal order process for solving finite element equations*, Math. Comp. 36 (1981), pp. 427–458.

[20] R. Bank, T. Dupont, and H. Yserentant, *The hierarchical basis multigrid method*, Numer. Math. 52 (1988), pp. 427–458.

[21] R. Blaheta, *Nested tetrahedral grids and strengthened C.B.S. inequality*, Num. Lin. Alg. Appl. 10 (2003), pp. 619–637.

[22] R. Blaheta, S. Margenov, and M. Neytcheva, *Uniform estimate of the constant in the strengthened CBS inequality for anisotropic non-conforming FEM systems*, Num. Lin. Alg. Appl. 11 (2004), pp. 309–326.

[23] ______, *Robust optimal multilevel preconditioners for non-conforming finite element systems*, Num. Lin. Alg. Appl. 12 (2005), pp. 495–514.

[24] ______, *Aggregation-based multilevel preconditioning of non-conforming FEM elasticity problems*, LNCS, vol. 3732, Springer, 2006, pp. 847–856.

[25] T. Dupont, R.P. Kendall, and H.H. Rachford, *An approximate factorization procedure for solving self-adjoint elliptic difference equations*, SIAM Journal on Numerical Analysis 5 (1968), pp. 559–573.

[26] V. Eijkhout, *On the existence problem of incomplete factorisation methods*, Tech. report, Computer Science Department, University of Tennessee, U.S.A., 1999, Lapack working note 144, UT-CS-99-435.

[27] V. Eijkhout and P.S. Vassilevski, *The role of the strengthened Cauchy-Bunyakowski-Schwarz inequality in multilevel methods*, SIAM Review 33 (1991), pp. 405–419.

[28] I. Georgiev, J. Kraus, and S. Margenov, *Multilevel preconditioning of 2D Rannacher-Turek FE problems; additive and multiplicative methods*, LNCS, Springer, to appear.

[29] ______, *Multilevel preconditioning of rotated bilinear non-conforming FEM problems*, Computers & Mathematics with Applications (to appear).

[30] J. Jones and P.S. Vassilevski, *AMGe based on element agglomeration*, SIAM J. Sci. Comput. 23 (2001), pp. 109–133.

[31] Tz. Kolev and S. Margenov, *Two-level preconditioning of pure displacement non-conforming FEM systems*, Num. Lin. Alg. Appl. 6 (1999), pp. 533–555.

[32] J. Kraus, *An algebraic preconditioning method for M-matrices: Linear versus nonlinear multilevel iteration*, Num. Lin. Alg. Appl. 9 (2002), pp. 599–618.

[33] ______, *Algebraic multilevel preconditioning of finite element matrices using local Schur complements*, Num. Lin. Alg. Appl. 13 (2006), pp. 49–70.

[34] J. Kraus and J. Synka, *An agglomeration-based multilevel-topology concept with application to 3D-FE meshes*, Tech. report, RICAM, Linz, 2004, No. 2004-08.

[35] J. Kraus, J. Synka, and S. Margenov, *On the multilevel preconditioning of Crouzeix-Raviart elliptic problems*, Num. Lin. Alg. Appl. (to appear).

[36] J. Kraus and S. Tomar, *Multilevel preconditioning of elliptic problems discretized by a class of discontinuous Galerkin methods*, SIAM J. Sci. Comput. (to appear).

[37] R.D. Lazarov and S. Margenov, *CBS constants for graph-Laplacians and application to multilevel methods for discontinuous Galerkin systems*, Journal of Complexity (to appear).

[38] R.D. Lazarov, P.S. Vassilevski, and S. Margenov, *Solving elliptic problems by the domain decomposition method using precondition matrices derived by multilevel splitting of finite element matrix*, Proceedings of the 1st Int. Conf. on Supercomputing (Greece), 1987, pp. 826–835.

[39] J.F. Maitre and F. Musy, *The contraction number of a class of two-level methods; an exact evaluation for some finite element subspaces and model problems*, Lect. Notes Math., vol. 960, 1982, pp. 535–54.

[40] S. Margenov, *Upper bound of the constant in the strengthened C.B.S. inequality for FEM 2D elasticity equations*, Num. Lin. Alg. Appl. 1 (1994), pp. 65–74.

[41] ———, *Semicoarsening AMLI algorithms for elasticity problems*, Num. Lin. Alg. Appl. 5 (1998), pp. 347–362.

[42] S. Margenov and J. Synka, *Generalized aggregation-based multilevel preconditioning of Crouzeix-Raviart FEM elliptic problems*, Tech. report, RICAM, Linz, 2006, No. 2006-23.

[43] S. Margenov and P.S. Vassilevski, *Algebraic multilevel preconditioning of anisotropic elliptic problems*, SIAM J. Sci. Comp. 15 (1994), pp. 1026–1037.

[44] ———, *Algebraic two-level preconditioning of non-conforming FEM systems*, Large-Scale Scientific Computations of Engineering and Environmental Problems, Notes on Numerical Fluid Mechanics, vol. 62, Vieweg, 1998, pp. 239–248.

[45] S. Margenov, L. Xanthis, and L. Zikatanov, *On the optimality of the semicoarsening AMLI algorithm*, Iterative Methods in Linear Algebra, II, Vol. 3 (New Jersey, USA), IMACS Series in Computational and Applied Mathematics, 1996, pp. 270–279.

[46] Y. Notay, *Using approximate inverses in algebraic multilevel methods*, Numer. Math. 80 (1998), pp. 397–417.

[47] ———, *A multilevel block incomplete factorization preconditioning*, Appl. Numer. Math. 31 (1999), pp. 209–225.

[48] I. Pultarova, *The strengthened CBS inequality constant for second order elliptic partial differential operator and for hierarchical bilinear finite element functions*, Applications of Mathematics 50 (2005), pp. 323–329.

[49] P.S. Vassilevski, *Nearly optimal iterative methods for solving finite element elliptic equations based on multilevel splitting of the matrix*, Tech. report, Institute for Scientific Computation, University of Wyoming, Laramie, U.S.A., 1989, No. 1989-09.

[50] ———, *Hybrid V-cycle algebraic multilevel preconditioners*, Math. Comp. 58 (1992), pp. 489–512.

[51] ———, *On two ways of stabilizing the hierarchical basis multilevel methods*, SIAM Rev. 39 (1997), pp. 18–53.

Author information

Johannes Kraus, Johann Radon Institute for Computational and Applied Mathematics, Austrian Academy of Sciences, Altenbergerstrasse 69, 4040 Linz, Austria.
Email: `johannes.kraus@oeaw.ac.at`

Svetozar Margenov, Institute for Parallel Processing, Bulgarian Academy of Sciences, Acad. G. Bonchev Bl. 25A, 1113 Sofia, Bulgaria.
Email: `margenov@parallel.bas.bg`

Domain decomposition methods

Sergey Nepomnyaschikh

Key words. Elliptic boundary value problems, finite element methods, iterative methods, preconditioning, domain decomposition methods, fictitious space methods

AMS classification. 65F10, 65N22, 65N55

1 Introduction

These notes aim at the construction of preconditioners B for solving systems of grid equations approximating elliptic boundary value problems in domains with complex geometry. A preconditioner B can be used, for example, in iterative processes of the following form:

$$B(u^{k+1} - u^k) = -\tau^k(Au^k - f), \tag{1.1}$$

where A is the stiffness matrix of the original system of grid equations. The convergence rate of the iterative process (1.1) depends on the constants c_1 and c_2 in the spectral equivalence inequalities

$$c_1(Bu,u) \leq (Au,u) \leq c_2(Bu,u), \tag{1.2}$$

which should be valid for any vector u. Here, we assume that A and B are symmetric positive definite matrices. In [8], a technique has been suggested for constructing the preconditioner B in the case of uniform grids of a rectangle. In addition, the constants c_1 and c_2 from (1.2) are independent of the mesh size, and, in order to perform the multiplication of B^{-1} with some vector, it is necessary to solve the system of grid equations corresponding to the five-point approximation of the Laplace operator on a uniform grid of a rectangle. The construction of a preconditioner B with similar characteristics in the case of boundary value problems in domains with complex geometry is of great interest.

The most efficient preconditioners for solving boundary value problems in domains with complex geometry can be constructed, as a rule, by 'simplifying' the geometry of the original domain. Here, we can point out two approaches. The first approach is to partition the original domain into simpler subdomains (domain decomposition methods), and the second approach is to embed the original domain in a domain of some canonical form, for example, a rectangle in the two-dimensional case and a parallelepiped in the three-dimensional case, by introducing additional equations (the fictitious domain method and its matrix counterparts) [9, 13, 16, 15, 17, 32, 36, 38].

Of the first group of methods, the so-called additive Schwarz method (ASM) is very effective. The classical overlapping domain decomposition method was first proposed by H. A. Schwarz in [40]. As a solution method for finite element equations, the ASM was suggested in [21]. Domain decomposition methods are the subject of the textbooks [34, 35, 41, 44]. In these notes, results from [21, 22, 23, 25, 26, 27, 29] are used.

In the second group of methods, major gains have been obtained for problems with natural boundary conditions [1]. Using the matrix counterpart of the fictitious domain method, we have managed to construct a preconditioner B such that the constants c_1 and c_2 from (1.2) are independent of h, and the main operation in performing multiplication of B^{-1} with a vector consists in solving the five-point finite difference counterpart of the Laplace operator in the rectangle. Later, in [18, 19], the so-called nonsymmetric augmentation of the original system of grid equations was proposed to solve the Dirichlet problem, as well as an iterative process for solving this augmented system of equations whose convergence rate is independent of h. Moreover, the main operation in one iteration step of the iterative process is the solution of the five-point finite difference counterpart of the Laplace operator in the rectangle. Some preconditioner using this idea (not optimal), has been constructed before. Finally, in [2, 14, 20, 24], the case of mixed boundary value problems was considered for second-order elliptic equations. The authors suggested an iterative process whose convergence rate is logarithmically dependent on h. It was the Dirichlet boundary condition that made it impossible to avoid the dependence on h. The most flexible approach to the construction of fictitious domain type preconditioning operators in domains with complex geometries is provided by the fictitious space lemma, which was presented in [28, 30]. This lemma gives the possibility to use "convenient" fictitious (auxiliary) spaces equipped with "convenient" norms. In particular, instead of a direct solver for the five-point approximation of the Laplace operator, BPX-like multilevel preconditioners can be used.

2 Domain decomposition

Let Ω be an L-shaped domain with the boundary $\Gamma = \partial\Omega$, and let Ω be decomposed into two rectangles Ω_1 and Ω_2 with the common boundary γ. We now consider the Dirichlet problem for the Poisson equation

$$-\Delta u = f \quad \text{in } \Omega \quad \text{and} \quad u = 0 \quad \text{on } \Gamma$$

as model problem. Let $H_0^1(\Omega) = \{u \in H^1(\Omega) \,|\, u = 0 \text{ on } \Gamma\}$ the subspace of all functions from the Sobolev space $H^1(\Omega)$ vanishing on the boundary Γ. Then the weak formulation reads as follows: Find $u \in H_0^1(\Omega)$ such that

$$\int_\Omega (\nabla u, \nabla v) d\Omega = \int_\Omega f v \, d\Omega \;\forall v \in H_0^1(\Omega).$$

We assume that Ω^h is an uniform triangulation of Ω and define the discrete space $H_h(\Omega) = \{u^h \in H_0^1(\Omega) \,|\, u^h = \text{piecewise-linear}\}$. The finite element function $u^h \in H_h(\Omega)$ can be identified with the vector $\bar{u} = [u_1, \ldots, u_N]^\top$, where $u_i = u^h(z_i)$ are the values at the nodes z_i. The finite element approximation to the weak formulation results in the finite element equations of the form

$$A\bar{u} = \bar{f}, \tag{2.1}$$

where the stiffness matrix A and the load vector $\bar{f}$ are defined by the identities

$$(A\bar{u}, \bar{v}) = \int_\Omega (\nabla u^h, \nabla v^h) d\Omega \quad \text{and} \quad (\bar{f}, \bar{v}) = \int_\Omega f v^h d\Omega,$$

respectively.

The vector $\bar{u}$ can be decomposed into three groups, that is, $\bar{u} = [\bar{u}_0, \bar{u}_1, \bar{u}_2]^\top = [\bar{u}_0^\top, \bar{u}_1^\top, \bar{u}_2^\top]^\top$, where $\bar{u}_0$, $\bar{u}_1$, and $\bar{u}_2$ are corresponding to γ, Ω_1, and Ω_2, respectively. Apparently, we get

$$A\bar{u} = \begin{bmatrix} A_0 & A_{01} & A_{02} \\ A_{10} & A_1 & 0 \\ A_{20} & 0 & A_2 \end{bmatrix} \begin{bmatrix} \bar{u}_0 \\ \bar{u}_1 \\ \bar{u}_2 \end{bmatrix} = \begin{bmatrix} \bar{f}_0 \\ \bar{f}_1 \\ \bar{f}_2 \end{bmatrix} = \bar{f}.$$

We observe that A_i corresponds to the Dirichlet problem in Ω_i, that is, $A_i \longleftrightarrow -\Delta_{\Omega_i}$, for $i = 1, 2$. From the second and the third equation, we obtain

$$\bar{u}_1 = A_1^{-1}\bar{f}_1 - A_1^{-1}A_{10}\bar{u}_0, \tag{2.2}$$

$$\bar{u}_2 = A_2^{-1}\bar{f}_2 - A_2^{-1}A_{20}\bar{u}_0. \tag{2.3}$$

Substituting (2.2) and (2.3) into the first equation, we get

$$(A_0 - A_{01}A_1^{-1}A_{10} - A_{02}A_2^{-1}A_{20})\bar{u}_0 = \bar{f}_0 - A_{01}A_1^{-1}\bar{f}_1 - A_{02}A_2^{-1}\bar{f}_2. \tag{2.4}$$

Introducing the Schur complement matrix

$$S = A_0 - A_{01}A_1^{-1}A_{10} - A_{02}A_2^{-1}A_{20}$$

and the vectors

$$\phi = \bar{u}_0 \quad \text{and} \quad \psi = \bar{f}_0 - A_{01}A_1^{-1}\bar{f}_1 - A_{02}A_2^{-1}\bar{f}_2,$$

we can rewrite system (2.4) in the abbreviated form

$$S\phi = \psi. \tag{2.5}$$

If we can find a good preconditioner Σ for S, then the solution ϕ can efficiently be approximated by an iterative method, e.g. by the Richardson iteration

$$\Sigma(\phi^{k+1} - \phi^k) = -\tau_k(S\phi^k - \psi),$$

where τ_k denotes some suitably chosen relaxation parameter. The correspondence between an approximate solution of the Schur complement problem (2.5) with the original problem (2.1) is given by the following lemma.

Lemma 2.1 *If $\|\phi^n - \phi\|_S = \epsilon$, then $\|\bar{u}^n - \bar{u}\|_A = \epsilon$, where the components of the vector $\bar{u}^n$ are given by $\bar{u}_0^n = \varphi^n$, $\bar{u}_1^n = A_1^{-1}(\bar{f}_1 - A_{10}\phi^n)$ and $\bar{u}_2^n = A_2^{-1}(\bar{f}_2 - A_{20}\phi^n)$.*

Proof. Using (2.2) and (2.3), we obtain the equations

$$\begin{aligned}
\|\bar{u}^n - \bar{u}\|_A^2 &= (A(\bar{u}^n - \bar{u}), \bar{u}^n - \bar{u}) \\
&= \left(\begin{bmatrix} A_0 & A_{01} & A_{02} \\ A_{10} & A_1 & 0 \\ A_{20} & 0 & A_2 \end{bmatrix} \begin{bmatrix} \bar{u}_0^n - \bar{u}_0 \\ \bar{u}_1^n - \bar{u}_1 \\ \bar{u}_1^n - \bar{u}_2 \end{bmatrix}, \begin{bmatrix} \bar{u}_0^n - \bar{u}_0 \\ \bar{u}_1^n - \bar{u}_1 \\ \bar{u}_1^n - \bar{u}_2 \end{bmatrix} \right) \\
&= \left(\begin{bmatrix} A_0\bar{u}_0^n + A_{01}\bar{u}_1^n + A_{02}\bar{u}_2^n - \bar{f}_0 \\ A_{10}\bar{u}_0^n + A_1\bar{u}_1^n - \bar{f}_1 \\ A_{20}\bar{u}_0^n + A_2\bar{u}_2^n - \bar{f}_2 \end{bmatrix}, \begin{bmatrix} \bar{u}_0^n - \bar{u}_0 \\ \bar{u}_1^n - \bar{u}_1 \\ \bar{u}_1^n - \bar{u}_2 \end{bmatrix} \right) \\
&= \left(\begin{bmatrix} S\bar{u}_0^n - \psi \\ 0 \\ 0 \end{bmatrix}, \begin{bmatrix} \bar{u}_0^n - \bar{u}_0 \\ \bar{u}_1^n - \bar{u}_1 \\ \bar{u}_1^n - \bar{u}_2 \end{bmatrix} \right) = \left(\begin{bmatrix} S\bar{u}_0^n - \psi \\ 0 \\ 0 \end{bmatrix}, \begin{bmatrix} \varphi^n - \varphi \\ 0 \\ 0 \end{bmatrix} \right) = \epsilon,
\end{aligned}$$

which proves the lemma. □

There are some interesting facts about S. The first is that though, S is an interface problem, it is closely related to the entire problem. The second is that the quadratic form $(S\phi, \phi)$ is equivalent to some trace norm. In order to obtain this, we consider

$$\begin{aligned}
(A\bar{u}, \bar{u}) &= \int_\Omega |\nabla u^h|^2 \, d\Omega \\
&= \int_{\Omega_1} |\nabla u^h|^2 \, d\Omega_1 + \int_{\Omega_2} |\nabla u^h|^2 \, d\Omega_2 \\
&= \left(A^{(1)} \begin{bmatrix} \bar{u}_0 \\ \bar{u}_1 \end{bmatrix}, \begin{bmatrix} \bar{u}_0 \\ \bar{u}_1 \end{bmatrix} \right) + \left(A^{(2)} \begin{bmatrix} \bar{u}_0 \\ \bar{u}_2 \end{bmatrix}, \begin{bmatrix} \bar{u}_0 \\ \bar{u}_2 \end{bmatrix} \right),
\end{aligned}$$

where $A^{(i)} = \begin{bmatrix} A_0^{(i)} & A_{0i} \\ A_{i0} & A_i \end{bmatrix}$ is just the discrete Laplacian $-\Delta_{\Omega_i}$, which satisfies the Dirichlet condition on $\partial\Omega_i \setminus \gamma$ and the Neumann condition on γ, $i = 1, 2$. Let $S_i = A_0^{(i)} - A_{i0}A_i^{-1}A_{i0}$. Then we have $S = S_1 + S_2$ and $A_0 = A_0^{(1)} + A_0^{(2)}$. Moreover,

$$\begin{aligned} \inf_{u_1}\left(A^{(1)}\begin{bmatrix}\phi \\ u_1\end{bmatrix}, \begin{bmatrix}\phi \\ u_1\end{bmatrix}\right) &= \inf_{u_1}((A_0^{(1)}\phi, \phi) + (A_1 u_1, u_1) + 2(A_{10}\phi, u_1)) \\ &= (A_0^{(1)}\phi, \phi) + \inf_{u_1}((A_1 u_1, u_1) - 2(-A_{10}\phi, u_1)). \end{aligned}$$

The quadratic form $(A_1 u_1, u_1) - 2(-A_{10}\phi, u_1)$ attains its minimum at $A_1 u_1 = -A_{10}\phi$, that is, $u_1 = -A_1^{-1}A_{10}\phi$. Thus,

$$\begin{aligned} \inf_{u_1}\left(A^{(1)}\begin{bmatrix}\phi \\ u_1\end{bmatrix}, \begin{bmatrix}\phi \\ u_1\end{bmatrix}\right) &= (A_0^{(1)}\phi, \phi) + (A_{10}\phi, A_1^{-1}A_{10}\phi) - 2(A_{10}\phi, A_1^{-1}A_{10}\phi) \\ &= (A_0^{(1)}\phi, \phi) - (A_{01}A_1^{-1}A_{10}\phi, \phi) \\ &= (S_1\phi, \phi). \end{aligned}$$

Hence we obtain the following relation:

$$(S\phi, \phi) = \inf_{u_h \in H_h(\Omega_1), u_h|_\gamma = \phi_h} |u^h|^2_{H^1(\Omega_1)} + \inf_{u_h \in H_h(\Omega_2), u_h|_\gamma = \phi_h} |u^h|^2_{H^1(\Omega_2)}.$$

In fact, the infimum occurs at u^h which solves the discrete Laplacian problem:

$$\begin{aligned} -\Delta_h u^h &= 0, \quad \text{in } \Omega_i, \\ u^h &= 0, \quad \text{on } \partial\Omega_i \setminus \gamma, \\ u^h &= \phi^h, \quad \text{on } \gamma. \end{aligned}$$

Let $\Gamma^h = \partial\Omega^h$, $\{z_i\} = \{x_i\}$ be the set of nodes contained in Γ^h, $I_i = [z_i, z_{i+1}]$, and $h_i = |z_{i+1} - z_i|$. Since Ω^h is shape regular, there exist $c_i \neq c_i(h)$ such that $c_1 \leq h_i/h_{i+1} \leq c_2$. Now, we define the discrete norms corresponding to the continuous trace norms by

$$\|\phi^h\|^2_{L_{2,h}(\Gamma^h)} := \sum_{z_i \in \Gamma^h} \phi^h(z_i)^2 h_i$$

and

$$|\phi^h|^2_{H_h^{1/2}(\Gamma^h)} := \sum_{z_i \in \Gamma^h} \sum_{z_j \in \Gamma^h, i \neq j} \frac{(\phi^h(z_i) - \phi^h(z_j))^2}{|z_i - z_j|^2} h_i h_j.$$

Lemma 2.2 *The above discrete norms are equivalent to* $\|\cdot\|_{L_2(\Gamma^h)}$ *and* $|\cdot|_{H^{1/2}(\Gamma^h)}$, *respectively, where the equivalence constants are independent of h.*

Proof. For the sake of simplicity we assume that Γ^h is the unit interval $[0,1]$ with the nodes $\{x_i\}$. For the $L_2(\Gamma^h)$ case, it is easy to prove the equivalence by observing

$$\|\phi^h\|_{L_2(\Gamma)} = \sum_{I_i} \|\phi^h\|^2_{L_2(I_i)}.$$

Let us consider the $H^{1/2}(\Gamma_h)$ case. Then, we have

$$\int_\Gamma \int_\Gamma \frac{(\phi^h(x) - \phi^h(y))^2}{|x-y|^2} dxdy = \sum_i \sum_j \int_{I_i} \int_{I_j} \frac{(\phi^h(x) - \phi^h(y))^2}{|x-y|^2} dxdy.$$

The above equation can be treated in the following three different cases:

- Case 1) Let $i = j$. On the interval $I_i = [x_i, x_{i+1}]$, ϕ^h is a linear function, in fact,

$$\phi^h(x) = \phi_i + \frac{\phi_{i+1} - \phi_i}{h_i} x.$$

Therefore,

$$\int_{I_i} \int_{I_j} \frac{(\phi^h(x) - \phi^h(y))^2}{|x-y|^2} dxdy$$
$$= \int_{x_i}^{x_{i+1}} \int_{x_i}^{x_{i+1}} \frac{(\phi_i + \frac{\phi_{i+1}-\phi_i}{h_i} x - \phi_i - \frac{\phi_{i+1}-\phi_i}{h_i} y)^2}{|x-y|^2} dxdy = (\phi_{i+1} - \phi_i)^2.$$

- Case 2) Let $i + 1 < j$. Then we have, $\forall x \in I_i, \forall y \in I_j$,

$$|x_{i+1} - x_j| \le |x - y| \le |x_{j+1} - x_j|,$$

and, therefore,

$$\int_{I_i} \int_{I_j} \frac{(\phi^h(x) - \phi^h(y))^2}{|x-y|^2} dxdy \simeq \frac{1}{|x_{i+1} - x_j|^2} \int_{x_j}^{x_{j+1}} \int_{x_i}^{x_{i+1}} (\phi^h(x) - \phi^h(y))^2 dxdy.$$

Let $x = x_i + h_i x'$ and $y = x_j + h_j y'$. Then the above equation is equal to

$$\frac{1}{|x_{i+1} - x_j|^2} h_i h_j \int_0^1 \int_0^1 (\tilde{\phi}^h(x') - \tilde{\phi}^h(y'))^2 dx' dy'.$$

We now define bilinear forms A and B by,

$$(A[\phi_i, \phi_{i+1}, \phi_j, \phi_{j+1}]^\top, [\phi_i, \phi_{i+1}, \phi_j, \phi_{j+1}]^\top)$$
$$= \int_0^1 \int_0^1 (\tilde{\phi}^h(x') - \tilde{\phi}^h(y'))^2 dx' dy',$$
$$(B[\phi_i, \phi_{i+1}, \phi_j, \phi_{j+1}]^\top, [\phi_i, \phi_{i+1}, \phi_j, \phi_{j+1}]^\top)$$
$$= (\phi_i - \phi_j)^2 + (\phi_i - \phi_{i+1})^2 + (\phi_{i+1} - \phi_j)^2 + (\phi_j - \phi_{j+1})^2.$$

By an easy calculation, we get

$$(A[\phi_i,\phi_{i+1},\phi_j,\phi_{j+1}]^\top,[\phi_i,\phi_{i+1},\phi_j,\phi_{j+1}]^\top)$$
$$= \frac{2}{3}(\phi_i^2+\phi_i\phi_{i+1}+\phi_{i+1}^2)+\frac{2}{3}(\phi_j^2+\phi_i\phi_{j+1}+\phi_{j+1}^2)-2(\phi_i+\phi_{i+1})(\phi_j+\phi_{j+1}).$$

We observe that Ker$A$ = Ker$B$ = $\{\phi \mid \phi_i=\phi_{i+1}=\phi_j=\phi_{j+1}\}$. Therefore, there exist constants $c_1, c_2 \neq c(h)$ such that

$$c_1(A\phi,\phi)\le(B\phi,\phi)\le c_2(A\phi,\phi).$$

- Case 3) Let $i+1=j$. Let $x=x_{i+1}-h_ix'$ and $y=x_{i+1}+h_{i+1}y'$. Then we have

$$\int_{I_i}\int_{I_j}\frac{(\phi^h(x)-\phi^h(y))^2}{|x-y|^2}dxdy = h_ih_{i+1}\int_0^1\int_0^1\frac{(\tilde{\phi}^h(-x')-\tilde{\phi}^h(y'))^2}{|h_ix'+h_{i+1}y'|^2}dx'dy'.$$

Assume that $h_{i+1}\ge h_i$. Then $\frac{h_{i+1}}{h_i}=a\ge 1$, so that $x'+y'\le x'+ay'\le a(x'+y')$. Therefore, the above equation is equal to

$$\begin{aligned}&\frac{h_{i+1}}{h_i}\int_0^1\int_0^1\frac{(\tilde{\phi}^h(-x')-\tilde{\phi}^h(y'))^2}{|x'+ay'|^2}dx'dy'\\ &\quad\simeq \frac{h_{i+1}}{h_i}\int_0^1\int_0^1\frac{(\tilde{\phi}^h(-x')-\tilde{\phi}^h(y'))^2}{|x'+y'|^2}dx'dy'\\ &\quad\simeq (\phi_i-\phi_{i+1})^2+(\phi_{i+1}-\phi_{i+2})^2.\end{aligned}$$

This completes the proof of our lemma. □

Then we obtain the following theorem.

Theorem 2.3 *Suppose that there exist constants c_1 and c_2 such that, for any $u_i^h\in H_h(\Omega_i)$ with $u_i^h=\phi^h$ on γ, the inequality*

$$\|\phi^h\|_{H_h^{1/2}(\partial\Omega_i)}\le c_1\|u_i^h\|_{H^1(\Omega_i)}$$

holds, and, for any $\phi^h\in H_h^{1/2}(\partial\Omega_i)$, there exist $u_i^h\in H_h(\Omega_i)$ with $u_i^h=\phi^h$ such that

$$\|u_i^h\|_{H^1(\Omega_i)}\le c_2\|\phi^h\|_{H_h^{1/2}(\partial\Omega_i)}.$$

Then

$$(S_i\phi,\phi)\simeq\|\phi^h\|^2_{H_h^{1/2}(\partial\Omega_i)}.$$

Here ϕ is the vector representation of ϕ^h.

We have just treated the Laplacian equation. For the general elliptic case, S is more complicated so that it is difficult to find a preconditioner for S. But the following lemma shows that it is enough to find a preconditioner for the Laplacian equation.

Lemma 2.4 *Let* $A = \begin{bmatrix} A_{11} & A_{12} \\ A_{21} & A_{22} \end{bmatrix}$ *and* $B = \begin{bmatrix} B_{11} & B_{12} \\ B_{21} & B_{22} \end{bmatrix}$. *Assume that* $A = A^\top \geq 0$, $B = B^\top \geq 0$, $A_{11} > 0$, *and* $B_{11} > 0$. *Let* $S_A = A_{22} - A_{21}A_{11}^{-1}A_{12}$ *and* $S_B = B_{22} - B_{21}B_{11}^{-1}B_{12}$. *If*

$$c_1(Au, u) \leq (Bu, u) \leq c_2(Au, u) \qquad \forall u$$

then

$$c_1(S_A u, u) \leq (S_B u, u) \leq c_2(S_A u, u) \qquad \forall u.$$

Proof. We start with the lower estimate. Indeed,

$$\begin{aligned} (S_B u_2, u_2) &= \inf_{u_1} \left(B \begin{bmatrix} u_1 \\ u_2 \end{bmatrix}, \begin{bmatrix} u_1 \\ u_2 \end{bmatrix} \right) \\ &= \left(B \begin{bmatrix} v_1 \\ u_2 \end{bmatrix}, \begin{bmatrix} v_1 \\ u_2 \end{bmatrix} \right) \qquad \text{(for some } v_1\text{)} \\ &\geq c_1 \left(A \begin{bmatrix} v_1 \\ u_2 \end{bmatrix}, \begin{bmatrix} v_1 \\ u_2 \end{bmatrix} \right) \\ &\geq c_1 \inf_{u_1} \left(A \begin{bmatrix} u_1 \\ u_2 \end{bmatrix}, \begin{bmatrix} u_1 \\ u_2 \end{bmatrix} \right) = c_1(S_A u_2, u_2). \end{aligned}$$

The upper estimate can be proved with the same arguments. □

Let

$$Lu = -\sum_{i,j=1}^{d} \frac{\partial}{\partial x_i}(a_{ij}(x)\frac{\partial u}{\partial x_j}) + a_0(x)u = f(x), \ x \in \Omega.$$

Assume that

$$(Lu, u) \simeq \|u\|^2_{H^1(\Omega)}.$$

Let S_L and S_1 be Schur complement matrices for L and -Δ, respectively. Lemma 2.4 implies that it is enough to construct a preconditioner for S_1 in place of S_L.

3 Finite element trace theorem

To construct effective preconditioners on interfaces we need the following result about the analytical characterization of finite element traces.

Theorem 3.1 *Let* Ω *be a bounded domain with the piecewise smooth boundary* Γ, *and let* Ω^h *(*Ω^h *is a polygonal approximation of* Ω *whose vertex may not lie on* Γ*) be a shape-regular triangulation of* Ω *such that*

i) we have

$$\frac{\operatorname{diam}\tau_i}{r_i} \le c \ne c(h),$$

where r_i denotes the radius of the largest ball inscribed in τ_i,

ii) there exists a mapping $T : \tau_i \to \tilde{\tau}_i$ such that $T(z_i) = \tilde{z}_i$ (z_i and $\tilde{z}_i$ are the vertices of τ_i and $\tilde{\tau}_i$, respectively) and

- *$T(\tau_i) = \tilde{\tau}_i$ is also shape regular,*
- *$z_i \in \Gamma^h \Longrightarrow \tilde{z}_i \in \Gamma$ (The map T moves $z_i \in \Gamma^h$ to $\tilde{z}_i \in \Gamma$.),*
- *$\exists c_1, c_2 \ne c(h),\ c_1|z_i - z_j| \le |\tilde{z}_i - \tilde{z}_j| \le c_2|z_i - z_j|$.*

Then

(1) There exists a constant $c_3 \ne c_3(h)$ such that

$$\|\varphi^h\|_{H_h^{1/2}(\Gamma^h)} \le c_3\|u^h\|_{H^1(\Omega^h)} \quad \forall u^h \in H_h(\Omega^h) \text{ with } u^h|_{\Gamma^h} = \varphi^h.$$

(2) There exists a constant $c_4 \ne c_4(h)$ such that, for any given $\varphi^h \in H_h(\Gamma^h)$, there exists a finite element function $u^h \in H_h(\Omega^h)$ satisfying the trace condition $u^h = \varphi^h$ on Γ_h and the inequality

$$\|u^h\|_{H^1(\Omega^h)} \le c_4\|\varphi^h\|_{H^{1/2}(\Gamma^h)}.$$

Remark 3.2 Such mappings T really exist if $\Gamma^h \approx \Gamma$ in $O(h^2)$ accuracy.

In the paper [12] by Korneev (1970), the special finite element function $\tilde{u}^h \in H_h(\widetilde{\Omega}^h)$ on the curvilinear triangulation $\widetilde{\Omega}^h$ was suggested such that $\tilde{u}^h(\tilde{z}_i) = u^h(z_i)$ where $\tilde{u}^h \in H_h(\widetilde{\Omega}^h)$ and $u^h \in H_h(\Omega^h)$. Moreover, the following lemma holds.

Lemma 3.3 *There exist constants c_5 and $c_6 \ne c(h)$ such that*

$$c_5\|\tilde{u}^h\|_{L^2(\tilde{\tau}_i)} \le \|u^h\|_{L^2(\tau_i)} \le c_6\|\tilde{u}^h\|_{L^2(\tilde{\tau}_i)},$$
$$c_5|\tilde{u}^h|_{H^1(\tilde{\tau}_i)} \le |u^h|_{H^1(\tau_i)} \le c_6|\tilde{u}^h|_{H^1(\tilde{\tau}_i)},$$
$$c_5\|\tilde{\varphi}^h\|_{L^2(\tilde{I}_i)} \le \|\varphi^h\|_{L^2(I_i)} \le c_6\|\tilde{\varphi}^h\|_{L^2(\tilde{I}_i)},$$

and

$$\begin{aligned} c_5\int_{\tilde{I}_i}\int_{\tilde{I}_j}\frac{(\tilde{\varphi}^h(x)-\tilde{\varphi}^h(y))^2}{|x-y|^2}\,dxdy &\le \int_{I_i}\int_{I_j}\frac{(\varphi^h(x)-\varphi^h(y))^2}{|x-y|^2}\,dxdy \\ &\le c_6\int_{\tilde{I}_i}\int_{\tilde{I}_j}\frac{(\tilde{\varphi}^h(x)-\tilde{\varphi}^h(y))^2}{|x-y|^2}\,dxdy, \end{aligned}$$

where $\Gamma^h = \bigcup_i I_i$ and $\Gamma = \bigcup_i \tilde{I}_i$.

The proof of Lemma 3.3 is based on the result from [12]. Now, we are able to prove Theorem 3.1.

Proof (Existence of c_3). There obviously exists a constant c_7 such that, for any given $u^h \in H_h(\Omega^h)$, there is a $\tilde{u}^h \in H^1(\tilde{\Omega}^h)$ satisfying the inequality $\|\tilde{u}^h\|_{H^1(\tilde{\Omega}^h)} \leq c_7\|u^h\|_{H^1(\Omega^h)}$. Setting $\tilde{\varphi}^h = \tilde{u}^h|_\Gamma \in H_h(\Gamma)$, we define $\varphi^h \in H_h(\Gamma_h)$ as a linear combinations of vertex values of $\tilde{\varphi}^h$. Now, we get the inequality

$$\|\tilde{\varphi}^h\|_{H^{1/2}(\Gamma)} \leq c_8\|\tilde{u}^h\|_{H^1(\Omega)}$$

from the usual trace theorem. By Lemma 3.3, it follows that

$$\|\varphi^h\|_{H_h^{1/2}(\Gamma^h)} \leq c_9\|\tilde{\varphi}^h\|_{H^{1/2}(\Gamma)}.$$

We remark that this is immediate in the case $\Omega^h = \overline{\Omega}$. □

Proof (Existence of c_4). For a given $\varphi^h \in H_h(\Gamma^h)$, let $\tilde{\varphi}^h \in H_h(\Gamma)$ be such that $\tilde{\varphi}^h(\tilde{z}_i) = \varphi^h(z_i)$. Then we have by Lemma 3.3

$$\|\tilde{\varphi}^h\|_{H^{1/2}(\Gamma)} \leq c\|\varphi^h\|_{H_h^{1/2}(\Gamma^h)}.$$

By the inverse trace theorem, there exists $u \in H^1(\Omega)$ such that $u|_\Gamma = \tilde{\varphi}^h$ and $\|u\|_{H^1(\Omega)} \leq c\|\tilde{\varphi}^h\|_{H^{1/2}(\Gamma)}$. But $u \notin H_h(\Omega)$. How can we construct $\tilde{u}^h \in H_h(\Omega)$? It is enough to have values at $\tilde{z}_i$. Let

$$\tilde{u}^h(\tilde{z}_i) = \begin{cases} \tilde{\varphi}^h(\tilde{z}_i), & \text{if } \tilde{z}_i \in \Gamma, \\ \dfrac{1}{\pi r_i}\displaystyle\int_{B(\tilde{z}_i, r_i)} u(x)\,dx, & \text{otherwise,} \end{cases}$$

where r_i is the radius of the largest ball $B(\tilde{z}_i, r_i)$ inscribed in the union of all elements sharing the vertex $\tilde{z}_i$ which is denoted by K_i. Then we take $u^h \in H_h(\Omega^h)$ with $u^h(z_i) = \tilde{u}^h(\tilde{z}_i)$. By Lemma 3.3 it follows that $\|u^h\|_{H^1(\Omega^h)} \leq c\|\tilde{u}^h\|_{H^1(\Omega)}$.

It remains to show that $\|\tilde{u}^h\|_{H^1(\Omega)} \leq c\|\varphi^h\|_{H^{1/2}(\Gamma)}$. We note that $\varphi^h \to \tilde{\varphi}^h \to u \to \tilde{u}^h \to u^h$. By Friedrich's inequality, we obtain

$$\|\tilde{u}^h\|_{L^2(\Omega)} \leq c(|\tilde{u}^h|_{H^1(\Omega)} + \|\tilde{\varphi}^h\|_{L^2(\Gamma)}),$$

and since $\|\tilde{\varphi}^h\|_{L^2(\Gamma)} \leq C\|\varphi\|_{H^{1/2}(\Gamma)}$ it is enough to estimate $|\tilde{u}^h|_{H^1(\Omega)}$. We note that

$$|\tilde{u}^h|^2_{H^1(\Omega)} \leq c \sum_{l_i \in \tilde{\Omega}^h} (\tilde{u}^h(\tilde{z}_{i_1}) - \tilde{u}^h(\tilde{z}_{i_2}))^2,$$

where $\tilde{z}_{i_1}$ and $\tilde{z}_{i_2}$ are the vertices of the edge l_i. Now we consider the following three cases separately:

- Case 1) $\tilde{z}_{i_1}, \tilde{z}_{i_2} \in \Gamma$. In this case, we immediately get

$$\begin{aligned} \sum(\tilde{u}^h(\tilde{z}_{i_1}) - \tilde{u}^h(\tilde{z}_{i_2}))^2 &= \sum(\tilde{\varphi}^h(\tilde{z}_{i_1}) - \tilde{\varphi}^h(\tilde{z}_{i_2}))^2 \\ &\leq \sum_{\tilde{z}_i}\sum_{\tilde{z}_j} \frac{(\tilde{\varphi}^h(\tilde{z}_i) - \tilde{\varphi}^h(\tilde{z}_i))^2}{|z_i - z_j|^2} h_i h_j \leq c|\tilde{\varphi}^h|^2_{H^{1/2}(\Gamma)}. \end{aligned}$$

- Case 2) $\tilde{z}_{i_1}, \tilde{z}_{i_2} \in \Omega$. Here, we use the following result.

Lemma 3.4 *Let* $0 < h_1 \leq h_2$. *Then we have the estimate*

$$\left(\frac{1}{\pi h_2^2} \int_{B(0,h_2)} u(x)\, dx - \frac{1}{\pi h_1^2} \int_{B(0,h_1)} u(x)\, dx \right)^2 \leq \frac{h_2}{\pi h_1} |u|^2_{H^1(B(0,h_2))}$$

that is valid for all $u \in H^1(B(0, h_2))$.

Proof. Let (r, θ) be the radial coordinate system given by the relation

$$x = (x_1, x_2) = (r \cos\theta, r \sin\theta).$$

Then we obtain

$$\begin{aligned}
&\left(\frac{1}{\pi h_2^2} \int_{B(0,h_2)} u(x)\, dx - \frac{1}{\pi h_1^2} \int_{B(0,h_1)} u(x)\, dx \right)^2 \\
&= \left(\frac{1}{\pi h_2^2} \int_0^{h_2} \int_0^{2\pi} u(r,\theta)\, r d\theta dr - \frac{1}{\pi h_1^2} \int_0^{h_1} \int_0^{2\pi} u(r,\theta)\, r d\theta dr \right)^2 \\
&= \left(\frac{1}{\pi h_2^2} \int_0^{h_2} \int_0^{2\pi} (u(r,\theta) - u(r/a,\theta))\, r d\theta dr \right)^2 \qquad (a = h_2/h_1 \geq 1) \\
&= \frac{1}{\pi^2 h_2^4} \left(\int_0^{h_2} \int_0^{2\pi} \left[\int_{\frac{r}{a}}^{r} r^{1/2} \frac{\partial u(t,\theta)}{\partial t}\, dt \right] r^{1/2} d\theta dr \right)^2 \\
&\overset{\text{C.B.}}{\leq} \frac{1}{\pi h_2} \int_0^{h_2} \int_0^{2\pi} \int_{\frac{r}{a}}^{r} \left(\frac{\partial u(t,\theta)}{\partial t} \right)^2 r\, dt d\theta dr \\
&\overset{r \leq at}{\leq} \frac{a}{\pi h_2} \int_0^{h_2} \int_0^{2\pi} \int_{\frac{r}{a}}^{r} \left(\frac{\partial u(t,\theta)}{\partial t} \right)^2 t\, dt d\theta dr \\
&\leq \frac{a}{\pi h_2} \int_0^{h_2} \int_0^{2\pi} \int_0^{h_2} \left(\frac{\partial u(t,\theta)}{\partial t} \right)^2 t\, dt d\theta dr \\
&= \frac{a}{\pi} \int_0^{2\pi} \int_0^{h_2} \left(\frac{\partial u(t,\theta)}{\partial t} \right)^2 t\, dt d\theta \\
&\leq \frac{a}{\pi} |u|^2_{H^1(B(0,h_2))}.
\end{aligned}$$

The last inequality follows from the fact that $(\frac{\partial u}{\partial r})^2 \leq (\frac{\partial u}{\partial x_1})^2 + (\frac{\partial u}{\partial x_2})^2$. This finishes the proof of Lemma 3.4. □

Now, we can continue the proof with case 2).

Can we find a constant c_4 such that for all $\phi_h \in H_h(\Gamma^h)$ there exists a finite element function $u^h \in H_h(\Omega^h)$ satisfying the trace condition $u^h(x) = \phi^h(x)$, $x \in \Gamma^h$ and the inequality $\|u^h\|_{H^1(\Omega^h)} \leq c_4 |\phi^h|_{H_h^{1/2}(\Gamma^h)}$? Using the construction $\phi^h \to \tilde{\phi}^h \in$

$\tilde{H}_h(\Gamma) \to u \in H^1(\Omega) \to \tilde{u}^h \in \tilde{H}_h(\Omega)$ with

$$\tilde{u}^h(\tilde{z}_i) = \frac{1}{\pi r_i^2} \int_{B(\tilde{z}_i, r_i)} u(x)dx, \tag{3.1}$$

we can proceed as follows. There are two cases:

$$1) \quad \tilde{z}_i, \tilde{z}_j \in \Gamma$$
$$2) \quad \tilde{z}_i, \tilde{z}_j \in \Omega$$

Let r denote the radius satisfying the following inclusion:

$$B(x, \sqrt{2}r) \subset K_{i_1} \cup K_{i_2}, \quad x \in l_i.$$

Now we estimate $(\tilde{u}^h(\tilde{z}_{i_2}) - \tilde{u}^h(\tilde{z}_{i_1}))^2$ as follows:

$$\begin{aligned}(\tilde{u}^h(\tilde{z}_{i_2}) - \tilde{u}^h(\tilde{z}_{i_1}))^2 \leq\ & 3\left(\left(\tilde{u}^h(\tilde{z}_{i_2}) - \frac{1}{\pi r^2}\int_{B(\tilde{z}_{i_2},r)} u(x)dx\right)^2\right. \\ & + \left(\frac{1}{\pi r^2}\int_{B(\tilde{z}_{i_1},r)} u(x)dx - \tilde{u}^h(\tilde{z}_{i_1})\right)^2 \\ & \left. + \left(\frac{1}{\pi r^2}\int_{B(\tilde{z}_{i_2},r)} u(x)dx - \frac{1}{\pi r^2}\int_{B(\tilde{z}_{i_1},r)} u(x)dx\right)^2\right).\end{aligned}$$

For the first two terms we can use Lemma 3.4. Let us now estimate the third term:

$$\begin{aligned}&\left(\frac{1}{\pi r^2}\int_{B(\tilde{z}_{i_2},r)} u(x)dx - \frac{1}{\pi r^2}\int_{B(\tilde{z}_{i_1},r)} u(x)dx\right)^2 \\ &= \frac{1}{\pi^2 r^4}\left(\int_{B(\tilde{z}_{i_1},r)} (u(x+y) - u(x)) \cdot 1dx\right)^2 \\ &\leq \frac{1}{\pi r^2}\int_{B(\tilde{z}_{i_1},r)} (u(x+y) - u(x))^2 dx \\ &\leq \frac{1}{\pi r^2}\int_{-r}^{r}\int_{-r}^{r} (u(s+h,t) - u(s,t))^2 dsdt \\ &\leq \frac{1}{\pi r^2}\int_{-r}^{r}\int_{-r}^{r}\int_{s}^{s+h} \left(\frac{\partial u(\xi,t)}{\partial \xi}\right)^2 d\xi dsdt \\ &\leq \frac{1}{\pi r^2}\int_{-r}^{r}\int_{-r}^{r}\int_{-r}^{r+h} \left(\frac{\partial u(\xi,t)}{\partial \xi}\right)^2 d\xi dsdt \\ &= \frac{2h}{\pi r^2}\int_{-r}^{r}\int_{-r}^{r+h} \left(\frac{\partial u(\xi,t)}{\partial \xi}\right)^2 d\xi dt \\ &\leq \frac{2h}{\pi r^2}|u|_{H^1(K_{i_1} \cup K_{i_2})}.\end{aligned}$$

• Case 3) $\tilde{z}_{i_1} \in \Gamma$, $\tilde{z}_{i_2} \in \Omega$. Next, let $\tilde{z}_{i_1} = (0,0)$, $\tilde{z}_{i_1}^{+} = (h_1, 0)$, $\tilde{z}_{i_1}^{-} = (-h_2, 0)$, $\tilde{z}_{i_2} = (0, h_3)$, and r such that $B(\tilde{z}_i, r) \subset S$, where $S = \{(s,h) | -h_2 \le s \le h_1, 0 \le h \le 2h_3\}$. Then, we have

$$
\begin{aligned}
(\tilde{u}^h(\tilde{z}_{i_2}) - \tilde{u}^h(\tilde{z}_{i_1}))^2 \ &\le \ 2\left(\tilde{u}^h(\tilde{z}_{i_2}) - \frac{1}{\pi r^2}\int_{B(\tilde{z}_{i_2}, r)} u(x)dx\right)^2 \\
&\quad + \left(\frac{1}{\pi r^2}\int_{B(\tilde{z}_{i_2}, r)} u(x)dx - \tilde{u}^h(\tilde{z}_{i_1})\right)^2 .
\end{aligned}
$$

The second term can be estimated as follows:

$$
\begin{aligned}
&\frac{1}{\pi^2 r^4}\left(\int_{B(\tilde{z}_{i_1}, r)} (u(x) - \tilde{u}^h(\tilde{z}_{i_2}))dx\right)^2 \overset{\text{C.B.}}{\le} \frac{1}{\pi r^2}\int_{B(\tilde{z}_{i_1}, r)} (u(x) - \tilde{u}^h(\tilde{z}_{i_2}))^2 dx \\
&\le \frac{1}{\pi r^2}\int_{-h_2}^{h_1}\int_0^{2h_3} (u(s,t) - \phi^h(0))^2 dtds \\
&\le \frac{2}{\pi r^2}\left(\int_{-h_2}^{h_1}\int_0^{2h_3} (u(s,t) - \phi^h(s))^2 dtds + \int_{-h_2}^{h_1}\int_0^{2h_3} (\phi(s) - \phi^h(0))^2 dtds\right) \\
&\le \frac{2}{\pi r^2}\left(\int_{-h_2}^{h_1}\int_0^{2h_3}\left(\int_0^t \frac{\partial u(s,\xi)}{\partial \xi} d\xi\right)^2 dtds\right. \\
&\quad \left. + 2\left(\int_0^{h_1}\left(\tilde{\phi}(s) - \tilde{\phi}^h(0)\right)^2 ds + \int_{-h_2}^{0}\left(\tilde{\phi}(s) - \tilde{\phi}^h(0)\right)^2 ds\right)\right) \\
&\le C\left(|u|_{H^1(S)}^2 + (\tilde{\phi}^h(z_{i_1}^{+}) - \tilde{\phi}^h(z_{i_1})^2 + (\tilde{\phi}^h(z_{i_1}^{-}) - \tilde{\phi}^h(z_{i_1})^2\right)^2 .
\end{aligned}
$$

Finally, we have

$$
\|u^h\|_{H^1(\Omega^h)} \le C_4 \|\phi^h\|_{H_h^{1/2}(\Gamma^h)} .
$$

□

In the following we need Sobolev's norm equivalence theorem, the proof of which can be found in [43].

Theorem 3.5 *Let $l\colon H^1(\Omega) \to \mathbb{R}$ be a linear bounded functional. If $l(c) = 0$ for some constant c yields $c = 0$, then $\|u\|_{H^1(\Omega)} \approx |u|_{H^1(\Omega)} + |l(u)|$.*

For instance, this theorem immediately yields the well-known Poincaré inequality

$$
\|u\|_{L^2(\Omega)}^2 \le C\left(|u|_{H^1(\Omega)}^2 + \left(\int_\Omega u(x)dx\right)^2\right).
$$

If $\int_\Omega u\, dx = 0$, then we have the usual Poincaré–Friedrich's inequality.

Lemma 3.6 (Poincaré inequality in $H^{1/2}(\Gamma)$)

$$\int_\Gamma \phi^2(x)dx \leq C\left(\int_\Gamma\int_\Gamma \frac{|\phi(x)-\phi(y)|^2}{|x-y|^2}dxdy + \left(\int_\Gamma \phi(x)dx\right)^2\right). \tag{3.2}$$

Proof. Let $x, y \in \Gamma$ and $x \neq y$ then

$$(\phi(x)-\phi(y))^2 \leq C_0\frac{(\phi(x)-\phi(y))^2}{|x-y|^2},$$

where $C_0 = \operatorname{diam}\Omega$. Thus, we have

$$\int_\Gamma\int_\Gamma (\phi(x)-\phi(y))^2 dxdy \leq C_0\int_\Gamma\int_\Gamma \frac{(\phi(x)-\phi(y))^2}{|x-y|^2}dydy.$$

Substituting

$$\begin{aligned}
&\int_\Gamma\int_\Gamma (\phi(x)-\phi(y))^2 dxdy \\
&= \int_\Gamma\int_\Gamma \phi(x)^2 dxdy - 2\int_\Gamma\int_\Gamma \phi(x)\phi(y)^2 dxdy + \int_\Gamma\int_\Gamma \phi(y)^2 dxdy \\
&= 2\cdot \text{meas}(\Gamma)\int_\Gamma \phi^2(x)dx - 2\left(\int_\Gamma \phi(x)dx\right)^2
\end{aligned}$$

into above equation, we arrive at (3.2). □

Theorem 3.7 (Trace theorem with semi-norm) *There are two positive constants C_1 and C_2 such that*

1) for all $u \in H^1(\Omega)$ its trace ϕ on Γ satisfies the inequality

$$|\phi|_{H^{1/2}(\Gamma)} \leq C_1|u|_{H^1(\Omega)},$$

and

2) for all $\phi \in H^{1/2}(\Gamma)$ there exists a function $u \in H^1(\Omega)$ such that $u = \phi$ on Γ and

$$|u|_{H^1(\Omega)} \leq C_2|\phi|_{H^{1/2}(\Gamma)}.$$

Proof. Let $u \in H^1(\Omega)$. Then, the function u can be split into two parts as follows:

$$u = u_0 + u_1, \quad u_0 = \text{ constant } = \frac{1}{\text{meas}(\Omega)}\int_\Omega u d\Omega, \quad \int_\Omega u_1 d\Omega = 0.$$

Now we split $\phi = \phi_0 + \phi_1$ into the traces $\phi_0 = u_0|_\Gamma$ and $\phi_1 = u_1|_\Gamma$ of the functions u_0 and u_1 on Γ. Then we have

$$|\phi|_{H^{1/2}(\Gamma)} = |\phi_1|_{H^{1/2}(\Gamma)} \leq C_3\|u_1\|_{H^1(\Omega)} \leq C_4|u_1|_{H^1(\Omega)} = C_4|u|_{H^1(\Omega)}.$$

This completes the proof of the first statement of the theorem.

The second statement can be proved as follows: Let $\phi \in H^{1/2}(\Gamma)$ be decomposed as

$$\phi = \phi_0 + \phi_1, \quad \phi_0 = \text{ constant } = u_0, \quad \int_\Gamma \phi_1 d\Gamma = 0.$$

By the standard trace theorem, there exists u_1 such that $u_1(x) = \phi_1(x)$ and

$$\|u_1\|_{H^1(\Omega)} \leq C_5 \|\phi_1\|_{H^{1/2}(\Gamma)}.$$

Set $u = u_0 + u_1$. Then $u(x) = \phi(x), x \in \Gamma$ and

$$|u|^2_{H^1(\Omega)} = |u_1|^2_{H^1(\Omega)} \leq C_5 \|\phi_1\|^2_{H^{1/2}(\Gamma)} \leq C_6 |\phi|^2_{H^{1/2}(\Gamma)}$$

where the last estimate follows from the Poincaré inequality . □

Remark 3.8 We have the same theorem for the finite element space because the FEM space contains the constant function.

Note that the definition of $\|\phi\|_{H^{1/2}(\Gamma)}$ is very complicated.

If $\phi \in H^1(-1,1)$ then we have

$$\|\phi\|^2_{H^1(-1,1)} = \|\phi\|^2_{H^1(-1,0)} + \|\phi\|^2_{H^1(0,1)}.$$

This kind of identity is only true for H^α with $0 < \alpha \leq 1$ and $\alpha \neq 1/2$, when the norm in H^α is additive with respect to the measure of the domain. In general cases we use the norm

$$\|\phi\|^2_{H^\alpha} = \|\phi\|^2_{L^2} + \int_{-1}^{1}\int_{-1}^{1} \frac{(\phi(x)-\phi(y))^2}{|x-y|^{1+2\alpha}} dxdy$$

and the following lemma.

Lemma 3.9 *There exist positive constants c_1 and c_2 such that*

$$\begin{aligned} c_1\|\phi\|^2_{H^{1/2}(-1,1)} &\leq \|\phi\|^2_{H^{1/2}(-1,0)} + \|\phi\|^2_{H^{1/2}(0,1)} + \int_0^1 \frac{(\phi(x)-\phi(-x))^2}{x} dx \\ &\leq c_2\|\phi\|^2_{H^{1/2}(-1,1)}. \end{aligned}$$

Proof. We have

$$\int_0^1 \frac{dy}{(x+y)^2} = \int_x^{1+x} \frac{dt}{t^2} = \frac{1}{x(1+x)}.$$

Thus,

$$\frac{1}{2x} \leq \int_0^1 \frac{dy}{(x+y)^2} \leq \frac{1}{x}, \quad x \in (0,1].$$

Then we can estimate the third term $I(\phi)$ in the first inequality as follows:

$$\begin{aligned} I(\phi) &= \int_0^1 \frac{(\phi(x)-\phi(-x))^2}{x}\,dx \\ &\le \int_0^1\int_0^1 \frac{(\phi(x)-\phi(-x))^2}{(x+y)^2}\,dydx \\ &\le 4\int_0^1\int_0^1 \frac{(\phi(x)-\phi(y))^2}{(x+y)^2}\,dydx + 4\int_0^1\int_0^1 \frac{(\phi(y)-\phi(-x))^2}{(x+y)^2}\,dydx \\ &\le 4(|\phi|^2_{H^{1/2}(0,1)} + |\phi|^2_{H^{1/2}(-1,1)}). \end{aligned}$$

Thus, the second inequality is proved. For the first inequality, we only need to consider the semi norm. First of all we have the representations

$$|\phi|^2_{H^{1/2}(0,1)} = \int_0^1\int_0^1 \frac{(\phi(x)-\phi(y))^2}{(x-y)^2}\,dydx$$

and

$$\begin{aligned} |\phi|^2_{H^{1/2}(-1,1)} &= \int_0^1\int_0^1 \frac{(\phi(x)-\phi(y))^2}{(x-y)^2}\,dydx + \int_0^1\int_{-1}^0 + \int_{-1}^0\int_0^1 + \int_{-1}^0\int_{-1}^0 \\ &= I + II + III + IV. \end{aligned}$$

Since $I = |\phi|^2_{H^{1/2}(0,1)}$ and $IV = |\phi|^2_{H^{1/2}(-1,0)}$, we have to consider only $II = III$. So, we estimate only one of them:

$$\begin{aligned} &\int_{-1}^0\int_0^1 \frac{(\phi(x)-\phi(y))^2}{(x-y)^2}\,dydx \\ &= 2\left(\int_{-1}^0\int_0^1 \frac{(\phi(x)-\phi(-x))^2}{(x-y)^2} dydx + \int_{-1}^0\int_0^1 \frac{(\phi(-x)-\phi(y))^2}{(x-y)^2}\right) dydx \\ &\le 2\int_0^1 (\phi(-x)-\phi(y))^2 \int_0^1 \frac{dy}{(x+y)^2} + \int_{-1}^0\int_{-1}^0 \frac{(\phi(x)-\phi(y))^2}{(x+y)^2}\,dydx. \end{aligned}$$

Here we have used the change of variables $(y \to -y')$, $(x \to -x')$. Now, we use the estimates of the integral

$$\frac{1}{2x} \le \int_0^1 \frac{dy}{(x+y)^2} \le \frac{1}{x}$$

to see that the third term is less than $|\phi|^2_{H^{1/2}(-1,0)}$. Then $I + II + III + IV$ can be estimated by

$$2(I(\phi) + |\phi|^2_{H^{1/2}(-1,0)}).$$

This proves the fist inequality in Lemma 3.9. □

Now divide the boundary (like a circle) by two points a and b on Γ, and the left-hand side is called Γ_1. Let us consider

$$H^{1/2}_{00}(\Gamma_1).$$

Let us assume that ϕ is equal to zero on $\Gamma_0 = \Gamma\backslash\Gamma_1$ and equivalent to the harmonic extension into the interior, i.e.,

$$\|\phi\|^2_{H^{1/2}(\Gamma)} \approx \|\phi\|^2_{H^{1/2}(\Gamma_1)} + \|\phi\|^2_{H^{1/2}(\Gamma_0)}(=0) + \int_{\Gamma_1} \frac{\phi^2}{|x-a|^2} + \int_{\Gamma_1} \frac{\phi^2}{|x-b|^2}.$$

With this motivation, we define

$$\|\phi\|^2_{H^{1/2}_{00}(\Gamma_1)} = \|\phi\|^2_{H^{1/2}(\Gamma_1)} + \int_{\Gamma_1} \frac{\phi^2}{|x-a|^2} + \int_{\Gamma_1} \frac{\phi^2}{|x-b|^2}. \tag{3.3}$$

Similarly, we introduce

$$\|\phi\|_{H^{1/2}_{00}(0,1)} = \|\phi\|^1_{H^{1/2}(0,1)} + \int_0^1 \frac{\phi^2}{x(1-x)}. \tag{3.4}$$

Meanwhile a function in $H^{1/2}(0,1)$ does not have anything to do with the value outside $(0,1)$.

Now, let us consider the FEM case. Let Ω be triangularized by Ω^h. Some part of its boundary is denoted by Γ_1^h some other by Γ_0^h. We now define the discrete counter part

$$\|\phi^h\|_{H^{1/2}_{00,h}(\Gamma_1^h)} = \|\phi^h\|^1_{H_h^{1/2}(\Gamma_1^h)} + \sum_{z_i\in\Gamma_1^h} \frac{(\phi^h(z_i))^2}{|z_i-a|} h_i + \sum_{z_i\in\Gamma_1^h} \frac{(\phi^h(z_i))^2}{|z_i-b|} h_i$$

of the norm. The second two terms correspond to an analog in the space $H^{1/2}$.

Let $\overset{0}{H}_h(\Gamma_1^h) = \{\phi^h \in H_h^{1/2}(\Gamma_1^h) |\ \phi^h(a) = \phi^h(b) = 0\}$ and $\phi^h \to \phi \in \mathbb{R}^n$. Then we have the following equivalences:

$$(S\phi,\phi) \approx \|\phi^h\|^2_{H^{1/2}(\Gamma^h)} \approx \|\phi^h\|^2_{H^{1/2}_{00}(\Gamma_1^h)} \approx \|\tilde{\phi}^h\|^2_{H^{1/2}_{00}(I)}.$$

Here I is the straightened boundary. Now, setting $\phi^h(z_i) = \tilde{\phi}(\tilde{z}_i)$ by mapping, extending it into the unit square and considering it on a uniform grid give

$$\|\tilde{\phi}^h\|_{H^{1/2}_{00}(I)} \approx (\tilde{S}\phi,\phi).$$

Finally, we have

$$(S\phi,\phi) \approx (\tilde{S}\phi,\phi).$$

Hence a preconditioner for $\tilde{S}$ suffices for the original problem. In summary, the Schur complement S is equivalent to the interface norm which is in turn equivalent to Schur complement $\tilde{S}$. On a good domain, the Schur complement $\tilde{S}$ can be found analytically.

A detailed study on the space $H^{1/2}_{00}(\Gamma_1)$

Let us start with a review on the Schur complement norm. First we recall that

$$(S\varphi,\varphi) = \inf_{w^h|_\Gamma = \varphi^h} \|w^h\|^2_{H^1(\Omega)} = \|u^h\|^2_{H^1(\Omega)}, \tag{3.5}$$

where u^h satisfying $u^h|_\Gamma = \varphi$ is the minimizer. Then, it is clear that

$$\|\varphi^h\|^2_{H^{1/2}(\Gamma)} \le C_3 \, \|u^h\|^2_{H^1(\Omega)} = C_3 \, (S\varphi, \varphi).$$

For φ^h, there exists a constant C_4 such that

$$\|v^h\|^2_{H^1(\Omega)} \le C_4 \|\varphi^h\|^2_{H^{1/2}(\Gamma)}.$$

Thus, we have

$$(S\varphi, \varphi) = \inf_{w^h|_\Gamma = \varphi^h} \|w^h\|^2_{H^1(\Omega)} = \|u^h\|^2_{H^1(\Omega)} \le \|v^h\|^2_{H^1(\Omega)} \le C_4 \|\varphi^h\|^2_{H^{1/2}(\Gamma)}.$$

We see that the Schur complement norm is equivalent to $H^{1/2}(\Gamma)$ norm. Let $\dot{C}^\infty(0,1)$ be the subspace of $C^\infty(0,1)$ with compact support. Then it is well known that

$$\overline{(C^\infty(0,1))}_{L^2} = L^2(0,1) \text{ and } \overline{(\dot{C}^\infty(0,1))}_{L^2} = L^2(0,1).$$

However, in the H^1 case, we have the closure relations

$$\overline{(C^\infty(0,1))}_{H^1(0,1)} = H^1(0,1) \text{ and } \overline{(\dot{C}^\infty(0,1))}_{H^1(0,1)} = H^1_0(0,1).$$

The definition of H^α for $\alpha \le 1/2$ follows from the L^2 case and for $\alpha > 1/2$ follows from the H^1 case:

$$\|\varphi\|^2_{H^\alpha(0,1)} = \|\varphi\|^2_{L^2(0,1)} + \int_0^1 \int_0^1 \frac{(\varphi(x) - \varphi(y))^2}{|x-y|^{1+2\alpha}} \, dx dy.$$

If $\alpha \le 1/2$ then

$$\overline{(C^\infty(0,1))}_{H^\alpha} = H^\alpha(0,1) \text{ and } \overline{(\dot{C}^\infty(0,1))} = H^\alpha(0,1). \tag{3.6}$$

If $\alpha < 1/2$ then, for $u \in H^\alpha(0,1)$, its extension by zero outside $(0,1)$ belongs to $H^\alpha(-1,2)$ like in the case of the L^2 space. However, for $\alpha = 1/2$, a function in $H^{1/2}(0,1)$ cannot be extended by zero (note that $H_0^{1/2} = H^{1/2}(0,1)$). If $\alpha > 1/2$ then

$$\overline{(C^\infty(0,1))}_{H^\alpha} = H^\alpha(0,1) \text{ and } \overline{(\dot{C}^\infty(0,1))}_{H^\alpha} = H^\alpha_0(0,1).$$

Let $\alpha = 1/2$. If we extend $\dot{C}^\infty(0,1)$ by the norm $\|\cdot\|_{H^{1/2}_{00}}$, then we obtain $H^{1/2}_{00}(0,1)$ and we can extend the function in $H^{1/2}_{00}(0,1)$ to a function in $H^{1/2}(-1,2)$ by zero. Hence, we have the proper inclusion

$$H^{1/2}(0,1) \supsetneq H^{1/2}_{00}(0,1).$$

We note that

$$\|\varphi\|^2_{H^{1/2}(-1,1)} \approx \|\varphi\|^2_{H^{1/2}(-1,0)} + \|\varphi\|^2_{H^{1/2}(0,1)} + \int_0^1 \frac{(\varphi(x) - \varphi(-x))^2}{x} \, dx.$$

For $\varphi \in H_{00}^{1/2}(0,1)$, we have

$$\|\varphi\|^2_{H_{00}^{1/2}(0,1)} = \|\varphi\|^2_{H^{1/2}(0,1)} + \int_0^1 \frac{(\varphi(x))^2}{x(1-x)}\,dx.$$

Hence, $\varphi \to 0$ as x tends to 0 and 1. Let us define the function $\tilde{\varphi} \in H^{1/2}(-1,2)$ by the formula

$$\tilde{\varphi}(x) = \begin{cases} 0 & x \in (-1,0), \\ \varphi(x) & x \in (0,1), \\ 0 & x \in (1,2). \end{cases}$$

Then we obtain the relations

$$\begin{aligned} &\|\tilde{\varphi}\|^2_{H^{1/2}(-1,2)} \\ &\approx \|\varphi\|^2_{H^{1/2}(0,1)} + \int_0^1 \frac{(\tilde{\varphi}(x)-\tilde{\varphi}(-x))^2}{x}\,dx + \int_0^1 \frac{(\tilde{\varphi}(x)-\tilde{\varphi}(2-x))^2}{1-x}\,dx \\ &= \|\varphi\|^2_{H^{1/2}(0,1)} + \int_0^1 \frac{\varphi(x)^2}{x} + \frac{\varphi(x)^2}{1-x}\,dx \\ &\approx \|\varphi\|^2_{H^{1/2}(0,1)} + \int_0^1 \frac{\varphi(x)^2}{x(1-x)}\,dx \\ &\approx \|\varphi\|^2_{H_{00}^{1/2}}(0,1) \end{aligned}$$

by simple calculations. In the first equivalence, we omitted both $\|\tilde{\varphi}\|^2_{H^{1/2}(-1,0)}$ and $\|\tilde{\varphi}\|^2_{H^{1/2}(1,2)}$ because they are zero by extension.

If $\alpha > 1/2$ then we have $H^\alpha \hookrightarrow C^0(0,1)$, i.e. if $\varphi \in H^\alpha(0,1)$, $\alpha > 1/2$, then $\lim_{x\to x_0} \varphi(x) = \varphi(x_0)$.

Example 3.10 Let the boundary Γ of Ω be divided by three points a, b and c into three pieces Γ_1, Γ_0 and $\tilde{\Gamma}_1$. Now let us consider the minimization problem

$$\inf_{w\in H^1(\Omega), w|_{\Gamma_1}=\varphi, w|_{\Gamma_0}=0} \|w\|^2_{H^1(\Omega)}.$$

The above minimization problem is obviously equivalent to the mixed boundary value problem

$$\begin{cases} -\Delta w + w = 0, \\ w|_{\Gamma_1} = \varphi, \\ w|_{\Gamma_0} = 0, \\ \dfrac{\partial w}{\partial n}|_{\tilde{\Gamma}_1} = 0. \end{cases}$$

The correct norm in $\breve{H}^{1/2}(\Gamma_1)$ is now given by the relation

$$\|\varphi\|^2_{\breve{H}^{1/2}(\Gamma_1)} = \|\varphi\|^2_{H^{1/2}(\Gamma_1)} + \int_{\Gamma_1} \frac{(\varphi(x))^2}{|x-a|}\,dx.$$

It is like in $H_{00}^{1/2}(\Gamma_1)$, but it is only an one-side norm because the integral near the point b is missing. For the FEM case, we use the following discrete norm

$$\|\varphi^h\|^2_{\breve{H}_h^{1/2}(\Gamma_1)} = \|\varphi^h\|^2_{H_h^{1/2}(\Gamma_1)} + \sum_{z_i \in \Gamma_1} \frac{(\varphi^h(z_i))^2}{|z_i - a|} h_i.$$

4 Domain decomposition method: the strip case

In this section, the decomposition of Ω into the subdomains Ω_i does not have any cross point. This constellation is called strip case which is characterized by a decomposition of the form

$$\bar{\Omega} = \bigcup_{i=1}^{n} \bar{\Omega}_i, \quad \gamma = \bigcup_{i=1}^{n} \partial\Omega_i \backslash \Gamma = \bigcup_{i=1}^{n-1} \gamma_i, \text{ with } \gamma_i \cap \gamma_j = \emptyset \text{ for } i \neq j,$$

where γ_i is an interface between the subdomains. We again consider the Dirichlet problem for the Poisson equation as model problem:

$$-\Delta u = f(x), \quad x \in \Omega,$$

$$u(x) = 0, \quad x \in \Gamma.$$

Then the finite element discretization yields the following system of linear algebraic equations

$$Au = f$$

which can be rewritten in block form as follows

$$\begin{pmatrix} A_0 & A_{01} & \cdot & \cdot & \cdot & A_{0n} \\ A_{10} & A_1 & & & & \\ \cdot & & \cdot & & 0 & \\ \cdot & & & \cdot & & \\ \cdot & & 0 & & \cdot & \\ A_{n0} & & & & & A_n \end{pmatrix} \begin{pmatrix} u_0 \\ u_1 \\ \cdot \\ \cdot \\ \cdot \\ u_n \end{pmatrix} = \begin{pmatrix} f_0 \\ f_1 \\ \cdot \\ \cdot \\ \cdot \\ f_n \end{pmatrix}.$$

Eliminating the vector u_i from the equation

$$A_{i0} u_0 + A_i u_i = f_i$$

and substituting

$$u_i = -A_i^{-1} A_{i0} u_0 + A_i^{-1} f_i$$

into the first (block) equation, we arrive at the Schur complement equations

$$S\varphi = \psi, \tag{4.1}$$

where $S = A_0 - \sum_{i=1}^{n} A_{0i}A_i^{-1}A_{i0}$, $\varphi = u_0$ and $f_0 - \sum_{i=1}^{n} A_{0i}A_i^{-1}f_i$. The Schur complement system (4.1) can be solved, for instance, by the Richardson iteration

$$\varphi^{k+1} = \varphi^k - \tau_k\Sigma^{-1}(S\varphi^k - \psi),$$

where Σ is a suitable preconditioner satisfying the spectral equivalence inequalities

$$c_1(\Sigma\varphi,\varphi) \le (S\varphi,\varphi) \le C_2(\Sigma\varphi,\varphi) \tag{4.2}$$

for all vectors φ, and $\hat{\tau}_k$ are appropriately chosen iteration parameters.

Now we arrange the vector u_0 in the form

$$u_0 = \begin{pmatrix} \varphi_1 \\ \vdots \\ \varphi_{n-1} \end{pmatrix},$$

where φ_i corresponds to γ_i. Then we have

$$S = S_1 + \cdots + S_{n-1},$$

where

$$\Omega_i \to S_i u_0 = \begin{pmatrix} 0 & 0 \\ S_{11}^{(i)} & S_{12}^{(i)} \\ S_{21}^{(i)} & S_{22}^{(i)} \\ 0 & 0 \end{pmatrix} \begin{pmatrix} 0 \\ \varphi_l \\ \varphi_m \\ 0 \end{pmatrix}.$$

Now, we look for Σ_l and Σ_m such that

$$\begin{pmatrix} \Sigma_l & 0 \\ 0 & \Sigma_m \end{pmatrix} \approx \tilde{S}_i = \begin{pmatrix} S_{11}^{(i)} & S_{12}^{(i)} \\ S_{21}^{(i)} & S_{22}^{(i)} \end{pmatrix},$$

i.e. $(\Sigma_l\varphi_l,\varphi_l) \approx \|\varphi_l\|^2_{H_{00}^{1/2}(\gamma_l)}$ and $(\Sigma_m\varphi_m,\varphi_m) \approx \|\varphi_m\|^2_{H_{00}^{1/2}(\gamma_m)}$. Figure 4.1 gives a typical situation of a subdomain Ω_i with interfaces γ_l and γ_m.

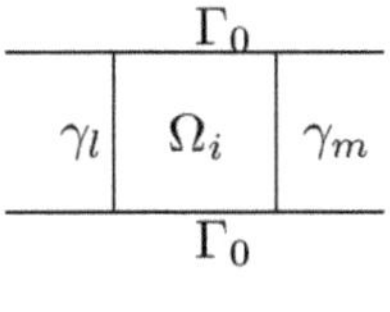

Figure 4.1 Subdomain Ω_i with interfaces γ_l and γ_m.

The corresponding Schur complement is equivalent to the following norm

$$\begin{aligned} \left(S_i\begin{pmatrix} \varphi_l \\ \varphi_m \end{pmatrix}, \begin{pmatrix} \varphi_l \\ \varphi_m \end{pmatrix}\right) &= \inf_{w^h \in H_h(\Omega_i), w^h|_{\gamma_l} = \varphi_l, w^h|_{\gamma_m} = \varphi_m, w^h|_{\Gamma\cap\partial\Omega_i} = 0} |w^h|^2_{H^1(\Omega_i)} \\ &\approx \|\varphi_l^h\|^2_{H_{00}^{1/2}(\gamma_l)} + \|\varphi_m^h\|^2_{H_{00}^{1/2}(\gamma_m)} \end{aligned}$$

due to the previous analysis. Hence, we have the preconditioner for the global Schur complement as a block diagonal matrix.

Let us now consider some γ_ℓ and let us omit the subindex ℓ. A norm which is equivalent to Schur complement norm has the following additive form.

$$\begin{aligned}
&\|\varphi^h\|^2_{H^{1/2}_{00}(\gamma)} \\
&= \sum_{z_i\in\gamma}(\varphi^h(z_i))^2\cdot h + \sum_{z_i,i\neq j}\sum_{z_j}\frac{(\varphi^h(z_i)-\varphi^h(z_j))^2}{|z_i-z_j|^2}h_ih_j + \sum_{z_i\in\gamma}\frac{(\varphi^2(z_i))^2}{(z_i-a)(z_j-b)}h_i \\
&\approx \sum_{z_i\in\gamma}(\tilde\varphi^h(z_i))^2\cdot h + \sum_{z_i,i\neq j}\sum_{z_j}\frac{(\tilde\varphi^h(z_i)-\tilde\varphi^h(z_j))^2}{|z_i-z_j|^2}h_ih_j + \sum_{z_i\in\gamma}\frac{(\tilde\varphi^2(z_i))^2}{(z_i-a)(z_j-b)}h_i \\
&= \|\tilde\varphi^h\|^2_{H^{1/2}_{00}(\tilde\gamma)},
\end{aligned}$$

where, in the second equation, we have everything replaced by its "tilde" (map it onto $[0,1]$) which is for a curved boundary.

Hence consider the square domain. For example, consider the domain with 4 subdomains (Ω_i, $i=1,2,3,4$) whose interfaces (γ_i, $i=1,2,3$) do not meet each other. Then, we have

$$S=\begin{bmatrix} S_1+S_2^{(1,1)} & S_2^{(1,2)} & \\ S_2^{(2,1)} & S_2^{(2,2)}+S_3^{(1,1)} & S_3^{(1,2)} \\ & S_3^{(2,1)} & S_3^{(2,2)}+S_4 \end{bmatrix}$$

where the submatrix S_i is the Schur complement (S-C) matrix corresponding to the subdomain Ω_i. For instance,

$$S_2=\begin{bmatrix} S_2^{(1,1)} & S_2^{(1,2)} \\ S_2^{(2,1)} & S_2^{(2,2)} \end{bmatrix}$$

and $S_2^{(i,j)}$ is the S-C matrix corresponding to Ω_2 and γ_i and γ_j. Here, we may write

$$S=\tilde S_1+\tilde S_2+\tilde S_3+\tilde S_4$$

where $\tilde S_i$ is just the extension of S_i by zero elements. Now, in terms of the spectral equivalence, we have

$$S_1\approx\Sigma_1,\qquad S_2\approx\begin{bmatrix}\Sigma_1 & \\ & \Sigma_2\end{bmatrix},\qquad S_3\approx\begin{bmatrix}\Sigma_2 & \\ & \Sigma_3\end{bmatrix}\qquad\text{and}\qquad S_4\approx\Sigma_3.$$

Note that $(\Sigma_i\phi_i,\phi_i)\approx\|\phi\|^2_{H^{1/2}_{00}(\gamma_i)}$. Hence, we have

$$S\approx\begin{bmatrix}\Sigma_1 & & \\ & \Sigma_2 & \\ & & \Sigma_3\end{bmatrix}.$$

Given a vertical interface-line segment, we introduce an artificial uniform domain and consider the problem with zero boundary condition on three side except Γ_1 on the left and consider the Schur complement of this problem, denote it by S.

For a given γ_i interface, we suppose that we have the mapping from γ_i onto one side of the rectangular domain with uniform mesh of size $h = 1/n$. Thus, we can now consider our interface problem arising from a rectangular model. The Schur complement of this model problem satisfies the spectral equivalence relation

$$(S\phi, \phi) \approx \|\phi^h\|^2_{H^{1/2}_{00}(\Gamma_1)}.$$

In the rectangular domain, we have

$$A_\Omega = \begin{bmatrix} A_0 + 2I & -I & & & \\ -I & A_0 + 2I & -I & & \\ & \ddots & \ddots & \ddots & \\ & & -I & A_0 + 2I & -I \\ & & & -I & \frac{1}{2}A_0 + I \end{bmatrix} := \begin{bmatrix} \bar{A}_{11} & \bar{A}_{12} \\ \bar{A}_{21} & \bar{A}_{22} \end{bmatrix},$$

where

$$A_0 = \begin{bmatrix} 2 & -1 & & & \\ -1 & 2 & -1 & & \\ & \ddots & \ddots & \ddots & \\ & & -1 & 2 & -1 \\ & & & -1 & 2 \end{bmatrix},$$

and

$$\bar{A}_{12} = \begin{bmatrix} 0 & 0 & 0 & \cdots & 0 & -I \end{bmatrix}^\top = (\bar{A}_{21})^\top, \qquad \bar{A}_{22} = \frac{1}{2}A_0 + I.$$

Now we have

$$S = \bar{A}_{22} - \bar{A}_{21}(\bar{A}_{11})^{-1}\bar{A}_{12}$$

and

$$(S\phi, \phi) = \inf_{u^h|_{\Gamma_1} = \phi^h, u^h|_{\partial\Omega\setminus\Gamma_1} = 0} \|u^h\|^2_{H^1(\Omega)} = \inf_{u^h|_{\Gamma_1} = \phi^h, u^h|_{\partial\Omega\setminus\Gamma_1} = 0} (A_\Omega u, u).$$

By the diagonalization, we decompose A_0 as

$$A_0 = Q\Lambda Q^\top,$$

where $Q = \begin{bmatrix} q_1 & q_2 & \cdots & q_{n-1} \end{bmatrix}$, Λ is the diagonal matrix with the diagonal entries $\lambda_1, \lambda_2, \ldots, \lambda_{n-1}$, and $A_0 q_i = \lambda_i q_i$. Note that it is well known that the eigenvalue $\lambda_i = 4\sin^2\frac{\pi}{2n}$, the jth component of the eigenvector q_i is $\sqrt{\frac{2}{n}}\sin(\frac{i\pi}{n}j)$, and $QQ^\top = I$.

Using this, we get

$$(\bar{A}_{11})^{-1} = \begin{bmatrix} Q & & & \\ & Q & & \\ & & \ddots & \\ & & & Q \end{bmatrix}$$

$$\times \begin{bmatrix} \Lambda + 2I & -I & & \\ -I & \Lambda + 2I & -I & \\ & \ddots & \ddots & \ddots \\ & & -I & \Lambda + 2I \end{bmatrix}^{-1} \times \begin{bmatrix} Q^\top & & & \\ & Q^\top & & \\ & & \ddots & \\ & & & Q^\top \end{bmatrix}$$

and

$$\begin{aligned} &\bar{A}_{21}(\bar{A}_{11})^{-1}\bar{A}_{12} \\ &= \begin{bmatrix} 0 & 0 & \cdots & -Q \end{bmatrix} \begin{bmatrix} \Lambda + 2I & -I & & \\ -I & \Lambda + 2I & -I & \\ & \ddots & \ddots & \ddots \\ & & -I & \Lambda + 2I \end{bmatrix}^{-1} \begin{bmatrix} 0 \\ 0 \\ \vdots \\ -Q^\top \end{bmatrix} \\ &= QB_{22}Q^\top, \end{aligned}$$

where

$$B := \begin{bmatrix} \Lambda + 2I & -I & & \\ -I & \Lambda + 2I & -I & \\ & \ddots & \ddots & \ddots \\ & & -I & \Lambda + 2I \end{bmatrix}^{-1} = \begin{bmatrix} B_{11} & B_{12} \\ B_{21} & B_{22} \end{bmatrix}.$$

Now let us compute the matrix B_{22}. Let $e_i = \begin{bmatrix} 0 & \cdots & 1 & \cdots & 0 \end{bmatrix}^\top$, where 1 is in the i-th position. Consider the following matrix equation.

$$\begin{bmatrix} \Lambda + 2I & -I & & \\ -I & \Lambda + 2I & -I & \\ & \ddots & \ddots & \ddots \\ & & -I & \Lambda + 2I \end{bmatrix} \begin{bmatrix} x_1^{(i)} \\ x_2^{(i)} \\ \vdots \\ x_{n-1}^{(i)} \end{bmatrix} = \begin{bmatrix} 0 \\ 0 \\ \vdots \\ e_i \end{bmatrix}.$$

Then the $(n-1)$-th solution vector of the above matrix equation is the i-th column of the matrix B_{22}, that is,

$$B_{22} = \begin{bmatrix} x_{n-1}^{(1)} & x_{n-1}^{(2)} & \cdots & x_{n-1}^{(n-1)} \end{bmatrix}.$$

We denote the vector $x_k^{(i)}$ by

$$x_k^{(i)} = \begin{bmatrix} x_k^{(i)}(1) \\ x_k^{(i)}(2) \\ \vdots \\ x_k^{(i)}(n-1) \end{bmatrix}.$$

Let us consider the j-th component. Then we obtain the following matrix equation

$$\begin{bmatrix} \lambda_j + 2 & -1 & & \\ -1 & \lambda_j + 2 & -1 & \\ & \ddots & \ddots & \ddots \\ & & -1 & \lambda_j + 2 \end{bmatrix} \begin{bmatrix} x_1^{(i)}(j) \\ x_2^{(i)}(j) \\ \vdots \\ x_{n-1}^{(i)}(j) \end{bmatrix} = \begin{bmatrix} 0 \\ 0 \\ \vdots \\ \delta_{ij} \end{bmatrix}.$$

To discuss some details we consider the vector equation for a fixed i, i.e.

$$\begin{array}{cccccc} (\Lambda + 2I)x_1^{(i)} & -x_2^{(i)} & & & = & 0 \\ -x_1^{(i)} & (\Lambda + 2I)x_2^{(i)} & -x_3^{(i)} & & = & 0 \\ & & \ddots & & \vdots & \\ & & -x_{n-3}^{(i)} & (\Lambda + 2I)x_{n-2}^{(i)} \quad -x_{n-1}^{(i)} & = & 0 \\ & & & -x_{n-2}^{(i)} \quad (\Lambda + 2I)x_{n-1}^{(i)} & = & e_i \end{array}$$

The first block corresponds to

$$\begin{array}{rrcl} (\lambda + 2)x_1^{(i)}(1) & -x_2^{(i)}(1) & = & 0 \\ +(\lambda + 2)x_1^{(i)}(2) & -x_2^{(i)}(2) & = & 0 \\ & & \vdots & \\ +(\lambda + 2)x_1^{(i)}(n-1) & -x_2^{(i)}(n-1) & = & 0 \end{array}$$

We collect j-th line. If $i = j$, we have

$$x_{n-1}^{(i)} = \begin{bmatrix} 0 \\ 0 \\ \vdots \\ x_{n-1}^{(i)}(i) \\ \vdots \\ 0 \end{bmatrix}.$$

Combining the vectors $x_{n-1}^{(i)}$, we can obtain the matrix B_{22} as

$$B_{22} = \begin{bmatrix} x_{n-1}^{(1)}(1) & 0 & \cdots & 0 \\ 0 & x_{n-1}^{(2)}(2) & \cdots & 0 \\ \vdots & \vdots & & \vdots \\ 0 & 0 & \cdots & x_{n-1}^{(n-1)}(n-1) \end{bmatrix}.$$

Then

$$\tilde{S} = \bar{A}_{22} - \bar{A}_{21}\bar{A}_{11}^{-1}\bar{A}_{12} = \frac{1}{2}A + I - QB_{22}Q^\top = Q(\frac{1}{2}\Lambda + I - B_{22})Q^\top$$

and the i-th eigenvalue of S is

$$\lambda_i(\tilde{S}) = \frac{1}{2}\lambda_i + 1 - x_{n-1}^{(i)}(i).$$

To compute $x_{n-1}^{(i)}(i)$, we have to solve the following system of algebraic equations:

$$\begin{bmatrix} \lambda_i + 2 & -1 & & \\ -1 & \lambda_i + 2 & -1 & \\ & \ddots & \ddots & \ddots \\ & & -1 & \lambda_i + 2 \end{bmatrix} \begin{bmatrix} x_1^{(i)}(i) \\ x_2^{(i)}(i) \\ \vdots \\ x_{n-1}^{(i)}(i) \end{bmatrix} = \begin{bmatrix} 0 \\ 0 \\ \vdots \\ 1 \end{bmatrix}.$$

Let $\alpha_i = \frac{1}{2}\lambda_i + 1$. By using Gauss-elimination technique (multiplying the j-th row by $2\alpha_i$ and adding the $(j-1)$-th row to j-th row) we obtain the triangular system

$$\begin{bmatrix} d_1 & -d_0 & & 0 \\ & d_2 & -d_1 & \\ & & \ddots & -d_{n-3} \\ 0 & & & d_{n-1} \end{bmatrix} \begin{bmatrix} x_1^{(i)}(i) \\ x_2^{(i)}(i) \\ \vdots \\ x_{n-1}^{(i)}(i) \end{bmatrix} = \begin{bmatrix} 0 \\ 0 \\ \vdots \\ d_{n-2} \end{bmatrix},$$

where $d_0 = 1$, $d_1 = 2\alpha_i$, and $d_{j+1} = 2\alpha_i d_j - d_{j-1}$, for $j = 1, 2, \ldots, n-2$. Let $U_n(x)$ be the second kind Chebyshev polynomial of degree n, that is,

$$U_n(x) = \frac{1}{2\sqrt{x^2-1}}\Big((x + \sqrt{x^2-1})^{n+1} - (x + \sqrt{x^2-1})^{-(n+1)}\Big).$$

Then $d_j = U_j(\alpha_i)$ [1] and

$$x_{n-1}^{(i)}(i) = \frac{d_{n-2}}{d_{n-1}} = \frac{U_{n-2}(\alpha_i)}{U_{n-1}(\alpha_i)}.$$

[1] We note that the first kind Chebyshev polynomials of degree n at α_j are determined by the same recursion scheme for d_j, but with the initial condition $d_0 = 1$ and $d_1 = \alpha_i$.

Hence,

$$\begin{aligned}\lambda_i(\tilde{S}) &= \alpha_i - \frac{d_{n-2}}{d_{n-1}} \\ &= \alpha_i - \frac{U_{n-2}(\alpha_i)}{U_{n-1}(\alpha_i)} \\ &= \alpha_i - \frac{(\alpha_i + \sqrt{\alpha_i^2 - 1})^{n-1} - (\alpha_i + \sqrt{\alpha_i^2 - 1})^{-n+1}}{(\alpha_i + \sqrt{\alpha_i^2 - 1})^{n} - (\alpha_i + \sqrt{\alpha_i^2 - 1})^{-n}} \\ &= \sqrt{\alpha_i^2 - 1}\, \frac{(\alpha_i + \sqrt{\alpha_i^2 - 1})^{n} + (\alpha_i + \sqrt{\alpha_i^2 - 1})^{-n}}{(\alpha_i + \sqrt{\alpha_i^2 - 1})^{n} - (\alpha_i + \sqrt{\alpha_i^2 - 1})^{-n}} \\ &= \sqrt{\alpha_i^2 - 1} f(x),\end{aligned}$$

where

$$f(x) = \frac{x + 1/x}{x - 1/x}, \quad x = \left(\alpha_i + \sqrt{\alpha_i^2 - 1}\right)^n.$$

Using $\sqrt{\alpha_i^2 - 1} = \sqrt{\lambda_i}\sqrt{1 + \frac{\lambda_i}{4}}$, we have the following estimates for $\lambda_i(\tilde{S})$:

$$\sqrt{\lambda_i} \le \lambda_i(\tilde{S}) \le \sqrt{\lambda_i}\, C(\lambda_{\min}, \lambda_{\max}),$$

where

$$C(\lambda_{\min}, \lambda_{\max}) = \sqrt{1 + \frac{\lambda_{\max}}{4}} \cdot \frac{\beta^n + \beta^{-n}}{\beta^n - \beta^{-n}},$$

with

$$\beta = 1 + \frac{1}{2}\lambda_{\min} + \sqrt{\lambda_{\min} + \frac{1}{4}\lambda_{\min}^2}.$$

Since $\lambda_{\min} = 4\sin^2(\frac{\pi}{2n}) \simeq \frac{1}{n^2}$ and $\lambda_{\max} \le 4$, we have

$$\beta^n \ge (1 + \sqrt{\lambda_{\min}})^n \simeq O(1).$$

Hence, by setting $\Sigma := A^{1/2} = Q\Lambda^{1/2}Q^\top$, we arrive at the following inequality:

$$(\Sigma\phi, \phi) \le (\tilde{S}\phi, \phi) \le C(\Sigma\phi, \phi).$$

Thus, we have

$$\tilde{S} = QJQ^\top \approx A^{1/2}, \quad \text{with } J = \text{diag}(\lambda_i(\tilde{S})),$$

and

$$\Sigma^{-1} = Q\Lambda^{-1/2}Q^\top,$$

where $Q = (q_1, \dots, q_{n-1}), q_i(j) = \sqrt{\frac{2}{n}} \sin\frac{i\pi j}{n}$. Since $\lambda_{\max} \le 4$, we have $C \le \frac{5\sqrt{2}}{3}$. If we use the Fast Fourier transform (FFT) algorithm, then the cost for computing $\Sigma^{-1}\phi$ is of order $h^{-1}\log(h^{-1})$. What is $\|\phi^h\|^2_{H^{1/2}}$? Since (as a discrete inner product on Γ)

$$(\phi^h, \psi^h)_{L_{2,h}} = h(\phi, \psi),$$

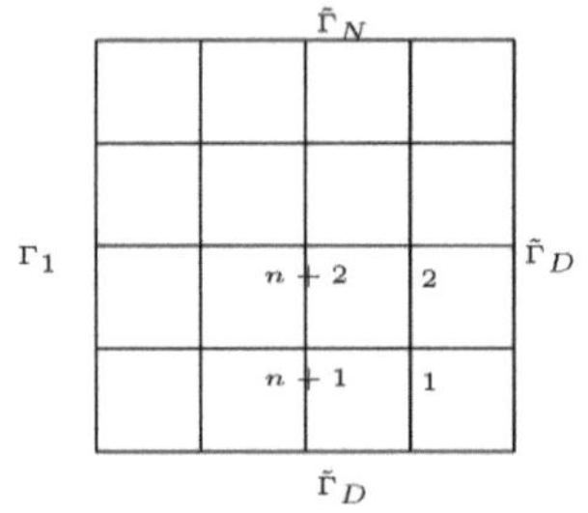

Figure 4.2 Domain and grid numbering.

where the right hand side is vector inner product and

$$\|\phi^h\|^2_{H^1} = (\frac{1}{h^2}A\phi, \phi)_{L_{2,h}} = \frac{1}{h}(A\phi, \phi),$$

we have

$$\|\phi^h\|^2_{H^{1/2}} \simeq \left(\left(\frac{1}{h^2}A \right)^{1/2} \phi, \phi \right)_{L_{2,h}} = (A^{1/2}\phi, \phi).$$

Thus, we are done with the case of Dirichlet boundary conditions. Next we consider mixed boundary conditions. Recall that, for Dirichlet boundary conditions, we have

$$(S\varphi, \varphi) \approx \sum_l (\Sigma_l \varphi, \varphi), \qquad \Sigma_l = \begin{bmatrix} 2 & -1 & & \\ -1 & 2 & -1 & \\ & \ddots & \ddots & \ddots \\ & & -1 & 2 \end{bmatrix}^{1/2}.$$

For mixed boundary conditions, we have

$$(S\varphi, \varphi) \approx \sum_l \|\varphi^h\|^2_{\breve{H}^{1/2}(\gamma_l)},$$

where a_l is the endpoint of the interface γ_l lying on the Dirichlet boundary, and

$$\|\varphi\|^2_{\breve{H}^{1/2}(\gamma_l)} = \|\varphi\|^2_{H^{1/2}(\gamma_l)} + \int_{\gamma_l} \frac{\varphi^2(x)}{|x - a_l|}\, dx.$$

Let

$$A_1 = \begin{bmatrix} 2 & -1 & & & \\ -1 & 2 & -1 & & \\ & \ddots & \ddots & \ddots & \\ & & -1 & 2 & -1 \\ & & & -1 & 1 \end{bmatrix}, \qquad D = \begin{bmatrix} 1 & & & \\ & \ddots & 0 & \\ & & \ddots & \\ & 0 & 1 & \\ & & & 1/2 \end{bmatrix},$$

and

$$A_\Omega = \begin{bmatrix} A_1 + 2D & -D & & & \\ -D & A_1 + 2D & -D & & \\ & \ddots & \ddots & \ddots & \\ & & -D & A_1 + 2D & -D \\ & & & -D & \frac{1}{2}A_1 + D \end{bmatrix}.$$

We note that A_1 corresponds to the first right vertical block. Now, we obtain

$$\begin{aligned}(A_\Omega u, u) &= \sum_{x(i,j)\in\Omega} \{(u_{i,j} - u_{i-1,j})^2 + (u_{i,j} - u_{i,j-1})^2\} \\ &\quad + \frac{1}{2}\sum_{j=1}^{n}(u_{n,j} - u_{n,j-1})^2 + \frac{1}{2}\sum_{i=1}^{n}(u_{i,n} - u_{i-1,n})^2 \\ &\approx (B_\Omega u, u),\end{aligned}$$

where

$$B_\Omega = \begin{bmatrix} A_1 + 2I & -I & & & \\ -I & A_1 + 2I & -I & & \\ & \ddots & \ddots & \ddots & \\ & & -I & A_1 + 2I & -I \\ & & & -I & \frac{1}{2}A_1 + I \end{bmatrix}.$$

The following lemma holds:

Lemma 4.1 *If $A \sim B$ then the corresponding Schur complements are also spectrally equivalent, i.e. $S_A \sim S_B$.*

Furthermore, we have $\lambda_{\min}(A_1) = O(h^2)$, $\lambda_{\max}(A_1) = O(1)$, and

$$S \approx \begin{bmatrix} 2 & -1 & & & \\ -1 & 2 & -1 & & \\ & \ddots & \ddots & \ddots & \\ & & -1 & 2 & -1 \\ & & & -1 & 1 \end{bmatrix}^{1/2}.$$

In this case the eigenvectors cannot be easily constructed. Thus, we consider the representation

$$\begin{aligned}
S &= \left(\frac{1}{2}A_1 + D\right) - \begin{bmatrix}0 & \cdots & 0 & -D\end{bmatrix} \begin{bmatrix} \ddots & \ddots & \ddots & & \\ & -D & A_1 + 2D & -D & \\ & & \ddots & \ddots & \ddots \end{bmatrix}^{-1} \begin{bmatrix} 0 \\ \vdots \\ 0 \\ -D \end{bmatrix} \\
&= D\left(\frac{1}{2}D^{-1}A_1 + I\right) \\
&\quad -D\begin{bmatrix}0 & \cdots & 0 & -I\end{bmatrix} \left(\begin{bmatrix} D & & \\ & \ddots & \\ & & D \end{bmatrix} \begin{bmatrix} D^{-1}(A_1 + 2I) & -I & & \\ -I & \ddots & \ddots & \\ & \ddots & \ddots & \end{bmatrix} \right)^{-1} \begin{bmatrix} 0 \\ \vdots \\ 0 \\ -D \end{bmatrix} \\
&= D\left(\frac{1}{2}D^{-1}A_1 + I\right) \\
&\quad -D\begin{bmatrix}0 & \cdots & 0 & -I\end{bmatrix} \begin{bmatrix} D^{-1}(A_1 + 2I) & -I & & \\ -I & \ddots & \ddots & \\ & \ddots & \ddots & \end{bmatrix}^{-1} \begin{bmatrix} 0 \\ \vdots \\ 0 \\ -I \end{bmatrix}.
\end{aligned}$$

The following matrix corresponds to the finite difference version for the one-side Neumann problem, see [37, 39]:

$$A_2 = D^{-1}A_1 = \begin{bmatrix} 2 & -1 & & & \\ -1 & \ddots & \ddots & & \\ & \ddots & \ddots & \ddots & \\ & & -1 & 2 & -1 \\ & & & -2 & 2 \end{bmatrix}.$$

We obtain

$$A_2 = Q_2 \Lambda_2 Q_2^{-1} = Q_2 \Lambda_2 Q_2^\top D,$$

where

$$Q_2 = \begin{bmatrix} q_1 & \cdots & q_n \end{bmatrix}, \qquad q_i(j) = \sqrt{\frac{2}{n}} \sin \frac{(2i-1)\pi j}{2n}, \qquad \lambda_i = 4 \sin^2 \frac{(2i-1)\pi}{2n},$$

for $i, j = 1, \dots, n$. Here $\tilde{\Lambda}_2$ is obtained from the Chebyshev polynomial. Therefore,

$$S = DQ_2 \tilde{\Lambda}_2 Q_2^\top D \approx DQ_2 \Lambda_2^{1/2} Q_2^\top D = \Sigma_{DN}, \qquad \Sigma_{DN}^{-1} = Q_2 \Lambda_2^{-1/2} Q_2^\top.$$

For the implementation, it is possible to use FFT for Q_2. In fact, we have a D-orthogonal basis such that the following properties hold:

- $A_2 q = D^{-1} A_1 q = \lambda q \implies A_1 q = \lambda D q$,
- $(Dq_i, q_j) = \delta_{ij}$,
- $(D^{-1/2} A_1 D^{-1/2}) D^{1/2} q = \lambda D^{1/2} q$,
- $(Dq_i, q_j) = (\tilde{q}_i, \tilde{q}_j) = \delta_{ij} \implies Q_2^\top D Q = I \implies Q_2^{-1} = Q^\top D$, with $\tilde{q} = D^{1/2} q$.

Neumann boundary conditions both on top and bottom of boundary

In this case, we have

$$A_3 = \begin{bmatrix} 1 & -1 & & & \\ -1 & 2 & -1 & & \\ & \ddots & \ddots & \ddots & \\ & & -1 & 2 & -1 \\ & & & -1 & 1 \end{bmatrix}, \qquad Q_3 = \begin{bmatrix} 1/2 & & & & \\ & 1 & & & \\ & & \ddots & & \\ & & & 1 & \\ & & & & 1/2 \end{bmatrix}.$$

Repeating the same analysis, we arrive at the following two possibilities (see Figure 4.3):

$$\Sigma_{NN} = A_3^{1/2} + \frac{1}{n} I, \qquad \Sigma_{NN}^{(N)} = A_3^{1/2}.$$

This problem generates the following Schur complements and their preconditioners:

$$\begin{aligned} \Omega_1 &\longrightarrow & S^{(1)} &\approx \Sigma_{DD}^{(1)}, \\ \Omega_2 &\longrightarrow & S^{(2)} &= \begin{bmatrix} S_{11}^{(2)} & S_{12}^{(2)} \\ S_{21}^{(2)} & S_{22}^{(2)} \end{bmatrix} \approx \begin{bmatrix} \Sigma_{DD}^{(1)} & \\ & \Sigma_{DD}^{(2)} \end{bmatrix}, \\ \Omega_3 &\longrightarrow & S^{(3)} &= \begin{bmatrix} S_{11}^{(3)} & S_{12}^{(3)} \\ S_{21}^{(3)} & S_{22}^{(3)} \end{bmatrix} \approx \begin{bmatrix} \Sigma_{DN}^{(1)} & \\ & \Sigma_{DN}^{(3)} \end{bmatrix}, \\ \Omega_4 &\longrightarrow & S^{(4)} &\approx \Sigma_{DN}^{(3)}, \end{aligned}$$

and, therefore,

$$\Sigma = \begin{bmatrix} \Sigma_{DD}^{(1)} & & \\ & \Sigma_{DD}^{(2)} + \Sigma_{DN}^{(2)} & \\ & & \Sigma_{DN}^{(3)} + \Sigma_{ND}^{(3)} \end{bmatrix} \approx \begin{bmatrix} \Sigma_{DD}^{(1)} & & \\ & \Sigma_{DD}^{(2)} & \\ & & \Sigma_{DD}^{(3)} \end{bmatrix}.$$

Here $\Sigma_{DN}^{(2)}$ is smaller than $\Sigma_{DD}^{(2)}$ and $\Sigma_{DN}^{(3)} + \Sigma_{ND}^{(3)} \approx \Sigma_{DD}^{(3)}$, i.e.

$$\begin{aligned} &\Sigma_{DD}^{(2)} \le \Sigma_{DD}^{(2)} + \Sigma_{DN}^{(2)} \le c\Sigma_{DD} \qquad \text{on } \gamma_2, \\ &\Sigma_{DN}^{(3)} + \Sigma_{ND}^{(3)} \approx \Sigma_{DD}^{(3)} \qquad \text{on } \gamma_3. \end{aligned}$$

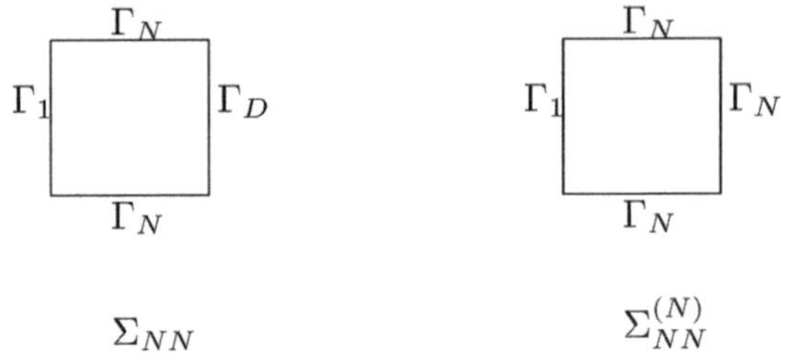

Figure 4.3 The two possibilities.

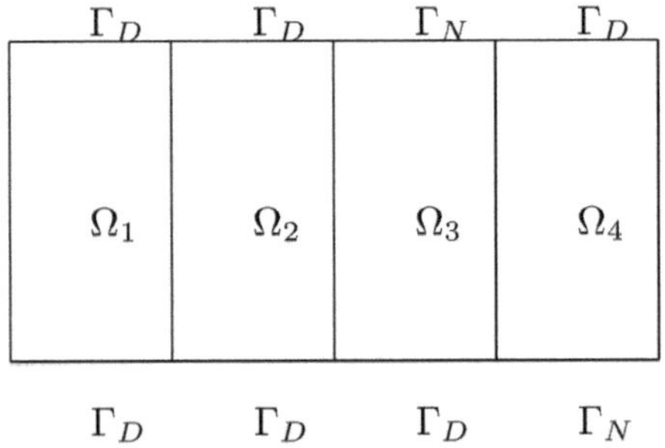

Figure 4.4 Domain partition with boundary conditions.

We note that $(\Sigma_{DN}^{(3)})^{-1} + (\Sigma_{ND}^{(3)})^{-1} \neq (\Sigma_{DN} + \Sigma_{ND})^{-1}$. Furthermore, we have

$$(\Sigma_{DN}\varphi, \varphi) \approx \|\varphi^h\|_{H^{1/2}}^2 + \int \frac{(\varphi^h(x))^2}{x - a_3}\, dx$$

and

$$\Sigma = \begin{bmatrix} \Sigma_{DD}^{(1)} & & \\ & \Sigma_{NN}^{(2)} & \\ & & \Sigma_{DD}^{(3)} \end{bmatrix} \not\approx S.$$

This is only related to the semi-norm. What should we do in this case? We use $-\Delta + I$ instead of $-\Delta$ and construct a preconditioner for $-\Delta u + u$. Hence, we have $\Sigma_{NN}^{(2)}$ in the second block of above expression.

Lemma 4.2 *Let*

$$A = \begin{bmatrix} A_{11} & A_{12} \\ A_{21} & A_{22} \end{bmatrix}, \qquad A^{-1} = \begin{bmatrix} B_{11} & B_{12} \\ B_{21} & B_{22} \end{bmatrix}.$$

Then $B_{11}^{-1} = A_{11} - A_{12}A_{22}^{-1}A_{21}$.

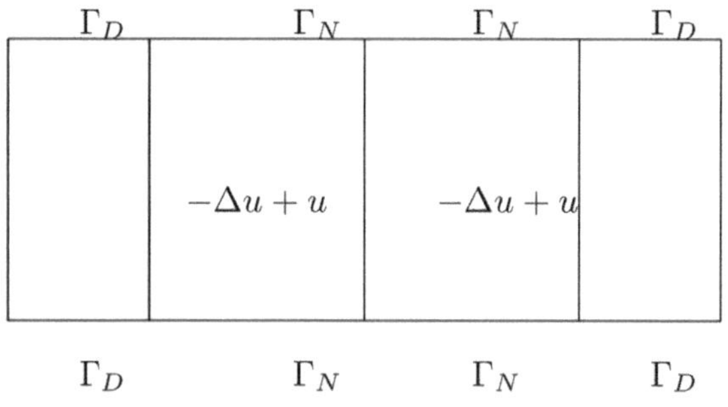

Figure 4.5 Other boundary conditions.

Proof. From the obvious relation

$$\begin{bmatrix} B_{11} & B_{12} \\ B_{21} & B_{22} \end{bmatrix} \begin{bmatrix} A_{11} & A_{12} \\ A_{21} & A_{22} \end{bmatrix} = \begin{bmatrix} I_1 & 0 \\ 0 & I_2 \end{bmatrix},$$

we immediately obtain the equations

$$\begin{aligned} B_{11}A_{11} + B_{12}A_{21} &= I_1, \quad & B_{11}A_{12} + B_{12}A_{22} &= 0, \\ B_{21}A_{11} + B_{22}A_{21} &= 0, \quad & B_{21}A_{12} + B_{22}A_{22} &= I_2, \end{aligned}$$

which lead to the relations

$$\begin{aligned} &\Longrightarrow \quad B_{12} = -B_{11}A_{12}A_{22}^{-1}, \quad B_{11}A_{11} - B_{11}A_{12}A_{22}^{-1}A_{21} = I_1, \\ &\Longrightarrow \quad B_{11}^{-1} = A_{11} - A_{12}A_{22}^{-1}A_{21}. \end{aligned}$$

This completes the proof of the lemma. □

We note that a similar result holds for B_{22}.

The cross-point case is more complicated. Let us consider some model boundary value problem in the domain Ω which consists of four subdomains as is shown in Figure 4.6. Let us construct a preconditioner $\Sigma^{(i)}$ for $S^{(i)}$ on each subdomain. Then we arrive at the spectral equivalence relation

$$S = S^{(1)} + S^{(2)} + S^{(3)} + S^{(4)} \approx \Sigma^{(1)} + \Sigma^{(2)} + \Sigma^{(3)} + \Sigma^{(4)} = \Sigma.$$

However, it is not immediately clear how one can efficiently solve the preconditioning system. The trace theorem is not enough. We need the so-called "Schwarz machinery", in particular, the theory of the additive Schwarz methods (ASM) will be very helpful.

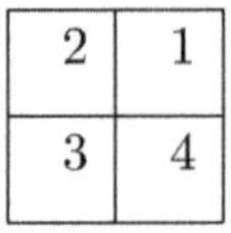

Figure 4.6 The cross point case.

5 The Schwarz alternating method

In 1869, H. A. Schwarz introduced an overlapping domain decomposition method [40]. He used this method, which is now called the Schwarz alternating method, for proving the existence of harmonic functions in complicated domains composed of simpler domains where the existence is known.

Let us consider again the Dirichlet problem for the Poisson equation

$$\begin{cases} -\Delta u = f & \text{in } \Omega, \\ u = 0 & \text{on } \Gamma, \end{cases} \tag{5.1}$$

in some domain $\overline{\Omega} = \overline{\Omega}_1 \cup \overline{\Omega}_2$ which is composed of two overlapping subdomains Ω_1 and Ω_2. Then the Schwarz alternating method is nothing but an iterative process, where we alternately solve Dirichlet problems in the two subdomains, i.e. starting with some initial guess u^0 that vanishes on Γ, for $k = 1, 2, \dots,$ we perform following iteration steps:

1. Solution on Ω_1: Determine u_{2k+1} such that

$$\begin{cases} -\Delta u_{2k+1} = f - \Delta u^{2k}, \\ u_{2k+1} = 0 \quad \text{in } \Gamma_1, \end{cases}$$

   and set $u^{2k+1} = u^{2k} + u_{2k+1}$.

2. Solution on Ω_2: Determine u_{2k+2} such that

$$\begin{cases} -\Delta u_{2k+2} = f - \Delta u^{2k+1}, \\ u_{2k+2} = 0 \quad \text{in } \Gamma_1, \end{cases}$$

   and set $u^{2k+2} = u^{2k+1} + u_{2k+2}$.

Let us recall the weak formulation of the boundary value problem (5.1): Find $u \in H_0^1(\Omega)$ such that

$$a(u, v) = l(v) \quad \forall v \in H_0^1(\Omega),$$

where

$$a(u, v) = \int_\Omega (\nabla u, \nabla v)\, d\Omega \quad \text{and} \quad l(v) = \int f v\, d\Omega.$$

The weak formulation of the Schwarz alternating method, which was first given by S. L. Sobolev in [42], reads as follows:

$$\begin{gathered} u_{2k+1} \in H_0^1(\Omega_1): \\ a(u^{2k} + u_{2k+1}, v) = l(v) \quad \forall v \in H_0^1(\Omega_1) \\ u^{2k+1} = u^{2k} + u_{2k+1} \end{gathered}$$

$$\begin{gathered} u_{2k+2} \in H_0^1(\Omega_2): \\ a(u^{2k+1} + u_{2k+2}, v) = l(v) \quad \forall v \in H_0^1(\Omega_2) \\ u^{2k+2} = u^{2k+1} + u_{2k+2}. \end{gathered}$$

Let us now introduce the abbreviations

$$H = H_0^1(\Omega), \quad H_1 = H_0^1(\Omega_1), \quad H_2 = H_0^1(\Omega_2)$$

and the orthogonal projection (the Ritz projection) $P_i : H \to H_i$ with respect to the bilinear form $a(u, v)$. Then we can do the following analysis. From the identity

$$a(u_{2k+1}, v) = l(v) - a(u^{2k}, v) = a(u, v) - a(u^{2k}, v) = a(u - u^{2k}, v) \quad \forall v \in H_1,$$

we get

$$\begin{cases} u_{2k+1} = P_1(u - u^{2k}) \\ u^{2k+1} = u^{2k} + P_1(u - u^{2k}) \\ u^{2k+2} = u^{2k+1} + P_2(u - u^{2k+1}) \end{cases} \Longleftarrow \begin{cases} u^{2k+1} - u \\ = u^{2k} - u + P_1(u - u^{2k}) \\ = (I - P_1)(u - u^{2k}). \end{cases}$$

Introducing the iteration error

$$\psi^k = u^k - u$$

and the co-projection

$$Q_i : H \to H_1^{\perp}, \quad Q_i = I - P_i,$$

we obtain the following recursions for the iteration error:

$$\begin{aligned} \psi^{2k+1} &= (I - P_1)\psi^{2k} = Q_1\psi^{2k}, \\ \psi^{2k+2} &= (I - P_2)\psi^{2k+1} = Q_2\psi^{2k+1}. \end{aligned}$$

Now, for $k \geq 1$, we have the relations

$$\begin{aligned} a(\psi^{2k+1}, \psi^{2k+1}) &= \|\psi^{2k+1}\|_a^2 = \|Q_1\psi^{2k}\|_a^2 \\ &= \|Q_1Q_2\psi^{2k}\|_a^2 = a(Q_1Q_2\psi^{2k}, Q_1Q_2\psi^{2k}) \\ &= a(Q_2Q_1Q_1Q_2\psi^{2k}, \psi^{2k}) = a(Q_2Q_1Q_2\psi^{2k}, \psi^{2k}) \\ &= a((I - (P_1 + P_2) + P_1P_2 + P_2P_1 - P_2P_1P_2)\psi^{2k}, \psi^{2k}) \\ &= a(\psi^{2k}, \psi^{2k}) - a((P_1 + P_2)\psi^{2k}, \psi^{2k}) + a(P_1P_2\psi^{2k}, \psi^{2k}) \\ &\quad + a(P_2P_1\psi^{2k}, \psi^{2k}) - a(P_2P_1P_2\psi^{2k}, \psi^{2k}) \\ &= a(\psi^{2k}, \psi^{2k}) - a((P_1 + P_2)\psi^{2k}, \psi^{2k}). \end{aligned}$$

Let us assume that there exists a positive constant $\alpha \leq 1$ such that

$$\alpha a(u,u) \leq a((P_1 + P_2)u, u) \quad \forall u \in H.$$

Then we easily get the estimates

$$\begin{cases} \|\psi^{2k+1}\|_a \leq (1-\alpha)^{1/2}\|\psi^{2k}\|_a \\ \|\psi^{2k+2}\|_a \leq (1-\alpha)^{1/2}\|\psi^{2k+1}\|_a \end{cases}$$

which finally yield the convergence rate estimate

$$\|\psi^{2k+2}\|_a \leq (1-\alpha)\|\psi^{2k}\|_a$$

for the Schwarz alternating method in the energy norm $\|\cdot\|_a$.

The error propagation operator of the Schwarz alternating method is multiplicative. Therefore, such types of methods are also called multiplicative Schwarz methods. Multiplicative Schwarz methods are not in parallel. To construct a parallel method we will consider an additive version of the Schwarz method in the next section.

6 Additive Schwarz method

The additive Schwarz method (ASM) was suggested by A. Matsokin and S. Nepomnyaschikh in 1985 [21].

In the case of two subdomains, the ASM is based on the following inequalities

$$\alpha\, a(u,u) \leq a((P_1 + P_2)u, u) \leq 2\, a(u,u) \quad \forall u \in H,$$

and can be written in the following form: Given initial guess $u^0 \in H$, find iteratively

$$u^{k+1} = u^k - \tau_k(P_1 + P_2)(u^k - u) \quad k = 0, 1, 2, \ldots$$

The general theory of ASM is given by the following abstract theorem:

Theorem 6.1 *Let H be a Hilbert space equipped with the inner product (u,v). Let us consider the decomposition of H into subspaces H_i, i.e. $H = H_1 + H_2 + \cdots + H_m$. Moreover, let $A : H \to H$ be a symmetric, bounded and positive definite operator on H and denote by $a(u,v) = (Au,v)$ the corresponding bilinear form. Finally, let $P_i : H \to H_i$ be the orthogonal projections with respect to $a(u,v)$. Then, the following two statements are equivalent:*

a) There exists an $\alpha > 0$ such that, for all $u \in H$, there exists a decomposition

$$u = u_1 + u_2 + \cdots + u_m, \quad \text{with } u_i = H_i,$$

satisfying the inequality

$$\alpha\,(a(u_1,u_1) + a(u_2 + u_2) + \cdots + a(u_m,u_m)) \leq a(u,u).$$

b) The inequality

$$\alpha\, a(u,u) \le a((P_1 + P_2 + \cdots + P_m)u, u)$$

holds for all $u \in H$.

Proof. For the proof of the implication b) $\Longrightarrow$ a), we introduce

$$P = P_1 + P_2 + \cdots + P_m.$$

The operator P is symmetric, bounded and positive definite. For any $u \in H$, there exists a $v \in H$ such that $u = Pv = \sum_{i=1}^m P_i v$. We set $u_i = P_i v$. Then, we get

$$\begin{aligned} \sum_{i=1}^m a(u_i, u_i) &= \sum_{i=1}^m a(P_i v, P_i v) = \sum_{i=1}^m a(P_i v, v) = a(\sum_{i=1}^m P_i v, v) \\ &= a(u, v) = a(u, P^{-1}u) \le \frac{1}{\alpha} a(u,u). \end{aligned}$$

This proves the first assertion.

In order to prove a) $\Longrightarrow$ b), we start with the equality

$$\|u\|_a = \sup_{v\in H} \frac{a(u,v)}{\|v\|_a}. \tag{6.1}$$

Indeed,

$$\sup_{v\in H} \frac{a(u,v)}{\|v\|_a} \overset{\text{C.B.}}{\le} \sup_{v\in H} \frac{\|u\|_a \|v\|_a}{\|v\|_a} = \|u\|_a.$$

With $v = u$, we get $\sup \frac{a(u,v)}{\|v\|_a} \ge \frac{a(u,u)}{\|u\|_a} = \|u\|_a$. This proves (6.1).

Now, for any $u \in H$, we can conclude

$$\begin{aligned} \|u\|_a &= \sup_{v\in H} \frac{a(u,v)}{\|v\|_a} = \sup_{v\in H} \frac{a(u, \sum_{i=1}^m v_i)}{\|v\|_a} \\ &= \sup_{v\in H} \sum_{i=1}^m \frac{a(u, P_i v_i)}{\|v\|_a} = \sup_{v\in H} \sum_{i=1}^m \frac{a(P_i u, v_i)}{\|v\|_a} \\ &\overset{\text{C.B.}}{\le} \sup_{v\in H} \frac{\sum_{i=1}^m \|P_i u\|_a \|v_i\|_a}{\|v\|_a} \\ &\overset{\text{C.B.}}{\le} \sup_{v\in H} \frac{\sqrt{\sum_{i=1}^m \|P_i u\|_a^2} \cdot \sqrt{\sum_{i=1}^m \|v_i\|_a^2}}{\|v\|_a} \\ &\le \frac{1}{\sqrt{\alpha}} \sqrt{\sum_{i=1}^m \|P_i u\|_a^2} \end{aligned}$$

using (6.1), the Cauchy–Bunjakowski (C.B.) inequality two times and (a). □

Let Ω be a union of two overlapping subdomains Ω_1 and Ω_2. We want to show that, for any $u \in H_0^1(\Omega)$, there exist $u_i \in H_0^1(\Omega)$ such that $u_1 + u_2 = u$ and

$$\|u_1\|^2_{H^1(\Omega_1)} + \|u_2\|^2_{H^1(\Omega_2)} \leq \frac{1}{\alpha}\|u\|^2_{H^1(\Omega)}.$$

Let

$$u_1(x) = \begin{cases} u(x), & x \in \Omega_1 \backslash \Omega_2 \\ \text{extension}, & x \in \Omega_1 \cap \Omega_2. \end{cases}$$

The extension implies the estimate $\|u_1\|_{H^1(\Omega_1)} \leq C\|u\|_{H^1(\Omega)}$ with $u_1 \in H_0^1(\Omega_1)$. Then, let $u_2 = u - u_1$ with $u_2 \in H_0^1(\Omega_2)$. This gives the estimates

$$\|u_2\|_{H^1(\Omega_2)} \leq \|u\|_{H^1(\Omega)} + \|u_1\|_{H^1(\Omega_1)} \leq (1 + C)\|u\|_{H^1(\Omega)}$$

where the constant depends on the extension.

In general, we have $a(Pu, u) \leq m \cdot a(u, u)$ where m is the number of subspaces. However, if m is large, a refined estimate is required.

Theorem 6.2 *The following two assertions are equivalent:*

a) $a(Pu, u) \leq \beta\, a(u, u)\ \forall u \in H$,

b) $a(u, u) \leq \beta \inf\limits_{u_1+\cdots+u_m=u, u_i \in H_i} \sum_{i=1}^{m} a(u_i, u_i)$.

Proof. Let $u \in H$ and put $u_i = P_i P^{-1} u$. Then, we have

$$u_1 + \cdots + u_m = P_1 P^{-1} u + \cdots + P_m P^{-1} u = u.$$

Let $u = v_1 + \cdots + v_m$ be another decomposition with $v_i \in H_i$ and $v_i = u_i + w_i$. Obviously, one obtains $\sum_{i=1}^{m} w_i = 0$. Moreover,

$$\begin{aligned}
\sum_{i=1}^{m} a(v_i, v_i) &= \sum_{i=1}^{m} a(u_i, u_i) + 2a(u_i, w_i) + a(w_i, w_i) \\
&= \sum_{i=1}^{m} a(u_i, u_i) + 2a(P_i P^{-1} u, w_i) + a(w_i, w_i) \\
&= \sum_{i=1}^{m} a(u_i, u_i) + 2a(P^{-1} u, \sum_{i=1}^{m} w_i) + \sum_{i=1}^{m} a(w_i, w_i).
\end{aligned}$$

Hence,

$$\begin{aligned}
\inf_{u=v_1+\cdots+v_m, v_i \in H_i} \sum_{i=1}^{m} a(v_i, v_i) &= \sum_{i=1}^{m} a(u_i, u_i) = \sum_{i=1}^{m} a(P_i P^{-1} u, P_i P^{-1} u) \\
&= \sum_{i=1}^{m} a(P^{-1} u, P_i P^{-1} u) = a(P^{-1} u, P P^{-1} u) \\
&= a(P^{-1} u, u).
\end{aligned}$$

Choosing $u = P^{1/2}v$, we can prove

$$a(Pu,u) \le \beta\, a(u,u)\ \forall u \in H \iff a(u,u) \le \beta\, a(P^{-1}u,u)\ \forall u \in H.$$

This proves the theorem. □

Lemma 6.3 *Let $a(u,v) = A(u,v)$. Moreover, let us define*

$$A_i : H_i \to H_i, \quad (A_i u_i, v_i) = A(u_i, v_i) \quad \forall u_i, v_i \in H_i.$$

Finally, let Q_i be the orthogonal projector with respect to $(\cdot,\cdot)$, i.e.

$$Q_i : H \to H_i \quad \textit{in } (\cdot,\cdot).$$

Then, we have $P_i = A_i^{-1} Q_i A$.

Proof. Let $u \in H$ and set $u_i = P_i u$, $w_i = A_i^{-1} Q_i u$. This gives $A w_i = Q_i A u$. For all $v_i \in H_i$, we can conclude that

$$a(w_i,v_i) = (Aw_i,v_i) = (A_i w_i, v_i) = (Q_i Au, v_i) = (Au, v_i) = a(u,v_i).$$

Hence, we have $w_i = u_i$. □

Let us recall that $a(u,v) = (u,v)_{H^1}$ and $(u,v) = (u,v)_{L_2}$ for our model problem.

Let us summarize the results obtained so far. The previous lemma has given the relation between the projection P_i corresponding to the given bilinear form $a(\cdot,\cdot)$ and the L_2 projection Q_i. That is, we began with

- the decomposition $H = H_1 + H_2 + \cdots + H_m$,
- the bilinear form $a(u,v) = (Au,v)$, and
- the energetic projections $P_i : H \to H_i$.

For $P = \sum_{i=1}^m P_i$, we have shown the inequalities

$$\alpha\, a(u,u) \le a(Pu,u) \le \beta\, a(u,u) \quad \forall u \in H.$$

Then, we have proved that

$$P_i = A_i^{-1} Q_i A$$

where $Q_i : H \to H_i$ are the projections in $(\cdot,\cdot)$. Now, we have

$$\alpha\,(Au,u) \le (A(\sum_{i=1}^m A_i^{-1} Q_i A)u, u) \le \beta\,(Au,u) \quad \forall u \in H$$

and this is equivalent to

$$\alpha\,(Au,u) \le (AB^{-1}Au, u) \le \beta\,(Au,u) \quad \forall u \in H,$$

where $B^{-1} = \sum_{i=1}^m Q_i A_i^{-1} Q_i$. Putting $Au = v$, one easily concludes that

$$\alpha\,(A^{-1}v,v) \le (B^{-1}v,v) \le \beta\,(A^{-1}v,v) \quad \forall v \in H,$$

or

$$\alpha\,(Bu,u) \leq (Au,u) \leq \beta\,(Bu,u) \quad \forall v \in H.$$

Thus, we have constructed a preconditioner B which is equivalent to A. Now, we can use B as a preconditioner in the Richardson iteration

$$u^{k+1} = u^k - \tau_k B^{-1}(Au^k - f),$$

or in conjugate gradient method for solving our original problem $Au = f$.

Example 6.4 (Simple one-dimensional example) We consider the equation $-u'' = f$ in $\Omega = (0,1)$ with the homogeneous Dirichlet boundary condition $u(0) = u(1) = 0$. Then we have $Au = f$ with

$$A = \begin{bmatrix} 2 & -1 & & & \\ -1 & 2 & -1 & & \\ & \ddots & \ddots & \ddots & \\ & & -1 & 2 & -1 \\ & & & -1 & 2 \end{bmatrix}.$$

Note that A is a $(n-1)\times(n-1)$ matrix and $H = \mathbb{R}^{n-1}$. As in Figure 6.1, we define $H = H_1 + H_2$, $\Omega = \Omega_1 \cup \Omega_2$ and

$$u = \begin{bmatrix} v_1 & v_2 & v_3 & v_4 & v_5 \end{bmatrix}^\top$$

where $H_1 = \{(v_1, v_2, v_3, 0, 0)^\top\}$ and $H_2 = \{(0, 0, v_3, v_4, v_5)^\top\}$. Here the vectors v_1, v_3 and v_5 correspond to the interior nodes inside the intervals, whereas the values u_2 and u_4 correspond to the interface nodes between the intervals. Now we can take $\beta = 2$ by the property of the projection. For $u \in H$, we want to find some α such that

$$\alpha \sum_{i=1}^{2} a(u_i, u_i) \leq a(u,u), \qquad u = u_1 + u_2,$$

where $u_i \in H_i$. From Figure 6.1, it can be observed

$$(Au_1, u_1) \leq c\,(Au, u),$$

where c is independent of h. Setting $u_2 = u - u_1$, we have such an α which is independent of h. In this case, we have

$$Q_1 = \begin{bmatrix} I_1 & 0 \\ 0 & 0 \end{bmatrix}, \quad Q_2 = \begin{bmatrix} 0 & 0 \\ 0 & I_2 \end{bmatrix}.$$

Moreover,

$$Q_1 \begin{bmatrix} v_1 \\ v_2 \\ v_3 \\ v_4 \\ v_5 \end{bmatrix} = \begin{bmatrix} v_1 \\ v_2 \\ v_3 \\ 0 \\ 0 \end{bmatrix}, \quad Q_2 \begin{bmatrix} v_1 \\ v_2 \\ v_3 \\ v_4 \\ v_5 \end{bmatrix} = \begin{bmatrix} 0 \\ 0 \\ v_3 \\ v_4 \\ v_5 \end{bmatrix}$$

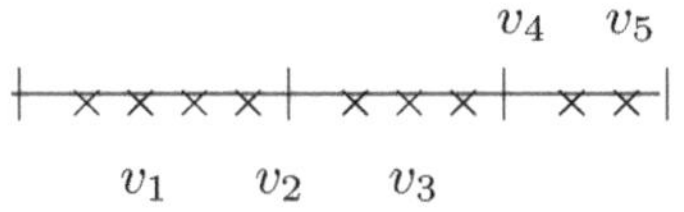

Figure 6.1 Numbering of (i, j) element and its subdivision.

and

$$A_1 = \begin{bmatrix} 2 & -1 & & & \\ -1 & 2 & -1 & & \\ & \ddots & \ddots & \ddots & \\ & & -1 & 2 & 0 \\ & & & 0 & 0 \end{bmatrix}, \quad A_2 = \begin{bmatrix} 0 & 0 & & & \\ 0 & 2 & -1 & & \\ & \ddots & \ddots & \ddots & \\ & & -1 & 2 & -1 \\ & & & -1 & 2 \end{bmatrix}.$$

Note that A_1 has zero entries that correspond to the multiplication with v_4 and v_5, whereas A_2 has zero entries corresponding to the v_1, v_2-multiplication. Now, we obtain the preconditioner in the form

$$B^{-1} = Q_1 A_1^+ Q_1 + Q_2 A_2^+ Q_2 = A_1^+ + A_2^+$$

where A_i^+ denotes the pseudo-inverse of A_i.

After finishing this simple one-dimensional example, we return to the general ASM theory. The following theorem covers the case where local preconditioners are used.

Theorem 6.5 *Let $H = H_1 + H_2 + \cdots + H_m$ and $a(u, v) = (Au, v)$. Let $P_i : H \to H_i$ be the orthogonal projection with respect to $a(\cdot, \cdot)$ and let A be symmetric and positive definite. Furthermore, let us assume that the following three conditions are satisfied:*

(1) $\alpha(a(u_1, u_1) + \cdots + a(u_m, u_m)) \le a(u, u)$ for $u_1 + \cdots + u_m = u$.

(2) $a(u, u) \le \beta \inf_{u_1 + \cdots + u_m = u} \sum_{i=1}^{m} a(u_i, u_i)$.

(3) There are local preconditioners $B_i : H \to H_i$, with $B_i = B_i^*$, such that there exist positive constants c_1 and c_2 such that the spectral equivalence inequalities

$$c_1 (B_i u, u) \le (Au, u) \le c_2 (B_i u, u) \quad \forall u \in H_i$$

are fulfilled.

Then, we have

$$\alpha c_1 (A^{-1} u, u) \le (B^{-1} u, u) \le \beta c_2 (A^{-1} u, u) \quad \forall u \in H,$$

where $B^{-1} = B_1^+ + \cdots + B_m^+$.

Proof. Note that $P_i = Q_i A_i^{-1} Q_i A$. We have a pseudo-inverse

$$(Q_i A Q_i)^+ = Q_i A_i^{-1} Q_i,$$

since $(Q_i A Q_i) Q_i A_i^{-1} Q_i = Q_i A Q_i A_i^{-1} Q_i = Q_i$. From (1) and (2) we have

$$\alpha(A^{-1}v, v) \le (((Q_1 A Q_1)^+ + \cdots + (Q_m A Q_m)^+)v, v) \le \beta(A^{-1}v, v),$$

and from (3)

$$c_1((Q_i A Q_i)^+ u, u) \le (B_i^+ u, u) \le c_2(c(Q_i A Q_i)^+ u, u) \quad \forall u \in H_i.$$

Combining the two above inequalities, we get the result of the theorem. □

Remark 6.6 Due to Theorem 6.1, condition (1) is equivalent to $\alpha\, a(u,u) \le a(Pu, u)$, with $P = P_1 + \cdots + P_m$ for all u. Using Theorem 6.2, condition (2) is equivalent to $a(Pu,u) \le \beta\, a(u,u)$ for all u.

In order to prove the stability of the next decompositions the following lemma is required.

Lemma 6.7 *Let $\varphi \in H^{1/2}(-1,0)$ and let us define an extension of φ to $[0,2]$ by the formula*

$$\varphi = \begin{cases} (1-x)\varphi(-x), & x \in [0,1], \\ 0, & x \in [1,2]. \end{cases}$$

Then there exists a constant C such that $\|\varphi\|_{H^{1/2}(-1,2)} \le C \|\varphi\|_{H^{1/2}(-1,0)}$.

Proof. By Lemma 3.9, we have

$$\|\varphi\|^2_{H^{1/2}(-1,2)} \le C_1(\|\varphi\|^2_{H^{1/2}(-1,0)} + \|\varphi\|^2_{H^{1/2}(0,1)} + \|\varphi\|^2_{H^{1/2}(1,2)} + I_1(\varphi) + I_2(\varphi))$$

Note that $\|\varphi\|_{H^{1/2}(1,2)} = 0$. It is trivial to prove

$$\|\varphi\|_{L^2(0,1)} \le \|\varphi\|_{L^2(-1,0)}.$$

Moreover, one obtains

$$\begin{aligned} |\varphi|^2_{H^{1/2}(0,1)} &= \int_0^1 \int_0^1 \frac{|\varphi(-x)(1-x) - \varphi(-y)(1-y)|^2}{|x-y|^2}\, dx dy \\ &\le 2 \int_0^1 \int_0^1 \frac{|\varphi(-x)(1-x) - \varphi(-x)(1-y)|^2}{|x-y|^2}\, dx dy \\ &\quad + 2 \int_0^1 \int_0^1 \frac{|\varphi(-x) - \varphi(-y)|^2 |(1-y)|^2}{|x-y|^2}\, dx dy \\ &\le 2 \left(\int_0^1 \int_0^1 \frac{|\varphi(-x)(x-y)|^2}{|x-y|^2}\, dx dy + \int_{-1}^0 \int_{-1}^0 \frac{|\varphi(x) - \varphi(y)|^2}{|x-y|^2}\, dx dy \right) \\ &= 2(|\varphi|^2_{H^{1/2}(-1,0)} + \|\varphi\|^2_{L^2(-1,0)}). \end{aligned}$$

On the other hand, we have the inequalities

$$I_1(\varphi) = \int_0^1 \frac{(\varphi(-x) - \varphi(-x)(1-x))^2}{x}\,dx \;\le\; \int_0^1 \varphi^2(-x)\frac{x^2}{x}\,dx \le \|\varphi\|^2_{L^2(-1,0)}$$

$$I_2(\varphi) = \int_0^1 \frac{(\varphi(-x)(1-x))^2}{1-x}\,dx \;\le\; \|\varphi\|^2_{L^2(-1,0)}.$$

Gathering all inequalities completes the proof. □

Our goal is to construct the Schur complement on the lines (sub-boundaries) of the interior of the given domain.

We consider a decomposition of Λ into substructures λ_i, i.e. $\Lambda = \bigcup_{i=1}^{n} \partial\Omega_i = \bigcup_{i=1}^{m} \lambda_i$. The substructures λ_i, $i = 1, \dots, m_1$, correspond to cross-points, whereas the substructures λ_i, $i = m_1 + 1, \dots, m$, correspond to usual lines.

Let us assume that there exists a constant r which is independent of h such that for all $p \in \Lambda$ there exists λ_i:

$$B(p, r) \cap \Lambda \subset \lambda_i,$$

where $B(p, r)$ denotes a ball with a center at p and with a radius r.

Let $H = H_h(\Lambda)$ and $H = H_1 + H_2 + \dots + H_m$ with

$$H_i = H_h(\lambda_i) = \{\varphi^h \in H_h(\Lambda) | \varphi(x) = 0, \exists x \notin \lambda_i\}.$$

By Lemma 6.7, we have $\forall \varphi^h \in H, \exists \varphi_i^h \in H_i$,

$$\|\varphi_1^h\|^2_{H_{00}^{1/2}(\lambda_1)} + \dots + \|\varphi_m^h\|^2_{H_{00}^{1/2}(\lambda_m)} \le C\|\varphi^h\|^2_{H^{1/2}(\Lambda)}.$$

Let us introduce

$$\tilde{H}_1 = H_1 + \dots + H_{m_1}, \quad \tilde{H}_2 = H_{m_1+1} + \dots + H_m.$$

Then we have $H = \tilde{H}_1 + \tilde{H}_2$. So far we have constructed the space satisfying the conditions (1) and (2) in the previous Theorem 6.5. Now, we will construct a preconditioner for the Schur complement by an additive form of pseudo-inverses such as

$$\Sigma^{-1} = \Sigma_1^+ + \dots + \Sigma_m^+$$

with

$$\Sigma_i = R_i \begin{bmatrix} 0 & 0 & 0 \\ 0 & X^{1/2} & 0 \\ 0 & 0 & 0 \end{bmatrix} R_i^\top, \quad i = m_1 + 1, \dots, m,$$

where R_i is the permutation matrix and X is the matrix corresponding to the one-dimensional Laplacian, i.e.

$$X = \begin{bmatrix} 2 & -1 & 0 & & \\ -1 & 2 & -1 & & \\ & \ddots & \ddots & \ddots & \\ & & -1 & 2 & -1 \\ & & & -1 & 2 \end{bmatrix}. \tag{6.2}$$

Hence, we get

$$\Sigma_i^+ = R_i \begin{bmatrix} 0 & 0 & 0 \\ 0 & X^{-1/2} & 0 \\ 0 & 0 & 0 \end{bmatrix} R_i^\top .$$

The preconditioners Σ_i, $i = 1, \ldots, m_1$, are defined later. First we prove the following lemma.

Lemma 6.8 *Let us assume that the symmetric and positive definite matrices*

$$\Sigma : \mathbb{R}^m \to \mathbb{R}^m \quad \textit{and} \quad S : \mathbb{R}^n \to \mathbb{R}^n$$

are given. Let $t : \mathbb{R}^m \to \mathbb{R}^n$ *such that*

$$\alpha\,(\varphi, \varphi)_\Sigma \le (t\varphi, t\varphi)_S \le \beta\,(\varphi, \varphi)_\Sigma \quad \forall \varphi \in \mathbb{R}^m,$$

and $(t^\top u, \varphi)_{\mathbb{R}^n} = (u, t\varphi)_{\mathbb{R}^n}$, *where* $(\cdot, \cdot)_{\mathbb{R}^i}$ *denotes the Euclidean inner product. Set* $C = t\Sigma^{-1}t^\top$. *Then we have*

$$\alpha\,(C^+ u, u) \le (u, u)_S \le \beta\,(C^+ u, u) \quad \forall u \in \mathrm{Im}(t).$$

Proof. By our assumptions, the matrix $(t^\top t)^{-1}$ exists. We note that

$$C^+ = t(t^\top t)^{-1}\Sigma(t^\top t)^{-1}t^\top,$$

which is easily verified from the following observation

$$C^+ C = t(t^\top t)^{-1}\Sigma(t^\top t)^{-1}t^\top(t\Sigma^{-1}t^\top) = t(t^\top t)^{-1}t^\top.$$

Now it is sufficient to check that $t(t^\top t)^{-1}t^\top$ is a projection. If $u \in \mathrm{Im}(t)$, then $u = t\varphi$ for some φ. Thus, we have

$$C^+ C u = t(t^\top t)^{-1}t^\top u = t\varphi = u.$$

Furthermore, for all $v_0 \in (\mathrm{Im}(t))^\perp$, we have

$$0 = (v_0, t\varphi) = (v_0, t(t^\top t)^{-1}t^\top t\varphi) = (t(tTt)^{-1}t^\top v_0, t\varphi) \quad \forall \varphi.$$

Hence, one obtains

$$C^+ C v_0 = 0 \quad \forall v_0 \in (\mathrm{Im}(t))^\perp.$$

Now, for all $u \in \mathrm{Im}(t)$, the relations

$$\begin{aligned} (C^+ u, u) &= (C^+ t\varphi, t\varphi) = (t(t^\top t)^{-1}\Sigma(t^\top t)^{-1}t^\top t\varphi, t\varphi) \\ &= (t^\top t(t^\top t)^{-1}\Sigma\varphi, t\varphi) \\ &= (\Sigma\varphi, \varphi) \end{aligned}$$

hold. Hence the proof is completed. □

Remark 6.9 In general, $m \le n$. The operator t can be interpreted as an extension operator.

In the additive Schwarz method, we need to define B_i^+. Now we will try to set

$$B_i^+ = (C_i^+)^+ = C_i = t\Sigma^{-1}t^\top,$$

where t is a proper extension operator. The space H_i is defined via $\mathrm{Im}(t) := H_i$. The stability of the decomposition depends on the choice of the extension operators.

7 Additive Schwarz method on interfaces

In this section we use the general framework of ASM in order to develop preconditioners on the interfaces.

Let z_0 be a fixed cross point. Let λ be the union of branches emerging from z_0. Let L_i, for $i = 1, \dots, m$ be each branch and let $L_{m+1} = L_1$ by definition. The trace norm on λ is defined via

$$\|\phi^h\|^2_{H_{00}^{1/2}(\lambda)} = \sum_{i=1}^{m} \|\phi^h\|^2_{H_{00}^{1/2}(L_i \cup L_{i+1})}.$$

Let $x_{i,j}$ be the point on the branch L_i which has the distance jh from z_0. We consider the space decomposition,

$$H_h(\lambda) = H_0 + H_1 + \cdots + H_m,$$

where $H_i = \{\phi^h \in H_h(\lambda) | \phi^h(x) = 0, x \notin L_i\}$ and $H_0 = \{\phi^h \in H^h(\lambda) | \phi^h(x_{1,j}) = \cdots = \phi^h(x_{m,j}), j = 1, 2, \dots, k\}$. Here we assume that each L_i has the same number k of nodes.

Lemma 7.1 *There exists a constant c independent of h such that, for each $\phi^h \in H^h(\lambda)$, there exist $\phi_i^h \in H_i$ with $\sum_{i=0}^m \phi_i^h = \phi^h$ which satisfies*

$$\|\phi_0^h\|^2_{H_{00}^{1/2}(\lambda)} + \|\phi_1^h\|^2_{H_{00}^{1/2}(\lambda)} + \cdots + \|\phi_m^h\|^2_{H_{00}^{1/2}(\lambda)} \le c\|\phi^h\|^2_{H_{00}^{1/2}(\lambda)}.$$

Proof. Let $\phi \in H_h(\lambda)$. First, we set $\phi_0^h(x_{i,j}) = \phi^h(x_{1,j})$, for $j = 1, \dots, k$ and $i = 1, \dots, m$, i.e. we take the values of the first branch on the another branches. Due to our definition, we have $\phi_0^h \in H_0$. Let $\psi^h = \phi^h|_{L_1}$. Since

$$\|\phi_0^h\|^2_{H_{00}^{1/2}(\lambda)} = m\|\phi^h\|^2_{H_{00}^{1/2}(L_1 \cup L_2)} \simeq \|\psi^h\|^2_{\tilde{H}^{1/2}(L_1)} \simeq (\Sigma_{ND}\psi, \psi),$$

there exists a constant c_1 which is independent of h such that

$$\|\phi_0^h\|^2_{H_{00}^{1/2}(\lambda)} \le c_1\|\phi^h\|^2_{H_{00}^{1/2}(\lambda)}.$$

Let $\xi^h = \phi^h - \phi_0^h$. Next, we define the hat functions ϕ_i^h, $i \ge 1$ via $\phi_i^h(x_{i,j}) = \xi^h(x_{i,j})$. Then, we have

$$\|\xi^h\|^2_{H_{00}^{1/2}(\lambda)} \le c_2\|\phi^h\|^2_{H_{00}^{1/2}(\lambda)}.$$

This implies the estimates

$$\|\phi_i^h\|^2_{H_{00}^{1/2}(\lambda)} \simeq \|\phi_i^h\|^2_{H_{00}^{1/2}(L_i)} \simeq (\Sigma_{DD}\phi_i, \phi_i).$$

Hence we obtain

$$\|\phi_i^h\|^2_{H_{00}^{1/2}(\lambda)} \le \|\xi^h\|^2_{H_{00}^{1/2}(\lambda)}.$$

Continuing the above processes, we can prove the lemma. □

The lemma shows that

$$\frac{1}{c}a(\phi^h, \phi^h) \le a((P_0 + \cdots + P_m)\phi^h, \phi^h) \le (m+1)a(\phi^h, \phi^h).$$

Let t be the extension operator such that, for each $\psi^h = \begin{bmatrix}\psi_0 & \psi_1 & \cdots & \psi_m\end{bmatrix}^\top$, the relation $t\phi^h = \begin{bmatrix}\psi_0 & \eta & \cdots & \eta\end{bmatrix}^\top$ holds, where $\eta = \begin{bmatrix}\psi_1 & \cdots & \psi_m\end{bmatrix}^\top$. Then, we have the relations $H_0 = t \cdot F$, $F = H_h(L_1)$ and

$$\|\psi^h\|_{\check{H}^{1/2}(L_1)} \le \|t\psi^h\|_{H_{00}^{1/2}(\lambda)} \le C\|\psi^h\|_{\check{H}^{1/2}(L_1)}.$$

This gives

$$B_i^+ = t\Sigma_{ND}^{-1}t^\top.$$

Now, the whole interface space $H_h(\Lambda)$ is decomposed into subspaces. Let

$$H_h(\Lambda) = H_1^{(N)} + \cdots + H_{m_1}^{(N)} + H_{m_1+1}^{(0)} + \cdots + H_m^{(0)}.$$

The spaces $H_i^{(N)}$, $i = 1, \dots, m_1$, are the subspaces corresponding to cross points, whereas the spaces

$$H_i^N = \{\varphi^h \in H_h(\Lambda) \mid \varphi^h(x) = t_i\psi^h(x), x \in \lambda_i, \varphi^h(x) = 0, x \notin \lambda_i\}$$

and $H_i^{(0)}$, $i = m_1 + 1, \dots, m$, are the subspaces corresponding to intervals between cross points, i.e.

$$H_i^0 = \{\varphi^h \in H_h(\Lambda) \mid \varphi^h(x) = 0, x \notin \lambda_i\}.$$

Let

$$B^{-1} = B_{N,1}^+ + \cdots + B_{N,m_1}^+ + B_{0,m_1+1}^+ + \cdots + B_{0,m}^+, \tag{7.1}$$

where

$$B_{0,i}^+ = \begin{bmatrix} 0 & 0 & 0 \\ 0 & X^{-1/2} & 0 \\ 0 & 0 & 0 \end{bmatrix}$$

with the one-dimensional Laplacian operator X, cf. (6.2) and

$$B_{N,i}^+ = t_i\Sigma_{ND}^{-1}t_i^\top.$$

Then, the matrix B (7.1) is a preconditioner for the Schur complement $S = A_0 - \sum_{i=1}^n A_{0i}A_i^{-1}A_{i0}$. However, the direct computation of A_i^{-1} is too expensive. In order to replace A_i^{-1} by a local preconditioner on Ω_i, we have to use ASM. This is presented in the following section.

7.1 Inexact solvers

In this section, we consider a nonoverlapping partition of Ω into subdomains Ω_i, i.e. $\overline{\Omega} = \bigcup_{i=1}^{n} \overline{\Omega_i}$, $\Omega_i \cap \Omega_j = \emptyset$, if $i \neq j$. The space decomposition $H_h(\Omega) = H_0 + H_1$ is considered, where

$$\begin{aligned} H_0 &= H_{h,0}(\Omega_1) \oplus \cdots \oplus H_{h,0}(\Omega_n) \quad \text{and} \\ H_{h,0}(\Omega_i) &= \{\phi^h \in H_h(\Omega) | \phi^h(x) = 0, x \notin \Omega_i\} \end{aligned}$$

correspond to the Dirichlet problems on the subdomains Ω_i. The space H_1 is defined below.

Theorem 7.2 *In addition to the above assumptions, let us assume the following:*

(A) There exist operators B_i which satisfy

$$c_1 \|u^h\|^2_{H^1(\Omega)} \leq (B_i u, u) \leq c_2 \|u^h\|^2_{H^1(\Omega)} \; \forall u^h \in H_{h,0}(\Omega_i), \quad i = 1, \ldots, n.$$

(B) There exist extension operators $t_i : H_h(\Gamma_i) \to H_h(\Omega_i)$ such that the inequalities

$$\|t_i \phi^h\|_{H_1(\Omega_i)} \leq c_3 \|\phi^h\|_{H^{1/2}(\Gamma_i)}$$

hold for $i = 1, \ldots, n$. Let $H_1 = tH_h(\lambda)$, where t now denotes the global extension operator composed of the local ones.

(C) There exists an operator Σ with

$$c_4 \|\phi^h\|^2_{H^{1/2}(\lambda)} \leq (\Sigma\phi, \phi) \leq c_5 \|\phi^h\|^2_{H^{1/2}(\Lambda)} \; \forall \phi^h \in H_h(\Lambda).$$

Set

$$B^{-1} = \begin{bmatrix} 0 & & & \\ & B_1^{-1} & & \\ & & \ddots & \\ & & & B_n^{-1} \end{bmatrix} + t\Sigma^{-1}t^{\top}. \tag{7.2}$$

Then there exist constants α and β which only depend on $c_1, c_2, \ldots, c_5$ such that

$$\alpha\,(Bu, u) \leq (Au, u) \leq \beta\,(Bu, u) \qquad \forall u \in H_h(\Omega).$$

Proof. Let $u^h \in H_h(\Omega)$ denote the trace function of u^h by $\phi^h \in H^h(\Lambda)$, i.e. $\phi^h(x) = u^h(x), x \in \Lambda$. By the trace theorem, there exists a constant c_6 independent of h such that

$$\|\phi^h\|_{H^{1/2}(\Lambda)} \leq c_6 \|u^h\|^2_{H^1(\Omega)}. \tag{7.3}$$

Let $u_1^h = t\phi^h$. Using (B) and (7.3), we have

$$\|u_1^h\|^2_{H^1(\Omega)} \leq c_3 \|\phi^h\|_{H^{1/2}(\Lambda)} \leq c_3 c_6 \|u^h\|^2_{H^1(\Omega)}.$$

Let $u_0^h = u^h - u_1^h$. Using the triangle inequality and ASM proves the assertion. □

7.2 Explicit extension operators

The definition of the preconditioner (7.2) requires an extension operator t which satisfies (B). This section is devoted to the construction of the extension operator t. The simplest choice of an extension operator which satisfies (B) is the harmonic extension. However, in every preconditioning step a problem for the Laplacian has to be solved. This is too expensive. Therefore, we have to find another one.

Let (s, n) be a near boundary coordinate system, where s denotes the tangential and n denotes the normal coordinate. Let ϕ be a given function defined on the boundary Γ of a domain Ω. For the continuous case, we can define the extended function $u = t\phi$ by

$$u(s,n) = \xi(n)\frac{1}{n}\int_s^{s+n} \phi(t)dt,$$

where $\xi(n) = 1 - \frac{n}{D}$, and D is the thickness of the near boundary strip. Let ϕ be some finite element function which is given by its nodal values $\phi(l)$, $l = 0, \ldots, i+j$. The function u is defined by its values at the nodal points z_{ij}, where the first index corresponds to the tangential component and the second to the normal component, i.e. z_{i0} are the nodes on the boundary. Moreover, we introduce D_{ij} as the cells of an auxiliary grid. Then, the extension u is defined by the following three steps.

1. $V(z_{ij}) = \sum_{l=0}^{j} \phi(i+l)$,

2. $U(z_{ij}) = \frac{1 - \frac{j}{n}}{j+1} V(z_{ij})$,

3. $u^h(z_l) = \begin{cases} U(z_{ij}), & \text{if } z_l \in D_{ij}, \\ 0, & \text{if } z_l \notin \bigcup_{ij} D_{ij}. \end{cases}$

In matrix vector-notation, we have $u = t\phi = P_3 P_2 P_1 \phi$, where the matrix P_3 is given by

$$P_3 = \begin{bmatrix} 0 & \cdots & 0 \\ \vdots & \ddots & \vdots \\ 0 & \cdots & 0 \\ & I & \end{bmatrix}.$$

Note that the lower identity matrix corresponds to D. Then we have

$$\|u^h\|_{H_h^1(\Omega^h)} \leq c\|V\|_{H_h^1(D^h)}.$$

Due to the second step of the extension, the matrix P_2 is a diagonal matrix, i.e. $P_2 =$

diag$\{\ldots, \frac{1-j/n}{j+1}, \ldots\}$. The matrix P_1 corresponds to the first step and has the form

$$P_1 = \begin{bmatrix} 1 & & & \\ & \ddots & & \\ & & \ddots & \\ & & & 1 \\ 1 & 1 & & \\ & \ddots & \ddots & \\ & & 1 & 1 \\ 1 & . & . & 1 \\ \vdots & \vdots & \vdots & \vdots \end{bmatrix}.$$

The identity block corresponds to $V(z_{i,0}) = \phi(i)$, $i = 0, \ldots, N-1$, whereas the values of the j-th layer in normal direction are given by $P_1\phi = V(z_{i,j+1}) = V(z_{i,j}) + \phi(i+j+1)$, $0 \le i \le N-1$, $0 \le j \le M$. The total computational cost for the multiplication with P_2 is $O(h^{-2})$.

Let $t^\top = P_1^\top P_2^\top P_3^\top$ be the adjoint operator. For any given function V, we define the function W by

$$W(z_{i,M}) = V(z_{i,M}), \quad i = 0, \ldots, M$$

and for $i = 0, \ldots, N$ and $j = M, M-1, \ldots, 1$, let

$$W(z_{i,j-1}) = W(z_{i,j}) + V(z_{i,j-1}).$$

If we set $\phi_i = \sum_{j=0}^{N} \sum_{l=i-j}^{i} V(z_{l,j})$, $W(z_{i,M}) = V(z_{i,M})$ and

$$W_1(t_{i,j-1}) = W_1(z_{i,j}) + V(z_{i,j-1})$$

then we arrive at $\phi_i = \sum_{j=0}^{M} W(z_{i-j})$. Therefore the total cost for a multiplication with $t^\top$ is again $O(h^{-2})$.

Remark 7.3 We note that multilevel (hierarchical) explicit extension operators were suggested in [10]. This multilevel decomposition on a boundary of some subdomain is based on the results of [45]. But this method is not asymptotically optimal. An optimal method (with respect to the arithmetical coast and the norms of the extension operators) of the multilevel explicit extension of functions was suggested in [31] and [11]. In this case the multilevel decomposition on the boundary of the subdomain is based on BPX-like decompositions.

8 Domain decomposition with many subdomains

In order to use parallel computers with many processors a decomposition of an original domain into many subdomains ($n \gg 1$) of small measure is required.

Let Ω be a domain of diameter $O(1)$ with boundary Γ, and set

$$\Omega_\varepsilon = \{(x,y) : x = \varepsilon s,\ y = \varepsilon t,\ (x,y) \in \Omega\}$$

with boundary Γ_ε. Here we present some results on the trace theory in general Sobolev spaces with small diameters which are characterized by a small parameter ε. For the FEM solution of elliptic and parabolic problems this parameter ε usually is equivalent to a mesh size of the coarse grid H or the average diameter H of the subdomains.

Lemma 8.1 *There exists $c_1 \neq c_1(\varepsilon)$ such that, for all $u \in H^1(\Omega_\varepsilon)$, we have*

$$\varphi(x) = u(x),\ x \in \Gamma_\varepsilon, \qquad |\varphi|_{H^{1/2}(\Gamma_\varepsilon)} \le c_1 |u|_{H^1(\Omega_\varepsilon)},$$

and vice versa there exists $c_2 \neq c_2(\varepsilon)$ such that for every $\varphi \in H^{1/2}(\Gamma_\varepsilon)$, there exists $u \in H^1(\Omega_\varepsilon)$ satisfying

$$\varphi(x) = u(x),\ x \in \Gamma_\varepsilon, \qquad |u|_{H^1(\Omega_\varepsilon)} \le c_2 |\varphi|_{H^{1/2}(\Gamma_\varepsilon)}.$$

Proof. Using the change of variables, i.e. $(x \to s), (y \to t)$, we have

$$|u|^2_{H^1(\Omega_\varepsilon)} = \int_{\Omega_\varepsilon} \left(\frac{\partial u}{\partial x}\right)^2 + \left(\frac{\partial u}{\partial y}\right)^2 = \int_{\Omega} \left(\frac{\partial \tilde{u}}{\partial s}\right)^2 + \left(\frac{\partial \tilde{u}}{\partial t}\right)^2 = |\tilde{u}|^2_{H^1(\Omega)}$$

and

$$\begin{aligned} |\varphi|^2_{H^{1/2}(\Gamma_\varepsilon)} &= \int_{\Gamma_\varepsilon}\int_{\Gamma_\varepsilon} \frac{(\varphi(x)-\varphi(y))^2}{|x-y|^2}\, dxdy \\ &= \int_{\Gamma}\int_{\Gamma} \frac{(\tilde{\varphi}(s)-\tilde{\varphi}(t))^2}{|s-t|^2}\, dsdt \ = \ |\tilde{\varphi}|^2_{H^{1/2}(\Gamma)} \end{aligned}$$

which prove the lemma. □

Now we define

$$\|\varphi\|^2_{H^{1/2}_\varepsilon(\Gamma_\varepsilon)} = \varepsilon \|\varphi\|^2_{L_2(\Gamma_\varepsilon)} + |\varphi|^2_{H^{1/2}(\Gamma_\varepsilon)}.$$

Lemma 8.2 *There exists $c_1 \neq c_1(\varepsilon)$ such that for all $u \in H^1(\Omega_\varepsilon)$,*

$$\varphi(x) = u(x),\ x \in \Gamma_\varepsilon, \qquad \|\varphi\|_{H^{1/2}_\varepsilon(\Gamma_\varepsilon)} \le c_1 \|u\|_{H^1(\Omega_\varepsilon)}.$$

There exists $c_2 \neq c_2(\varepsilon)$ such that for every $\varphi \in H^{1/2}(\Gamma_\varepsilon)$, there exists $u \in H^1(\Omega_\varepsilon)$ satisfying

$$\varphi(x) = u(x),\ x \in \Gamma_\varepsilon, \qquad \|u\|_{H^1(\Omega_\varepsilon)} \le c_2 \|\varphi\|_{H^{1/2}_\varepsilon(\Gamma_\varepsilon)}.$$

Proof. Using evident transformations we have

$$\|u\|^2_{H^1(\Omega_\varepsilon)} = \int_{\Omega_\varepsilon} u^2 + \int_{\Omega_\varepsilon} |\nabla u|^2 = \varepsilon^2 \|\tilde{u}\|^2_{L_2(\Omega)} + |\tilde{u}|^2_{H^1(\Omega)} \approx \varepsilon^2 \|\tilde{\varphi}\|^2_{L_2(\Gamma)} + |\tilde{u}|^2_{H^1(\Omega)},$$

$$\varepsilon \|\varphi\|^2_{L_2(\Gamma_\varepsilon)} + |\varphi|^2_{H^{1/2}(\Gamma_\varepsilon)} = \varepsilon^2 \|\tilde{\varphi}\|^2_{L_2(\Gamma)} + |\varphi|^2_{H^{1/2}(\Gamma)}.$$

□

Lemma 8.3 *There exists* $c_1 \neq c_1(\varepsilon)$ *such that if* $\int_{\Gamma'_\varepsilon} \varphi(x)\,dx = 0$, $\text{meas}(\Gamma'_\varepsilon) \approx \varepsilon$, *then*

$$\frac{1}{\varepsilon}\|\varphi\|^2_{L_2(\Gamma_\varepsilon)} + |\varphi|^2_{H^{1/2}(\Gamma_\varepsilon)} \leq c_1|\varphi|^2_{H^{1/2}(\Gamma_\varepsilon)}.$$

Proof. In order to estimate $\frac{1}{\varepsilon}\|\varphi\|^2_{L_2(\Gamma_\varepsilon)}$, we use the following simple manipulations

$$\begin{aligned}\frac{1}{\varepsilon}\|\varphi\|^2_{L_2(\Gamma_\varepsilon)} + |\varphi|^2_{H^{1/2}(\Gamma_\varepsilon)} &= \|\tilde{\varphi}\|^2_{L_2(\Gamma)} + |\tilde{\varphi}|^2_{H^{1/2}(\Gamma)} \\ &\leq c_2\|\tilde{u}\|^2_{H^1(\Omega)} \quad \text{(Theorem 3.5)} \\ &\leq c_3|\tilde{u}|^2_{H^1(\Omega)} \\ &\leq c_1|\varphi|^2_{H^{1/2}(\Gamma_\varepsilon)}.\end{aligned}$$

This completes the proof of the lemma. □

Let $t : H^{1/2}(\Gamma_\varepsilon) \to H^1(\Omega_\varepsilon)$ be given by

$$u = t\varphi = \xi v, \qquad \xi(n) = 1 - \frac{n}{D},$$

where $\xi(n)$ is defined above, $v \in H^1(\Omega_\varepsilon)$ is norm preserving, but does not satisfy the homogeneous Dirichlet boundary condition. The cut-off function ξ gives to u the homogeneous Dirichlet boundary condition. Then

$$|u|^2_{H^1(\Omega_\varepsilon)} \cong |\xi'|^2\|v\|^2_{L_2(\Omega_\varepsilon)} + |\xi|\,|v|^2_{H^1(\Omega_\varepsilon)}, \qquad |\xi'| = \frac{1}{\varepsilon}.$$

Since the $|\xi'|$ can be big, we suggest the following construction of the extension operator. For $\varphi \in H^{1/2}(\Gamma_\varepsilon)$, let $\varphi = \varphi_0 + \varphi_1$, where $\varphi_0 \equiv$ constant and $\int_{\Gamma_\varepsilon} \varphi_1(x)\,dx = 0$. Now we define $u_0 \equiv$ constant $= \varphi_0$ and $u_1 = t\varphi_1 = \xi v$. Then we have the inequalities

$$\begin{aligned}\|u_0\|_{L_2(\Omega_\varepsilon)} &\leq c_2\varepsilon\|\varphi_0\|_{L_2(\Gamma_\varepsilon)}, \\ \left(\frac{1}{\varepsilon}\right)^2\|v\|^2_{L_2(\Omega_\varepsilon)} + |v|^2_{H^1(\Omega_\varepsilon)} &\leq c_3|\varphi_1|^2_{H^{1/2}(\Gamma_\varepsilon)} = c_3|\varphi|^2_{H^{1/2}(\Gamma_\varepsilon)}.\end{aligned}$$

Using Lemma 6.7 we can easily prove the following lemma.

Lemma 8.4 *For a given function* $\varphi : [-1, 0] \to \mathbb{R}$, *let*

$$\varphi(x) = \begin{cases}(1-x)\varphi(-x), & x \in [0,1], \\ 0, & x \in [1,2].\end{cases}$$

Then there exists a constant c *such that* $\|\varphi\|_{H^{1/2}(-1,2)} \leq c\|\varphi\|_{H^{1/2}(-1,0)}$.

Lemma 8.5 *For $\varphi \in H^{1/2}(0,3\varepsilon)$, we have ($c \neq c(\varepsilon)$)*

$$\frac{1}{\varepsilon}\|\varphi\|^2_{L_2(0,3\varepsilon)} + |\varphi|^2_{H^{1/2}(0,3\varepsilon)} \leq c\|\varphi\|^2_{H^{1/2}(0,3\varepsilon)}.$$

Let $\varphi = \varphi_1 + \varphi_2$ with

$$\varphi(x) = \begin{cases} \varphi_1(x), & x \in (0,\varepsilon), \\ \varphi_2(x), & x \in (2\varepsilon, 3\varepsilon), \end{cases}$$

where φ_1 and φ_2 are defined on $[0,3\varepsilon]$ according to Lemma 8.4. Then

$$\frac{1}{\varepsilon}\|\varphi_1\|^2_{L_2(0,3\varepsilon)} + |\varphi_1|^2_{H^{1/2}(0,3\varepsilon)} + \frac{1}{\varepsilon}\|\varphi_2\|^2_{L_2(0,3\varepsilon)} + |\varphi_2|^2_{H^{1/2}(0,3\varepsilon)} \leq c_1\|\varphi\|^2_{H^{1/2}(0,3\varepsilon)}.$$

Proof. It is easy to see that

$$\begin{aligned} \int_0^{3\varepsilon}\int_0^{3\varepsilon} \frac{(\varphi(x)-\varphi(y))^2}{|x-y|^2}\,dxdy &= \int_0^3\int_0^3 \frac{(\tilde{\varphi}(s)-\tilde{\varphi}(t))^2}{|s-t|^2}\,dsdt \\ \frac{1}{\varepsilon}\int_0^{3\varepsilon} \varphi^2(x)\,dx &= \int_0^3 \tilde{\varphi}(s)^2\,ds. \end{aligned}$$

□

Lemma 8.6 *Let $\varphi^\varepsilon \in H^{1/2}(0,3\epsilon)$ be a continuous and piecewise linear function with $\varphi^\varepsilon(i\varepsilon) = \varphi_i$, and linear on the intervals $[i\varepsilon,(i+1)\varepsilon]$, $i = 0,\ldots,3$. Then*

$$\|\varphi^\varepsilon\|^2_{H^{1/2}_\varepsilon(0,3\varepsilon)} \approx \sum_{i=0}^3 \varepsilon^2\varphi_i^2 + \sum_{i=0}^3\sum_{j=0}^3(\varphi_i-\varphi_j)^2.$$

Proof. The assertion immediately follows from the relations

$$\varepsilon\|\varphi^\varepsilon\|^2_{L_2(0,3\varepsilon)} \approx \sum_{i=0}^3 \varepsilon^2\varphi_i^2, \quad \text{and} \quad |\varphi^\varepsilon|^2_{H^{1/2}(0,3\varepsilon)} \approx \sum_{i=0}^3\sum_{j=0}^3(\varphi_i-\varphi_j)^2.$$

□

In order to construct asymptotically optimal decompositions in the case of many subdomains, we have to consider the so-called coarse subspace beside the local subspaces. To do it, we use the following lemma.

Lemma 8.7 *There exists $c \neq c(h,\varepsilon)$ such that, for every $\varphi^h \in H_h(0,3\varepsilon)$, there are $\varphi^\varepsilon, \varphi_1^h, \varphi_2^h$ satisfying*

$$\begin{aligned} &\varphi^h = \varphi^\varepsilon + \varphi_1^h + \varphi_2^h, \qquad \varphi^\varepsilon \text{ piecewise linear}, \\ &\varphi_1^h(x) = 0, \qquad x \in (2\varepsilon, 3\varepsilon), \\ &\varphi_2^h(x) = 0, \qquad x \in (0,\varepsilon), \end{aligned}$$

and

$$\|\varphi^\varepsilon\|^2_{H^{1/2}(0,3\varepsilon)} + \|\varphi_1^h\|^2_{H^{1/2}(0,3\varepsilon)} + \|\varphi_2^h\|^2_{H^{1/2}(0,3\varepsilon)} \leq c\|\varphi^h\|^2_{H^{1/2}(0,3\varepsilon)}.$$

Proof. We define the piecewise linear function φ^ε by the values

$$\varphi_0 = \varphi_1 = \frac{1}{\varepsilon}\int_0^\varepsilon \varphi^h(x)\,dx,$$
$$\varphi_2 = \varphi_3 = \frac{1}{\varepsilon}\int_{2\varepsilon}^{3\varepsilon} \varphi^h(x)\,dx.$$

Then we arrive at the following estimates:

$$\begin{aligned}
(\varphi_i)^2 &= \left(\frac{1}{\varepsilon}\int_{x_i}^{x_{i+1}} \varphi(x)\,dx\right)^2 \le \frac{1}{\varepsilon^2}\varepsilon\int_{x_i}^{x_{i+1}} \varphi^2(x)\,dx,\\
\sum_{i=0}^{3}\varepsilon^2(\varphi_i)^2 &\le \varepsilon\|\varphi\|^2_{L_2(0,3\varepsilon)},\\
(\varphi_i-\varphi_j)^2 &= \left(\frac{1}{\varepsilon}\int_{x_i}^{x_{i+1}} \varphi^h(x)\,dx - \frac{1}{\varepsilon}\int_{x_j}^{x_{j+1}} \varphi^h(x)\,dx\right)^2\\
&= \frac{1}{\varepsilon^2}\left(\frac{1}{\varepsilon}\int_{x_i}^{x_{i+1}}\int_{x_j}^{x_{j+1}} \varphi^h(x)\,dydx - \frac{1}{\varepsilon}\int_{x_i}^{x_{i+1}}\int_{x_j}^{x_{j+1}} \varphi^h(x)\,dxdy\right)^2\\
&\le \frac{4}{\varepsilon^2}\left(\int_{x_i}^{x_{i+1}}\int_{x_j}^{x_{j+1}} \frac{\varphi^h(x)-\varphi^h(y)}{|x-y|}\,dxdy\right)^2\\
&\le 4\int_{x_i}^{x_{i+1}}\int_{x_j}^{x_{j+1}} \frac{(\varphi^h(x)-\varphi^h(y))^2}{|x-y|^2}\,dxdy,\\
\psi^h &= \varphi^h - \varphi^\varepsilon,\\
\int_0^\varepsilon \psi^h(x)\,dx &= \int_{2\varepsilon}^{3\varepsilon} \psi^h(x)\,dx = 0.
\end{aligned}$$

This completes the proof of the lemma. □

Let us define the substructure as above. Then the following lemma holds.

Lemma 8.8 *Let* $\overline{\Omega} = \bigcup_{i=1}^n \overline{\Omega_i}$, *where* Ω_i *is polygonal and* diam $\Omega_i = O(H)$, *and let* $\Lambda = \bigcup_{i=1}^m \lambda_i$. *Then there exists* $c \ne c(h,H)$ *such that, for every* $\varphi^h \in H_h(\Lambda)$, *there are* $\varphi^H, \varphi_1^h, \dots, \varphi_m^h$ *satisfying*

(i) φ^H *piecewise linear on the coarse grid* $\bigcup_{i=1}^n \partial\Omega_i$, *and*

(ii) $\varphi_i^h(x) = 0$, *if* x *is a cross point of* λ_i, $i = 1,\dots,m$.

Then we have

$$\|\varphi^H\|_{H^{1/2}(\Lambda)} \le c_1\|\varphi^h\|_{H^{1/2}(\Lambda)},$$
$$\sum_{i=1}^m \|\varphi_i^h\|^2_{H^{1/2}(\Lambda)} \le C_1\|\varphi^h\|^2_{H^{1/2}(\Lambda)},$$

$$\Sigma^{-1} = \Sigma_H^+ + \Sigma_1^+ + \cdots + \Sigma_m^+, \qquad (\Sigma_i\varphi,\varphi) \simeq \|\varphi^h\|_{H_{00}^{1/2}(\lambda_i)},$$

$$(\Sigma_H\varphi,\varphi) = H^2\sum \varphi_i^2 + \sum_i\sum_j(\varphi_i - \varphi_j)^2,$$

$$(\Sigma\varphi,\varphi) \simeq \|\varphi^h\|_{H^{1/2}(\Lambda)}.$$

Proof. The proof of this lemma follows from the general theory of ASM and the previous results. □

9 Additive Schwarz method and multilevel decomposition

The ASM preconditioner also requires preconditioners for the subdomains. This section presents a possible preconditioner. Let Ω be a domain in $\mathbb{R}^2$, $\Omega_i, i = 1,\dots,n$ be a disjoint subdomain of Ω and $\Lambda = \bigcup_{i=1}^n \partial\Omega_i$. In addition, we introduce matrices B_i which define equivalent norms for $-\Delta_{\Omega_i}$ and a matrix Σ which generates an equivalent norm on the space $H^{1/2}(\Lambda)$, i.e. $B_i \longleftrightarrow -\Delta_{\Omega_i}$ and $\Sigma \longleftrightarrow H^{1/2}(\Lambda)$, respectively.

Let $t\colon H^{1/2}(\Lambda) \to H(\Omega)$ be an extension operator. Then the inexact ASM preconditioner (7.2) has the form

$$B^{-1} = \begin{bmatrix} 0 & & & \\ & B_1^{-1} & & \\ & & \ddots & \\ & & & B_n^{-1} \end{bmatrix} + t\Sigma^{-1}t^\top,$$

where the first block corresponds to Λ and the block $i+1$ to Ω_i. Let us fix the subdomain Ω_i and omit the index i. First we consider the case when Ω is polygonal. Let $\Omega_0^h, \Omega_1^h, \dots, \Omega_J^h$ be a sequence of grids on Ω and $W_0 \subset W_1 \subset \cdots \subset W_J = W$ be a sequence of nested spaces, respectively. We denote the nodal basis in W_k by $\{\phi_i^{(k)}\}_{i=1,2,\dots,n_k}$, and $\Phi_i^{(k)} = \{\alpha\cdot\phi_i^{(k)} | \alpha \in \mathbb{R}\}$ the vector space spanned by the basis function $\phi_i^{(k)}$. Then we obviously have the representations

$$W_k = \Phi_1^{(k)} + \cdots + \Phi_{n_k}^{(k)} \quad \text{and} \quad W = \sum_{k=0}^{J}\sum_{i=1}^{n_k}\Phi_i^{(k)}. \tag{9.1}$$

Let $P_i^{(k)} : W \to \Phi_i^{(k)}$ be an orthogonal projection with respect to $a(\cdot,\cdot)$.

The so-called BPX preconditioner (multilevel preconditioner), which corresponds to the space decomposition (9.1), was proposed by J. H. Bramble, J. E. Pasciak and J. Xu in [5]. The investigation of the optimality of the BPX preconditioner is due to P. Oswald [33], see also [3, 7, 34, 46]. The BPX preconditioner can also be considered as ASM for a special decomposition of the original finite element space into subspaces.

Due to the space decomposition (9.1), the following result can be shown.

Theorem 9.1 *There exist two constants α and β, which are independent of h, such that*

(1) For every $u^h \in W$, there exist $u_i^{(k)} \in \Phi_i^{(k)}$ such that

$$\sum_{k=0}^{J}\sum_{i=1}^{n_k} u_i^{(k)} = u^h \tag{9.2}$$

with

$$\sum_{k=0}^{J}\sum_{i=1}^{n_k} \|u_i^{(k)}\|^2_{H^1(\Omega)} \le \frac{1}{\alpha}\|u^h\|_{H^1(\Omega)}. \tag{9.3}$$

(2) Moreover, the inequality

$$\|u^h\|^2_{H^1(\Omega)} \le \beta \inf_{\sum_{k=0}^{J}\sum_{i=1}^{n_k} v_i^{(k)}=u^h} \|v_i^{(k)}\|^2_{H^1(\Omega)}, \quad v_i^{(k)} \in \Phi_i^{(k)} \tag{9.4}$$

holds.

Remark 9.2 Note that relation (9.3) implies the estimate

$$\alpha\,\|u^h\|^2_{H^1(\Omega} \le a\left(\sum_{k=0}^{J}\sum_{i=1}^{n_k} P_i^{(k)}u^h, u^h\right) \le \beta\,(u^h, u^h).$$

The proof uses the following fundamental result that was already suggested by P. L. Butzer and K. Scherer in [6].

Lemma 9.3 *Let $Q_k : W \to W_k$ be the orthogonal projection in $L_2(\Omega)$. Then there exist two constants C_1 and C_2, which are independent of h and J, such that*

$$\begin{aligned} C_1\|u^h\|^2_{H^1(\Omega)} \;\le\; |||h^h||| \;&:=\; \|Q_0u^h\|^2_{L_2(\Omega)} + \sum_{k=1}^{J} h_k^{-2}\|(Q_k - Q_{k-1})u^h\|^2_{L_2(\Omega)} \\ &\le\; C_2\|u^h\|^2_{H^1(\Omega)} \end{aligned}$$

and

$$C_1|||u^h||| \le \inf_{\substack{u^h=u_0^h+\cdots+u_J^h \\ u_k^h\in W_k}} \sum_{k=0}^{J} h_k^{-2}\|u_k^h\|^2_{L_2(\Omega)} \le C_2|||u^h|||.$$

For the finite element case, the rigorous proof of the lemma can be found in [34].

Proof (Theorem 9.1). The proof of (9.3) considers the decomposition

$$u^h = Q_0u^h + \sum_{k=1}^{J}(Q_k - Q_{k-1})u^h = v_0^h + v_1^h + \cdots + v_J^h,$$

where $v_k^h \in W_k$. Note that $Q_J u^h = u^h$. On the other hand, the fact that $v_k^h \in W_k$ implies the relation

$$v_k^h = \sum_{i=1}^{n_k} \alpha_i^{(k)} \phi_i^{(k)} = \sum_{i=1}^{n_k} v_i^{(k)},$$

where $v_i^{(k)} \in \Phi_i^{(k)}$. Thus, we have

$$u^h = \sum_{k=0}^{J} \sum_{i=1}^{n_k} v_i^{(k)}.$$

Using the inverse inequality, Lemma 9.3 and the fact $v_k^h = (Q_k - Q_{k-1})u^h$, we obtain

$$\begin{aligned}
\sum_{k=0}^{J} \sum_{i=0}^{n_k} \|v_i^{(k)}\|^2_{H^1(\Omega)} &\approx \sum_{k=0}^{J} \sum_{i=0}^{n_k} h_k^{-2} \|v_i^{(k)}\|^2_{L_2(\Omega)} \\
&\approx \sum_{k=0}^{J} h_k^{-2} \|v_k^h\|^2_{L_2(\Omega)} \\
&\leq \|u^h\|^2_{H^1(\Omega)}.
\end{aligned}$$

This proves (9.3). In order to prove (9.4), we start with

$$\|u^h\|^2_{H^1(\Omega)} \leq \beta \inf \sum_{k=0}^{J} \sum_{i=1}^{n_k} h_k^{-2} \|v_i^{(k)}\|^2_{L_2(\Omega)}.$$

Moreover,

$$\begin{aligned}
\inf_{\substack{v_i^{(k)} \in \Phi_i^{(k)} \\ \sum_{k=0}^{J} \sum_{i=1}^{n_k} v_i^{(k)} = u^h}} \sum_{k=0}^{J} \sum_{i=1}^{n_k} h_k^{-2} \|v_i^{(k)}\|^2_{L_2(\Omega)} &= \inf_{\alpha_i^{(k)}} \sum_{k=0}^{J} \sum_{i=1}^{n_k} h_k^{-2} \|\alpha_i^{(k)} \phi_i^{(k)}\|^2_{L_2(\Omega)} \\
&\geq C \inf_{\alpha_i^{(k)}} \sum_{k=0}^{J} h_k^{-2} \|v_k^h\|^2_{L_2(\Omega)} \\
&= C \inf_{v_k^h \in W_k} \sum_{k=0}^{J} \|v_k^h\|^2_{L_2(\Omega)} \\
&\geq C \cdot C_1 \|u^h\|^2_{H^1(\Omega)}.
\end{aligned}$$

This completes the proof of Theorem 9.1. □

Let us give an example of the above theorem. Let $A_i^{(k)} : \Phi_i^{(k)} \to \Phi_i^{(k)}$. Let us define the L_2 orthogonal projection $Q_i^{(k)} : W \to \Phi_i^{(k)}$ as follows:

$$Q_i^{(k)} u^h = \frac{(u^h, \phi_i^{(k)})_{L_2(\Omega)}}{(\phi_i^{(k)}, \phi_i^{(k)})_{L_2(\Omega)}} \phi_i^{(k)}.$$

We define $P_i^{(k)} : W \to \Phi_i^{(k)}$ by setting

$$P_i^{(k)} = (A_i^{(k)})^{-1} Q_i^{(k)}$$

and $a(\cdot,\cdot)$ by

$$a(u^h, v^h) = (Au, v).$$

Then

$$(A_i^{(k)} \phi_i^{(k)}, \phi_i^{(k)}) = (A\phi_i^{(k)}, \phi_i^{(k)}) = a(\phi_i^{(k)}, \phi_i^{(k)}) = (\alpha_i^{(k)} \phi_i^{(k)}, \phi_i^{(k)})_{L_2(\Omega)},$$

where $\alpha_i^{(k)} = \dfrac{a(\phi_i^{(k)}, \phi_i^{(k)})}{(\phi_i^{(k)}, \phi_i^{(k)})_{L_2(\Omega)}}$. We have the following equalities:

$$\begin{aligned} A_i^{(k)} \phi_i^{(k)} &= \frac{a(\phi_i^{(k)}, \phi_i^{(k)})}{(\phi_i^{(k)}, \phi_i^{(k)})_{L_2(\Omega)}} \phi_i^{(k)}, \\ (A_i^{(k)})^{-1} \phi_i^{(k)} &= \frac{(\phi_i^{(k)}, \phi_i^{(k)})_{L_2(\Omega)}}{a(\phi_i^{(k)}, \phi_i^{(k)})} \phi_i^{(k)}. \end{aligned}$$

Hence we have the following relations for the preconditioner B

$$B^{-1} u^h = \sum_{k=0}^{J} \sum_{i=0}^{n_k} (A_i^{(k)})^{-1} Q_i^{(k)} u^h = \sum_{k=0}^{J} \sum_{i=0}^{n_k} \frac{(u^h, \phi_i^{(k)})_{L_2(\Omega)}}{a(\phi_i^{(k)}, \phi_i^{(k)})} \phi_i^{(k)}.$$

Remark 9.4 If $a(\phi^h, \phi^h) = O(1)$, it is possible to replace the above preconditioner by the original BPX preconditioner

$$B_{BPX}^{-1} u^h = \sum_{k=0}^{J} \sum_{i=0}^{n_k} (u^h, \phi_i^{(k)})_{L_2(\Omega)} \phi_i^{(k)}.$$

10 The fictitious space method

The ASM decomposes an original problem into subproblems. But in some cases (subproblems with unstructured grids) a construction of effective preconditioners is still a difficult problem. To solve an original problem we will use the combination of ASM and fictitious space method (FSM). FSM is the generalization of fictitious domain method. Some references on the fictitious domain methods are given in the introduction.

The following abstract theorem is the basis for the FSM.

Theorem 10.1 *Let H_0 and H be two Hilbert spaces with the scalar products $(\cdot,\cdot)_{H_0}$, and $(\cdot,\cdot)_H$, respectively. Let $A : H_0 \to H_0$ and $B : H \to H$ be some self adjoint positive definite operators, i.e., $A^* = A > 0$ and $B^* = B > 0$. Assume that there exists an operator $R : H \to H_0$ such that*

$$(ARv, Rv)_{H_0} \leq C_R (Bv, v)_H \qquad \forall v \in H,$$

and an operator $T : H_0 \to H$ such that

$$RTu_0 = u_0 \qquad \forall u_0 \in H_0,$$

and

$$C_T (BTu_0, Tu_0)_H \leq (Au_0, u_0)_{H_0} \qquad \forall u_0 \in H_0.$$

Set $C^{-1} = RB^{-1}R^*$, where $R^* : H \to H_0$ and $(R^*u_0, v)_H = (u_0, Rv)_{H_0}$. Then the spectral equivalence inequalities

$$C_T (A^{-1}u_0, u_0)_{H_0} \leq (C^{-1}u_0, u_0) \leq C_R (A^{-1}u_0, u_0) \qquad \forall u_0 \in H$$

are valid.

The proof of this theorem uses the following result.

Lemma 10.2 *Let $A = A^* > 0$ in a Hilbert space with scalar product $(\cdot,\cdot)$. Then the identity*

$$(A^{-1}u, u)^{1/2} = \sup_{v \in H} \frac{(u, v)}{(Av, v)^{1/2}}$$

holds.

Proof (Lemma 10.2). By the Cauchy–Bunjakowski inequality, one obtains

$$(u, v) = (A^{-1/2}u, A^{1/2}v) \overset{\text{C.B.}}{\leq} \|A^{-1/2}u\| \|A^{1/2}v\| = (A^{-1}u, u)^{1/2}(Av, v)^{1/2}.$$

With $v = A^{-1}u$, one can conclude that

$$(A^{-1}u, u)^{1/2} = \sup_{v \in H} \frac{(u, v)}{(Av, v)^{1/2}},$$

which proves the lemma. □

Proof (Theorem 10.1). In order to prove the lower estimate, we use the above assumptions about R and T and Lemma 10.2. This gives

$$\begin{aligned}
(RB^{-1}R^*u_0, u_0)^{1/2})_{H_0} &= (B^{-1}R^*u_0, R^*u_0)_H = \sup_{v \in H} \frac{(R^*u_0, v)_H}{(Bv, v)_H^{1/2}} \\
&\geq \sup_{v_0 \in H_0} \frac{(R^*u_0, Tv_0)_H}{(BTv_0, Tv_0)_H^{1/2}} \geq \sqrt{C_T} \sup_{v_0 \in H_0} \frac{(R^*u_0, Tv_0)_H}{(Av_0, v_0)^{1/2}} \\
&= \sqrt{C_T} \sup_{v_0 \in H_0} \frac{(u_0, v_0)_{H_0}}{(Av_0, v_0)^{1/2}} = \sqrt{C_T}(A^{-1}u_0, u_0)^{1/2}.
\end{aligned}$$

For the upper estimate, we have

$$\begin{aligned}(RB^{-1}R^*u_0,u_0)^{1/2})_{H_0} &= \sup_{v\in H}\frac{(u_0,Rv)_{H_0}}{(Bv,v)_H^{1/2}} = \sup_{v\in H}\frac{(A^{-1/2}u_0,A^{1/2}Rv)_{H_0}}{(Bv,v)_H^{1/2}}\\ &\overset{\text{C.B.}}{\leq} (A^{-1}u_0,u_0)_{H_0}^{1/2}\sup_{v\in H}\frac{(ARu_0,Rv)^{1/2}}{(Bv,v)^{1/2}}\\ &\leq \sqrt{C_R}(A^{-1}u_0,u_0)^{1/2}.\end{aligned}$$

This completes the proof of Theorem 10.1. □

11 Application to the fictitious domain method

In this section we show how the general framework of FSM can be used for the analysis of the classical fictitious domain methods. However, we have to distinguish between different cases of the boundary conditions.

11.1 Neumann boundary condition

The simplest case is the case of Neumann boundary conditions. Let us consider the following model problem:

$$\begin{cases}-\Delta u+u=f & \text{in } \Omega,\\ \dfrac{\partial u}{\partial n}=0 & \text{on } \Gamma,\end{cases}$$

where Ω is not regular (not polygonal) and Γ is its boundary. Let Π be a domain of much simpler form which includes the domain Ω. A possible candidate for Π is a cube.

Let $H_0 = H^1(\Omega)$ and $H = H_0^1(\Pi)$. Let A and B be the differential operators according to the domain Ω and Π, i.e.,

$$A \longleftrightarrow -\Delta_\Omega + I, \qquad B \longleftrightarrow -\Delta_\Pi.$$

Let $R : H_0^1(\Pi) \to H^1(\Omega)$ be a restriction operator. In this case, we define it by $R = I_\Omega$. Then we have

$$(Ru,Ru)_{H^1(\Omega)} \leq C_R(\nabla u,\nabla u)_{L_2(\Pi)}.$$

Let $T : H^1(\Omega) \to H_0^1(\Pi)$ be an extension operator. For any $u \in H^1(\Omega)$, we have

$$\|u\|_{H^1(\Omega)} \geq C_1\|\phi\|_{H^{1/2}(\Gamma)} \geq C_2\|Tu\|_{H^1(\Pi)}$$

with

$$RTu_0 = u_0 \qquad \forall u_0 \in H^1(\Omega).$$

Hence, we obtain the preconditioner for the domain Ω by setting

$$C^{-1} = RB^{-1}R^*.$$

In matrix notation, we choose

$$C^{-1} = \begin{bmatrix} I & 0 \end{bmatrix} (-\Delta_{\Pi}^{-1}) \begin{bmatrix} I \\ 0 \end{bmatrix} \quad \text{and} \quad Ru = u_{\bar{\Omega}},$$

where

$$R = \begin{bmatrix} I & 0 \end{bmatrix} \quad \text{and} \quad u = \begin{bmatrix} u_{\bar{\Omega}} \\ u_{\Pi \setminus \Omega} \end{bmatrix}$$

and I is an identity block.

11.2 Dirichlet boundary condition (1-D case)

The situation for Dirichlet boundary conditions is much more difficult. In order to understand the difficulties, we start with the one-dimensional case and investigate the boundary value problem

$$\begin{cases} -\dfrac{d^2u}{dx^2} = f \text{ in } (a,b) \subset (0,1), \\ u(a) = u(b) = 0. \end{cases}$$

Let $H_0 = H_0^1(a,b)$ and $H = H_0^1(0,1)$ with $\Pi = (0,1)$ and $\Omega = (a,b)$. Let $A = -\Delta_\Omega$ and $B = -\Delta_\Pi$. In order to extend u from Ω to Π, we define an extension operator $T : H_0^1(a,b) \to H_0^1(0,1)$ by

$$Tu = \begin{cases} u(x), & x \in (a,b), \\ 0, & x \in \Pi \setminus (a,b). \end{cases}$$

Then the relation $(Tu_0, Tu_0)_{H^1(\Pi)} = (u_0, u_0)_{H^1(\Omega)}$ implies $C_T = 1$. Next, we consider the restriction operator $R : H_{0,h}^1(\Pi) \to H_{0,h}^1(\Omega)$. There are many ways to define R. Here we investigate two different definitions of R which we will compare.

(1) The first definition is as follows:

$$Ru^h = \begin{cases} u^h(x), & x_i \in (a,b), \\ 0, & x_i = a \text{ or } b. \end{cases}$$

Note that $\|R\| \to \infty$ as $h \to 0$ which implies $C_T \to \infty$. This is not a good choice for a possible restriction operator.

(2) Thus, we introduce another restriction operator. Let $I_\Omega : H_0^1(\Omega) \to H^1(\Omega)$ be the natural restriction defined as follows:

$$(I_\Omega u)(x) = u(x), x \in \Omega \quad \forall u \in H_0^1(\Omega),$$

and $I_\Gamma : H_0^1(\Pi) \to \mathbb{R}^2$ be the trace operator defined by

$$I_\Gamma u = \begin{bmatrix} u(a) \\ u(b) \end{bmatrix} \qquad \forall u \in H_0^1(\Pi).$$

Let $t : \mathbb{R}^2 \to H^1(\Omega)$ be the extension operator defined by

$$t\left(\begin{bmatrix} u(a) \\ u(b) \end{bmatrix}\right) = u(a) + \frac{u(b) - u(a)}{b - a}(x - a).$$

Now we define the restriction operator $R : H_0^1(\Pi) \to H_0^1(\Omega)$ by

$$R = I_\Omega - tI_\Gamma. \tag{11.1}$$

Clearly, we have the estimates

$$|u(a)| \leq C\|u\|_{H^1(\Pi)}, \qquad |u(b)| \leq C\|u\|_{H^1(\Pi)},$$

and

$$|tI_\Gamma u|_{H^1}^2 = \int_a^b \frac{(u(b) - u(a))^2}{(b - a)^2} dx \leq C\|u\|_{H^1(\Pi)}^2.$$

This implies

$$\|Ru\|_{H^1(\Omega)} \leq \|I_\Omega u\|_{H^1(\Omega)} + \|tI_\Gamma u\|_{H^1(\Omega)} \leq C_R\|u\|_{H^1(\Pi)}$$

since $\|I_\Omega u\|_{H^1(\Omega)} \leq \|I_\Omega u\|_{H^1(\Pi)}$. Note that the constant C_R is independent of h in this case (FEM). It is easy to see that $RTu_0 = u_0 - 0 = u_0$ for all $u_0 \in H_0^1(\Omega)$.

Summarizing, only the definition of R in (11.1) leads to a restriction operator with constant c_R bounded independently of the meshsize h.

11.3 Dirichlet boundary condition (2-D case)

Next, we consider the two-dimensional case. Let $H_0 = H_0^1(\Omega)$, $H = H_0^1(\Pi)$, $A = -\Delta_\Omega$ and $B = -\Delta_\Pi$. The operator T is defined as in the one-dimensional case, i.e. let $T : H_0^1(\Omega) \to H_0^1(\Pi)$ with

$$Tu = \begin{cases} u(x), & x \in \Omega, \\ 0, & x \in \Pi \setminus \Omega. \end{cases}$$

This gives $C_T = 1$. For the definition of R, we have to generalize the definition in (11.1) and introduce $R = I_\Omega - tI_\Gamma$, where t is the extension operator from Section 7.2. Then we obtain a constant C_R (independent of h).

11.4 Mixed boundary condition (2-D case)

In the case of mixed boundary conditions, the ideas for the Dirichlet and the Neumann case have to be combined. Let $\breve{H}^1(\Omega) = \{u \in H^1(\Omega) | u(x) = 0, x \in \Gamma_D\}$. Let $H = H_0^1(\Pi)$, $A = -\Delta_\Omega$ and $B = -\Delta_\Pi$. We consider the subdomains G_N and G_D in the vicinity of Γ_N and Γ_D, respectively, such that

$$\overline{\Pi \setminus \Omega} = \overline{G}_N \cup \overline{G}_D$$

and

$$\partial G_D \cap \Gamma_N = \emptyset, \quad \partial G_N \cap \overset{\circ}{\Gamma}_D = \emptyset.$$

Let $T_{ND} u_0 = T_N T_D u_0$ where the operator $T_D : \breve{H}^1(\Omega) \to H^1(\bar{\Omega} \cup G_D)$ for the Dirichlet data is defined by

$$T_D u_0 = \begin{cases} u_0(x), & x \in \Omega, \\ 0, & x \in G_D. \end{cases}$$

Next, by the trace theorem, there exists an operator $T_N : H^1(\bar{\Omega} \cup G_D) \to H^1(\Pi)$. Now, we define a restriction operator R by $R = I_\Omega - t_\Gamma t_N \cdot I_D$ where $I_\Omega : H_0^1(\Pi) \to H^1(\Omega)$ and $I_D : H_0^1(\Pi) \to H^{1/2}(\Gamma_D)$. We define $t_N : H^{1/2}(\Gamma_D) \to H^{1/2}(\Gamma)$ by

$$(t_N \phi)(-s) = (1 - \frac{s}{D})\phi(s) \quad \text{for } \phi(s) \in H^{1/2}(\Gamma_D).$$

Here $(1 - s/D)$ is a linear cut-off function. We note that D is independent of h. For t_N, the estimate

$$\|t_N \phi\|_{H^{1/2}(\Gamma)} \le C_1 \|\phi\|_{H^{1/2}(\Gamma_D)}.$$

can easily be proved. Let $t_\Gamma : H^{1/2}(\Gamma) \to H^1(\Omega)$ be the extension operator of Section 7.2. Then we obtain the estimates

$$\|Ru\| \le \|I_\Omega\| + \|t_\Gamma\| \cdot \|t_N\| \cdot \|I_D u\| \le C_R \|u\|_{H^1(\Pi)},$$

where C_R is independent of h in the FEM case.

11.5 Unstructured and nonuniform grid (2-D case)

In the previous sections, we have investigated the case of a structured grid. Here we consider the case of an unstructured and nonuniform grid for Ω. In other words, we can design a preconditioner for the differential operator on Ω from that on Π. In the case when Ω is not polygonal, though, we want to design a preconditioner from the uniform grid differential operator on Π. Let $Q^{h_\sharp}$ denote the uniform grid on Π and $h_\sharp$ be the mesh size of $Q^{h_\sharp}$ satisfying

$$h_\sharp < \frac{1}{\sqrt{2}} r_{\min} \quad \text{where} \quad r_{\min} = \min_{z_l \in \Omega^h} r_l.$$

Here r_l is the radius of the largest ball $B(z_l, r_l)$ inscribed in the union of all elements of the triangulation Ω^h sharing the vertex z_l.

Let $H_0 = H_h(\Pi^h)$ and $H = H_h(Q^{h_\sharp})$. Let $A = -\Delta_{\Pi^h}$ and $B \approx -\Delta_{Q^{h_\sharp}}$ be defined as in the previous sections. Now, we introduce a restriction operator

$$R_Q : H_h(Q^{h_\sharp}) \to H_h(\Pi^h), \tag{11.2}$$

i.e. for any $U^{h_\sharp} \in H_h(Q^{h_\sharp})$ the values $u^h \in H_h(\Pi^h)$ have to be defined. Let z_l denote the nodal point of Π^h and Z_{ij} is the node of some Q_{ij}. We set $u^h(z_l) = U^{h_\sharp}(Z_{i,j})$, i.e., $RU^{h_\sharp} = u^h$ is a simple restriction. Next, we define the extension operator $T : H_h(\Pi^h) \to H_h(Q^{h_\sharp})$ by the following way:

$$\begin{cases} U^{h_\sharp}(z_{ij}) = u^h(z_l), & \text{if } z_l \text{ belongs to some } Q_{ij}, \\ U^{h_\sharp}(z_{ij}) = \dfrac{1}{3}(u^h(z_1) + u^h(z_2) + u^h(z_3)), & \text{otherwise}. \end{cases}$$

With the condition on the mesh size there are only two cases. That is, there is a one-to-one correspondence between Π^h and some subset $\tilde{Q}^{h_\sharp}$ of $Q^{h_\sharp}$. Then we can see that

$$RTu^h = u^h \qquad \forall u^h \in H_h(\Pi^h).$$

Lemma 11.1 *Let us assume that $c_1 r_{\min} \le h_\sharp$, i.e., $h_\sharp$ is of order h. Then, there exist two constants C_R^Q and C_T^Q (independent of h) such that*

$$\|R_Q U^{h_\sharp}\|_{H^1(\Pi)} \le C_R^Q \|U^{h_\sharp}\|_{H^1(\Pi)} \quad \textit{and} \quad \|T_Q u^h\|_{H^1(\Pi)} \le C_T^Q \|u^h\|_{H^1(\Pi)}.$$

Proof. Using $u^h = RU^{h_\sharp}$, we can estimate

$$\begin{aligned} &\|u^h\|_{H^1(\Pi)} \\ &\quad\approx \sum_{\tau_i \subset \Pi^h} (h^2\{(u^h(z_{i_1}))^2 + (u^h(z_{i_2}))^2 + (u^h(z_{i_3}))^2\} \\ &\qquad + (u^h(z_{i_1}) - u^h(z_{i_2}))^2 + (u^h(z_{i_2}) - u^h(z_{i_3}))^2 + (u^h(z_{i_3}) - u^h(z_{i_1}))^2) \\ &\quad= \sum_{\tau_i \subset \Pi^h} h^2((U^{h_\sharp}_{i_1,j_1})^2 + \cdots + (U^{h_\sharp}_{i_k,j_k})^2) \\ &\qquad + \sum_{\tau_i \subset \Pi^h} ((U^{h_\sharp}_{i_1,j_1} - U^{h_\sharp}_{i_2,j_2})^2 + (U^{h_\sharp}_{i_2,j_2} - U^{h_\sharp}_{i_3,j_3})^2 + (U^{h_\sharp}_{i_3,j_3} - U^{h_\sharp}_{i_1,j_1})^2). \end{aligned}$$

Clearly,

$$\sum_{\tau_i \subset \Pi^h} h^2((U^{h_\sharp}_{i_1,j_1})^2 + \cdots + (U^{h_\sharp}_{i_k,j_k})^2) \le \|U^{h_\sharp}\|_{L_2,h}(Q^{h_\sharp}).$$

Next,

$$\begin{aligned} (U^{h_\sharp}_{i_1,j_1} - U^{h_\sharp}_{i_2,j_2})^2 &\le \text{some differences of neighbors} \\ &\le (U^{h_\sharp}_{i_1,j_1} - U^{h_\sharp}_{i'_2,j'_2})^2 + \cdots + (U^{h_\sharp}_{i'_k,j'_k} - U^{h_\sharp}_{i_2,j_2})^2. \end{aligned}$$

Thus, there exists a constant C_1 such that

$$\sum_{\tau_i \subset \Pi^h} ((U^{h_\sharp}_{i_1,j_1} - U^{h_\sharp}_{i_2,j_2})^2 + (U^{h_\sharp}_{i_2,j_2} - U^{h_\sharp}_{i_3,j_3})^2 + (U^{h_\sharp}_{i_3,j_3} - U^{h_\sharp}_{i_1,j_1})^2) \leq C_1 |U^{h_\sharp}|^2_{H^1(Q^{h_\sharp})}.$$

This completes the proof of the existence of C_R. The proof of the existence of C_T is the same as the case of C_R. □

Let

$$C^{-1}_{\text{FSM},u} = RR_Q(-\Delta_{Q^{h_\sharp}})^{-1} R_Q^\top R^\top \tag{11.3}$$

be the FSM preconditioner for unstructured grids. Using Theorem 10.1 twice and Lemma 11.1, we have proved the following final result about the fictitious space method.

Theorem 11.2 *Let* $C_{\text{FSM},u}$ *be defined via* (11.3). *Then the spectral equivalence relation* $\Delta\Omega_h \sim C_{\text{FSM},u}$ *holds.*

12 Fictitious space method and multilevel additive Schwarz method

In the previous section, the FSM-preconditioner $C_{\text{FSM},u}$ (11.3) has been developed. For this preconditioner, we have to solve a potential problem on a much simpler geometry. In this section, we consider the following mixed boundary value problem:

$$\begin{aligned} -\sum_{i,j=1}^{2} \frac{\partial}{\partial x_i} a_{ij} \frac{\partial u}{\partial x_j} + a_0(x)u &= f(x), \qquad x \in \Omega, \\ u(x) &= 0, \qquad x \in \Gamma_D, \\ \frac{\partial u}{\partial n} + \sigma(x)u &= 0, \qquad x \in \Gamma_N, \end{aligned}$$

where the diffusion matrix $(a_{ij})^2_{i,j=1}$ is symmetric and positive definite, and the coefficients a_0 and σ are uniformly positive. Then, we introduce the corresponding bilinear form

$$a(u,v) = \int_\Omega \left(\sum_{i,j=1}^{2} \frac{\partial u}{\partial x_i}\frac{\partial v}{\partial x_j} + a_0(x)uv \right) dx + \int_{\Gamma_N} \sigma(x)uv\, ds,$$

which satisfies the relations

$$a(u,v) = a(v,u) \quad \text{and} \quad a(u,u) \approx \|u\|^2_{H^1_\Omega}. \tag{12.1}$$

A flexible domain decomposition method was suggested in [4]. This method is not optimal with respect to convergence rate, but it is simple to implement and can be very effective for parabolic problems. To suggest an optimal algorithm we consider the following approach. We assume that the triangulation $\overline{\Omega}^h = \bigcup_{i=1}^M \overline{\tau}_i$ is quasi-uniform and shape regular with $\partial\Omega^h = \Gamma_D^h \cup \Gamma_N^h$ and $\Gamma_D^h \subset \overline{\Omega}, \Gamma_N^h \subset \overline{(\mathbb{R}^2 \setminus \Omega)}$. Also suppose that $h_\sharp \leq r_{\min}/2\sqrt{2}$. We define an auxiliary mesh Q^h by the minimum collection of Q_{ij} enclosing Ω^h. Let $\partial Q^h = S^h$ with $S^h = S_D^h \cup S_N^h$ such that, if $\overline{Q}_{ij} \cap \Gamma^h \neq \emptyset$, then

$$S^h \cap \overline{Q}_{ij} \in S_D^h \quad \text{and} \quad S_N^h = S^h \setminus S_D^h.$$

The operator induced by (12.1) is defined as

$$(Au, v) = a(u^h, v^h).$$

Moreover, let B be an operator which satisfies

$$(BU, U) \approx \|U^h\|^2_{H^1(Q^h)} \quad \forall U^h \in H_h(Q^h)$$

and

$$R_Q : H_h(Q^h) \to H_h(\Omega^h)$$

be defined via (11.2).

Theorem 12.1 *There exist two constants C_1 and C_2, independent of h, such that*

$$C_1\,(A^{-1}u, v) \leq (RB^{-1}R^\top u, v) \leq C_2\,(A^{-1}u, u) \quad \forall u.$$

Proof. The proof is similar to the proof of Theorem 11.2. □

Remark 12.2 The condition $h_\sharp \leq r_{\min}/2\sqrt{2}$ instead of $h_\sharp < r_{\min}/\sqrt{2}$ is needed in the mixed boundary case.

First we consider $\Gamma_D = \emptyset$, i.e. the case of pure Neumann boundary conditions.

Let us assume that $h_\sharp = l \cdot 2^{-J}$, $h_k = l \cdot 2^{-k}$, $k = 0, 1, \ldots, J$, and $h_J = h_\sharp$, and that we have a sequence of triangulations

$$\Pi_0^h, \Pi_1^h, \ldots, \Pi_J^h$$

and spaces

$$W_0^h \subset W_1^h \subset \cdots \subset W_J^h = \breve{H}_h(Q^h),$$

where $W_k^h = \{u_k^h = \sum_i \alpha_i^{(k)} \phi_i^{(k)}\}$ and $\phi_i^{(k)}$ is a nodal basis. Let $S^h = S_N^h$. Now, we introduce the preconditioner in Π via

$$B_N^{-1} U^h = \sum_{k=0}^{J} \sum_{\{\mathrm{supp}\phi_i^{(k)} \cap Q^h \neq \phi\}} (\tilde{U}^h, \phi_i^{(k)})_{L_2(Q^h)} \tilde{\phi}_i^{(k)}, \tag{12.2}$$

where $\tilde{U}^h = U^h(Z_{i,j})$ for $Z_{i,j} \in Q^h$ and $\tilde{U}^h = 0$ otherwise.

Theorem 12.3 *Let B_N be defined via (12.2). Then there exist two constants C_1 and C_2, which are independent of h, such that*

$$C_1 \|U^h\|^2_{H^1(Q^h)} \leq (B_N U, U) \leq C_2 \|U^h\|^2_{H^1(Q^h)} \quad \forall U^h \in H_h(Q^h).$$

Proof. Note that

$$B_\Pi^{-1}(U_\Pi^h) = \sum_{k=0}^{J} (U_\Pi^h, \Phi_i^{(k)})_{L_2(\Pi)} \Phi_i^{(k)} \quad \text{(BPX in } \Pi),$$

$$R_N U_\Pi^h = \begin{cases} U_\Pi^h(Z_{i,j}), & Z_{i,j} \in Q^h, \\ 0, & \text{otherwise,} \end{cases}$$

and

$$R_N = \begin{bmatrix} I & 0 \end{bmatrix}.$$

Then we have $R_N B_N^{-1} R_N^\top = B_N^{-1}$ which proves the theorem. □

Next, we consider the case of Dirichlet boundary conditions ($S^h = S_D^h$), i.e. $\Gamma_N = \emptyset$. Here, we define

$$B^{-1} U^h = \sum_{k=0}^{J} \sum_{\mathrm{supp}\Phi_i^{(k)} \subset Q^h} (U_\Pi^h, \Phi_i^{(k)})_{L_2(\Pi)} \Phi_i^{(k)}.$$

Theorem 12.4 *There exist two constants C_1 and C_2, which are independent of h, such that*

$$C_1 \|U^h\|^2_{H^1(Q^h)} \leq (B_D U, U) \leq C_2 \|U^h\|^2_{H^1(Q^h)} \quad \forall U \in H_h(Q^h).$$

Proof. The proof is moved to the end of this section. Let us assume temporarily that the theorem was proved already. □

At last we consider the case of mixed boundary conditions ($S_D^h \neq \emptyset, S_N^h \neq \emptyset$) and define

$$B_M^{-1} U = \sum_{k=0}^{J} \sum (U^h, \Phi_i^{(k)})_{L_2(\Pi)} \Phi_i^{(k)}$$

as the BPX preconditioner. Note that the second summation is taken on the set $\mathrm{supp}(\Phi_i^{(k)} \cap Q^h) \neq \emptyset$ and $\mathrm{supp}(\Phi_i^{(k)} \cap S_D^h) = \emptyset$.

Theorem 12.5 *There exist positive constants C_1 and C_2 such that*

$$C_1 \|U^h\|^2_{H^1(Q^h)} \leq (B_M U, U) \leq C_2 \|U^h\|^2_{H^1(Q^h)} \; \forall U \in H_h(Q^h).$$

Proof. Note that $\Pi^h \backslash Q^h = \bar{G}_D^h \cup \bar{G}_N^h$ and $G_D^h \cap G_N^h = \emptyset$. We have that $\partial G_D^h \cap S^h = S_D^h$ and $\bar{G}^h = \bar{Q}^h \cup \bar{G}_N^h$. Now we define

$$\dot{H}_h(G^h) = \{u^h | u^h(x) = 0, x \in \partial G^h\}.$$

Then we have by the previous case

$$B_{D,G}^{-1}U_G^h = \sum_{k=0}^{J} \sum_{\mathrm{supp}\Phi_i^{(k)} \subset G^h} (U_G^h, \Phi_i^{(k)})_{L_2(G)} \Phi_i^{(k)}$$

and

$$C_1 \|G_G^h\|_{H^1(G)}^2 \le (B_{D,G}U_G, U_G) \le C_2 \|U_G^h\|_{H^1(G)}^2.$$

Furthermore, let us define $R_{N,G} : \dot{H}_h(G^h) \to \breve{H}_h(Q^h)$ by

$$R_{N,G}U_G^h(Z_{i,j}) = \begin{cases} U_G^h(Z_{i,j}), & Z_{i,j} \in Q^h, \\ 0, & \text{otherwise.} \end{cases}$$

Then we finally have $R_{N,G}B_{D,G}^{-1}R_{N,G}^{-1} = B_M^{-1}$ with $R_{N,G} = \begin{bmatrix} I & 0 \end{bmatrix}$. □

Now we are in a position to show Theorem 12.4 for the case of Dirichlet boundary conditions. We define $\dot{W}_k = W_k \cap \dot{H}_h(Q^h)$. Then the proof of the theorem is completed if the following conditions are satisfied:

(a) For all $u^h \in \dot{W}_J$, there exists $u_i^{(k)} = \alpha_i^{(k)} \Phi_i^{(k)}$ such that

$$\sum_{k=0}^{J} \sum_{\mathrm{supp}(u_i^{(k)}) \subset Q^h} u_i^{(k)} = u^h \quad \text{and} \quad \alpha \sum_{\mathrm{supp}(u_i^{(k)}) \subset Q^h} \|u_i^{(k)}\|_{H^1(Q^h)}^2 \le \|u^h\|_{H^1(\Omega)}^2,$$

and,

(b) for all $u^h \in \dot{W}_J$, the inequality

$$\|u^h\|_{H^1(Q^h)}^2 \le \beta \inf \sum_{\mathrm{supp}(u_i^{(k)}) \subset Q^h} \|u_i^{(k)}\|_{H^1(Q^h)}^2$$

is valid, where the infimum is taken over all decompositions satisfying

$$\sum_{k=0}^{J} \sum_{\mathrm{supp}(u_i^{(k)}) \subset Q^h} u_i^{(k)} = u^h.$$

The positive constants are supposed to be independent of h.

Now, in order to prove the above conditions (a) and (b), we need three lemmas. The proof of each lemma is easy, so omitted. The first and second lemma imply the condition (b) and the last lemma with the BPX preconditioner in Π implies the condition (a). Finally, we will now state the three lemmas.

Lemma 12.6 *There exists C, independent of h, such that*

$$(\nabla v^h, \nabla w^h)_{L_2(\tau_i)} \le C(1/\sqrt{2})^{l-k} |v|_{H^1(\tau_i)} 2^l \|w\|_{L_2(\tau_i)}$$

for all triangles τ_i of the triangulation $\Pi_k^h \cap \mathrm{supp} W_k$, $v^h \in \dot{W}_k$, $w^h \in \dot{W}_l$ $(l > k)$.

Lemma 12.7 *For all* $u^h = u_0^h + \sum_{k=1}^J u_k^h$, $u_k^h \in \dot{W}_k$, *we have*

$$|u^h|^2 \leq C(|u_0^h|^2_{H^1(Q^k)} + \sum_{k=1}^J 4^k \|u_k^h\|^2_{L_2(Q^h)}).$$

Lemma 12.8 *For given* $u^h \in \dot{W}_J$, *we define* $\tilde{u}^h(x) = u^h(x)$ *if* $x \in Q^h$, *otherwise* $\tilde{u}^h = 0$. *Then, for a given decomposition*

$$\tilde{u}^h = \tilde{u}_0 + \sum_{k=1}^J \tilde{u}_k, \quad \tilde{u}_k \in W_k,$$

there exists a decomposition

$$u^h = u_0 + \sum_{k=1}^J u_k, \quad u_k \in \dot{W}_k$$

such that

$$4^k \|u_k\|^2_{L_2(Q^h)} \leq C(|\tilde{u}_0| + \sum_{k=1}^J \|\tilde{u}_k\|^2_{L_2(\Pi)})$$

for some constant C, independent of h.

Acknowledgments

The author gratefully acknowledges the support by the Austrian Academy of Sciences during the Special Semester on Computational Mechanics. Special thanks are due to Robert Burger (Linz) for improving the LaTeX. Last but not least, the author also wishes to thank Sven Beuchler (Linz) and Ulrich Langer (Linz) for helpful hints and for their work in order to improve the final manuscript.

Bibliography

[1] G.P. Astrakhantsev, *Method of fictitious domains for a second-order elliptic equation with natural boundary conditions*, U.S.S.R. Computational Math. and Math. Phys. 18 (1978), pp. 114–121.

[2] ———, *On numerical solution of mixed boundary value problems for second-order elliptic equations in an arbitrary domain.*, U.S.S.R. Computational Math. and Math. Phys. 25 (1985), pp. 200–209.

[3] F.A. Bornemann and H. Yserentant, *A basic norm equivalence for the theory of multilevel methods*, Numerische Mathematik 64 (1993), pp. 455–476.

[4] J.H. Bramble, J.E. Pasciak, and A.T. Vassiliev, *Analysis of non-overlapping domain decomposition algorithms with inexact solvers*, Math. Comp. 66 (1997), pp. 179–196.

[5] J.H. Bramble, J.E. Pasciak, and J. Xu, *Parallel multilevel preconditioners*, Math. Comp. 55 (1990), pp. 1–22.

[6] P.L. Butzer and K. Scherer, *Approximationsprozesse und Interpolationsmethoden*, BI-Hochschulskripten 826/826a, Mannheim, 1968.

[7] W. Dahmen and A. Kunoth, *Multilevel preconditioning*, Numerische Mathematik 63 (1992), pp. 315–344.

[8] E.G. D'yakonov, *The construction of iterative methods based on the use of spectrally equavalent operators*, USSR Comp. Math. and Math. Phys. 6 (1966), pp. 14–46.

[9] R. Glowinski, G.H. Golub, G. A. Meurant, and J. Périaux (eds.), *Domain Decomposition Methods for Partial Differential Equations*, Philadelphia, PA, USA, SIAM, 1988, Proceedings of the First International Symposium on Domain Decomposition Methods for Partial Differential Equations, Paris, France, January 1987.

[10] G. Haase, U. Langer, A. Meyer, and S.V. Nepomnyaschikh, *Hierarchical extension operators and local multigrid methods in domain decomposition preconditioners*, East-West J. Numer. Math. 2 (1994), pp. 172–193.

[11] G. Haase and S.V. Nepomnyaschikh, *Explicit extension operators on hierarchical grids*, East-West J. Numer. Math. 5 (1997), pp. 231–248.

[12] V.G. Korneev, *On construction of high-order accuracy variational difference methods*, Lecture Notes of the Leningrad State University. Series: Mathematics, Mechanics and Astronomy 4 (1970), pp. 28–40 (in Russian).

[13] Yu.A. Kuznetsov, *Iterative methods in subspaces for eigenvalue problems*, Vistas in Applied Math., Numerical Analysis, Atmospheric Sciences, Immunology (A.V. Balakrishinan, A.A. Dorodnitsyn, and J.L. Lions, eds.), Optimization Software, Inc., New York, 1986, pp. 96–113.

[14] Yu.A. Kuznetsov and S.A. Finogenov, *Two–stage fictitious component methods for solving the Dirichlet boundary value problem*, Sov. J. Numer. Anal. Math. Modeling 3 (1988), pp. 301–323.

[15] V. I. Lebedev, *Composition methods*, USSRAcademy of Sciences, Moscow, USSR, 1986 (in Russian).

[16] V.I. Lebedev and V.I. Agoshkov, *Poincaré–Steklov operators and domain decomposition methods in variational problems*, Vychisl. Protsessy i Sistemy 2 (1985), pp. 173–227.

[17] G.I. Marchuk, *Methods of numerical mathematics*, Springer-Verlag, New York, 1975.

[18] A.M. Matsokin, *On one method for solving systems of grid equations*, Metody Reshenya Sistem Variats.-Raznost (1979), pp. 136–138.

[19] ______, *Fictitious component method and the modified difference analogue of the Schwarz method*, Vychislitel'nye Metody Lineinoi Algebry (G.I. Marchuk, ed.), Vychisl. Tsentr. Sib. Otdel. Akad. Nauk. SSR, Novosibirsk, 1980, Sov. J. Numer. Anal. Math. Modeling, 4 (1989), pp. 66–77.

[20] ______, *Fictitious components method and alternating method in subspace.*, Ph.D. thesis, Novosibirsk, 1988.

[21] A.M. Matsokin and S.V. Nepomnyaschikh, *A Schwarz alternating method in a subspace*, Soviet Mathematics 29 (1985), pp. 78–84.

[22] ______, *Norms in the space of traces of mesh functions*, Sov. J. Numer. Anal. Math. Modeling 3 (1988), pp. 199–216.

[23] ______, *On using the bordering method for solving systems of mesh equations*, Sov. J. Numer. Anal. Math. Modeling 4 (1989), pp. 487–492.

[24] A.M. Matsokin and I.N. Skripko, *Fictitious components method and mixed boundary conditions.*, Vychisl. Algoritmy v Zadachakh Mat. Fiz., Comp. Cent. Sib. Branch, USSR Acad. Sci., Novosibirsk (1983), pp. 110–119.

[25] S.V. Nepomnyaschikh, *On the application of the bordering method to the mixed boundary value problem for elliptic equations and on mesh norms*, Sov. J. Numer. Anal. Math. Modeling 4 (1989), pp. 493–506.

[26] ______, *Schwarz alternating method for solving the singular Neumann problem.*, Soviet J. Numer. Anal. and Math. Model. 5 (1990), pp. 69–78.

[27] ______, *Application of Domain Decomposition to Elliptic Problems on with Discontinuous Coefficients*, Fourth International Symposium on Domain Decomposition Methods for Partial Differential Equations (Philadelphia, PA, USA) (R. Glowinski, Yu.A. Kuznetsov, G.A. Meurant, J. Périaux, and O. Widlund, eds.), SIAM, 1991.

[28] ______, *Mesh theorems of traces, normalizations of function traces and their inversion*, Sov. J. Numer. Anal. Math. Modeling 6 (1991), pp. 151–168.

[29] ______, *Method of splitting into subspaces for solving elliptic boundary value problems in complex form domains*, Sov. J. Numer. Anal. Math. Modeling 6 (1991), pp. 1–25.

[30] ______, *Decomposition and Fictious Domains for Elliptic Boundary Value Problems*, Fifth International Symposium on Domain Decomposition Methods for Partial Differential Equations (Philadelphia, PA) (D.E. Keyes, T.F. Chan, G.A. Meurant, J.S. Scroggs, and R.G. Voigt, eds.), SIAM, 1992, pp. 62–72.

[31] ______, *Optimal multilevel extension operators*, Preprint SPC Preprint SPC 95-3, Technische Universität Chemnitz-Zwickau, 1995.

[32] L.A. Oganesian and L.A. Rukhovets, *Variational difference methods for solving equations*, Akad. Nauk Arm. SSR, Erevan, 1979.

[33] P. Oswald, *On discrete norm estimates related to multilevel preconditioners in the finite element method*, Constructive Theory of Functions (K.G. Ivanov, P. Petrushev, and B. Sendov, eds.), Bulg. Acad. Sci., 1992, Proc. Int. Conf., Varna, 1991, pp. 203–241.

[34] ______, *Multilevel Finite Element Approximation. Theory and Applications*, Teubner Skripten zur Numerik, Teubner, Stuttgart, 1994.

[35] A. Quarteroni and A. Valli, *Domain decomposition methods for partial differential equations*, Oxford University Press, Oxford, 1999.

[36] V.S. Ryaben'ky, *Difference Potential Method for Some Problems of Continuum Mechanics*, Nauka, Moscow, 1987.

[37] A.A. Samarskii, *Theory of Difference Schemes*, Nauka, Moscow, 1977 (in Russian).

[38] A.A. Samarskii, I.E. Kaporin, A.B. Kucherov, and E.S. Nikolaev, *Some modern methods for solving grid equations*, Izv. Vuzov. Mat. 7 (1983), pp. 3–12.

[39] A.A. Samarskii and E.S. Nikolaev, *Numerical Methods for Grid Equations*, Nauka, Moscow, 1978 (in Russian).

[40] H.A. Schwarz, *Über einige Abbildungsaufgaben*, Ges. Math. Abh. 11 (1869), pp. 65–83.

[41] B.F. Smith, P.E. Bjørstad, and W.D. Gropp, *Domain Decomposition: Parallel Multilevel Methods for Elliptic Partial Differential Equations*, Cambridge University Press, 1996.

[42] S.L. Sobolev, *Schwarz's algorithm in the theory of elasticity*, Doklady Academy of Sciences of SSSR 4 (1936), pp. 235–238.

[43] ———, *Applications of Functional Analysis in Mathematical Physics*, SO AN USSR, Novosibirsk, 1962.

[44] A. Toselli and O. Widlund, *Domain Decomposition Methods – Algorithms and Theory*, Springer Series in Computational Mathematics, vol. 34, Springer, Heidelberg, 2004.

[45] H. Yserentant, *On the multi-level splitting of finite element spaces*, Numerische Mathematik 49 (1986), pp. 379–412.

[46] X. Zhang, *Multilevel Schwarz methods*, Numerische Mathematik 63 (1992), pp. 521–539.

Author information

Sergey Nepomnyaschikh, Institute for Computational Mathematics and Computational Geophysics, Sibirian Department of the Russian Academy of Sciences, Novosibirsk, Russia.
Email: `svnep@oapmg.sscc.ru`

Radon Series Comp. Appl. Math **1**, 161–226 © de Gruyter 2007

A posteriori error estimation methods for partial differential equations

Sergey Repin

Key words. A posteriori error control, partial differential equations, functional a posteriori estimate, Stokes problem, linear elasticity, evolutionary problems, modeling error

AMS classification. 65N15, 65N30, 65M15

Supported by Radon Institute of Computational and Applied Mathematics, Linz and FIM Institute in ETH, Zürich.

1 Introduction

Partial differential equations (PDE's) were discovered as mathematical models of various physical phenomena. Over the 20th century, mathematical models based on PDE's have been intensively investigating as a part of new rapidly developing science *Mathematical Modeling*. Modern applied mathematics considers the creation of *fully reliable* mathematical modeling methods as one of the most challenging scientific tasks whose development is highly important for natural sciences and engineering.

Reliable modeling consists of (a) efficient computing of approximate solutions that must converge to the desired exact solution and (b) reliable verification of the accuracy of approximate solutions. These two fundamental tasks started receiving serious attention in the 60s and 70s when the first systematic results in the error control theory for PDE's were obtained. Classical convergence theory created at that time (see, e.g., Ph. Ciarlét [99], G. Strang and G. Fix [289]) was focused on *asymptotic estimates* of approximation errors. The aim of the asymptotic error control theory is to prove that the difference between an exact solution u and its approximation u_n (found in a finite dimensional subspace V_n with $\dim V_n = n$) tends to zero as n tends to $+\infty$. In a sense, proving such a fact can be considered as a formal justification of the approximation method used. More exact estimates show the convergence rate, i.e., establish the relation

$$\|u - u_n\| \leq C(1/n)^k, \tag{1.1}$$

where C and k are some positive numbers. The value of the constant C depends on the solution u and properties of the approximations used. Estimates of the type (1.1) are often called *a priori* error estimates. In the majority of cases, such estimates are expressed in terms of the mesh-size parameter h, which is proportional to a certain negative power of n.

Usually, a priori error estimates establish asymptotical behavior of an upper bound of the approximation error for the *whole set of solutions and their approximations of a certain type*. However, they are generally unable to reliably (efficiently) evaluate the error related to a particular approximate solution computed on a particular mesh. Besides, they possess other features that make their practical exploitation rather difficult. First of all, they are valid *only for Galerkin approximations*. In practice, it is often difficult to guarantee that an approximate solution computed by a numerical procedure is indeed the exact solution of the respective finite dimensional problem. Secondary, such estimates require *extra regularity* of the exact solution, which may not hold in many practically important cases.

These drawbacks stimulated efforts focused on new error estimation methods able to explicitly characterize the accuracy of approximate solutions. In the late 70s and early 80s, it had become clear that successive numerical methods for PDE's should be based on the so-called *mesh-adaptive* procedures that modify finite dimensional spaces (and meshes) using the information contained in the approximate solution computed on the previous step[1]. At that time it was understood that efficient numerical methods for PDE's should be based on a posteriori error estimates.

At present, *a posteriori* error control methods have formed a new important part of the numerical analysis, whose purpose is to obtain computable error estimates in various norms and indicators of local errors. The necessity of them is motivated not only by standard arguments (such as getting efficient adaptive procedures and reliable results in a particular numerical experiment). A more fundamental task is *Validation of mathematical models* (see, e.g., I. Babuška [18]). Of course, proper validation of a mathematical model can be performed only by joint efforts of mathematicians and specialists in the respective natural science. However, a posteriori error control methods that guarantee the reliability of computer solutions must play an essential role in such a research.

The present paper is aimed to present an overview of a posteriori error estimation methods for partial differential equations. The amount of publications devoted to a posteriori estimates is vast, so that it is almost impossible to review all of them. Therefore, our goal is to discuss the basic ideas and main approaches to posteriori error control and relate them to a certain amount of distinctive publications. A reader will find more references in the books and articles cited.

The paper is organized as follows. Section 2 is devoted to a posteriori methods developed in the first half of the 20th century. Here, we discuss Runge's heuristic rule, estimates of Prager, Synge and Mikhlin. Also, we consider estimates of Ostrowski for iteration schemes and a posteriori methods based on monotonicity. A posteriori error estimation methods developed for finite element approximations are presented in Section 3. In it, we discuss mathematical foundations and specific features of several main methods and give references for a reader interested in a deeper investigation of them. A considerable part of the paper is devoted to the so-called *functional a posteriori estimates*. Functional approach to a posteriori error control arose about ten years ago. In Section 4, the basic ideas of this approach are discussed on the paradigm of a simple elliptic problem. An overview of functional a posteriori estimates for linear and some nonlinear problems is exposed in Sections 5 and 6, respectively. An important feature of these estimates is their validity for the whole class of conforming approximations (in other words, they are applicable for any function from the admissible energy class). Therefore, such estimates can be also used for the evaluation of *modeling errors* arising if a complicated model is replaced by a simplified one. A somewhat close topic concerns errors caused by *uncertainty* of the data. By the same reasons, functional a posteriori estimates open a way of evaluating such errors. These questions are considered in Sections 7 and 8.

[1] An interesting discussion on the reliability of different computer simulation methods in nonlinear mechanical problems is contained in [13]. It reflects early observations of the fact that approximations of strongly nonlinear problems (and not only of them) may seriously depend on the structure of the finite dimensional space (mesh) used.

2 First approaches to a posteriori error control

2.1 Runge's heuristic rule

First attempts to formulate accuracy criteria for approximate solutions date back to the very beginning of the 20th century when C. Runge suggested a heuristic error indication rule. Originally, it was applied to quadrature formulas, but later the rule was also adapted to integration procedures that he developed for ordinary differential equations. In essence, Runge's rule is based on comparing "coarse" and "refined" solutions.

Assume that an approximate solution u_h has been found on a mesh $\mathcal{T}_h$ with the mesh size h and $u_{h\text{REF}}$ is a solution on a refined mesh $\mathcal{T}_{h\text{REF}}$. The method is based on the assumption that h is small enough and the difference between u_h and exact solution u is subject to the asymptotical relation

$$u_h - u \cong \phi h^m + O(h^{m+p}), \quad m, p > 0, \tag{2.1}$$

where ϕ is a certain (unknown) function. Then, the same relation holds for $h_{\text{REF}} \le h$:

$$u_{\text{REF}} - u \cong \phi h^m_{\text{REF}} + O(h^{m+p}_{\text{REF}})$$

and, consequently,

$$u_h - u_{h\text{REF}} \cong \phi(h^m - h^m_{\text{REF}}) + O(h^{m+p}) + O(h^{m+p}_{\text{REF}}).$$

From here, we conclude that up to higher order terms

$$u_h - u \cong h^m \frac{u_h - u_{h\text{REF}}}{h^m - h^m_{\text{REF}}}. \tag{2.2}$$

If $h_{\text{REF}} = \kappa h$ with $\kappa \in (0, 1)$, then the above relation reads

$$\boxed{u_h - u \cong \frac{1}{1 - \kappa^m}\,(u_h - u_{h\text{REF}}).} \tag{2.3}$$

Certainly, the estimate adequately presents the error only if h is small enough (so that the higher order terms are negligibly small) and the error satisfies the asymptotic relation (2.1) with a certain ϕ. In general, it is impossible to guarantee such strong a priori requirements and (2.3) should be viewed as a conditional estimate. Anyway, (2.3) partially justifies the following heuristic rule often used in engineering computations:

> **if $u_h - u_{h\text{REF}}$ is small, then both approximations $u_{h\text{REF}}$ and u_h are probably close to the exact solution u.**

Using modern terminology, we can say that Runge's rule suggests to use $u_h - u_{h\text{REF}}$ as *an error indicator* and a certain norm of $u_h - u_{h\text{REF}}$ as *a stopping criteria*. Runge's heuristic rule is very simple. It is widely used in practical computations. In many cases, it gives quite acceptable (and even good) results. Regrettably, in general such a simple rule is not correct. Numerous examples show that two subsequent elements of an approximation sequence may be close to each other, but far from a desired limit function (in these examples the asymptotic assumptions are of course violated). Besides, in practice, it is important to have guaranteed upper and lower bounds of the error $\|u_h - u\|$ that does not follow from the above heuristic arguments. In Sections 4 and 5, we present the quantities computed by u_h and $u_{h\mathrm{REF}}$ that *indeed provide a guaranteed upper bound of the approximation error.* However, their derivation requires a special mathematical theory, which did not exist at the time of C. Runge.

2.2 Estimate of Prager and Synge

In 1947, W. Prager and J. L. Synge [228] presented an a posteriori estimate valid for approximations of linear elliptic problems. Originally, their proof was motivated by purely geometrical arguments (therefore, this approach is often called *the hypercircle method* [292]). Using modern notation, the Prager–Synge estimate for the problem

$$\Delta u + f = 0 \quad \text{in } \Omega, \tag{2.4}$$

$$u = 0 \quad \text{on } \partial\Omega, \tag{2.5}$$

where Ω is a bounded connected domain in $\mathbb{R}^d$, is as follows:

$$\|\nabla(u - v)\|^2 + \|\nabla u - q\|^2 = \|\nabla v - q\|^2. \tag{2.6}$$

Here, $q \in Q_f$ and Q_f is a subset of $H(\Omega, \operatorname{div})$ that contains the functions satisfying the equations $\operatorname{div} q + f = 0$, and $\| \; \|$ denotes the norm in $L^2(\Omega)$. By the symbols $\cdot$ and $:$ we denote the products in the spaces of vectors and tensors, respectively. The Euclidean norms of vectors and tensors associated with the above products are denoted by $| \; |$. To simplify the notation, we denote the norm in the space of vector-valued functions $L^2(\Omega, \mathbb{R}^d)$ by the same symbol. The derivation of (2.6) is based upon the *orthogonal decomposition* of the space of square summable vector-valued functions $L^2(\Omega, \mathbb{R}^d)$ into the gradients of $\overset{\circ}{H}{}^1(\Omega)$-functions and the solenoidal functions from the set Q_0. Since $\nabla u - q \in Q_0$, we have the orthogonality relation

$$\int_\Omega \nabla(u - v) \cdot (\nabla u - q)\, dx = 0,$$

which implies (2.6). From (2.6) it also follows that

$$\boxed{\|\nabla(u - v)\| \leq \|\nabla v - q\| \qquad \forall q \in Q_f.} \tag{2.7}$$

2.3 Estimate of Mikhlin

A similar estimate was derived by S. Mikhlin [201] with the help of *variational arguments*. The main idea of his approach can be explained on the paradigm of the problem (2.4)–(2.5). It is not difficult to show that

$$\frac{1}{2}\|\nabla(u-v)\|^2 = J(v) - J(u), \quad J(u) := \inf_{v \in \overset{\circ}{H}{}^1(\Omega)} J(v), \tag{2.8}$$

where

$$J(v) := \frac{1}{2}\|\nabla v\|^2 - \int_\Omega f v dx$$

is the variational functional associated with the problem in question. In the framework of the variational approach, the value of $J(u)$ is estimated by means of the so-called *dual variational problem* (see, e.g., [112, 209]), namely

$$J(u) = \sup_{q \in Q_f} \left\{ -\frac{1}{2}\|q\|^2 \right\},$$

From this relation, we find that for any $q \in Q_f$

$$\frac{1}{2}\|\nabla(u-v)\|^2 \leq J(v) + \frac{1}{2}\|q\|^2 = \frac{1}{2}\|\nabla v\|^2 - \int_\Omega f v \, dx + \frac{1}{2}\|q\|^2 =$$

$$= \frac{1}{2}\|\nabla v\|^2 - \int_\Omega q \cdot \nabla v \, dx + \frac{1}{2}\|q\|^2 = \frac{1}{2}\|\nabla v - q\|^2$$

and the estimate (2.7) follows. Later (see, e.g., H. Gajewski, K. Gröger and K. Zacharias [135], S. Mikhlin [202], P. Mosolov and V. Myasnikov [205]), similar estimates were derived for some classes of convex variational problems. Estimates based on complementary principles were also studied in A. M. Arthurth [12].

It should be noted that from the practical point of view (2.7) has an essential drawback: *it uses a function q in the set Q_f defined by the differential relation, which is difficult to exactly satisfy*. In [201], a strategy close to the known *orthogonal projections method* (see M. Vishik [304], H. Weil [310], S. Zaremba [315]) is discussed. It is based on constructing an approximating sequence for the dual problem. However, practical realization of fully equilibrated approximations within the framework of locally supported trial functions may be faced with serious difficulties. The so-called equilibrated finite element approximations are either complicated (see, e.g., C. Johnson and B. Mercier [166], where such approximations were constructed using macroelements) or may not exactly satisfy the differential relation. In the latter case, the reliability of the upper bound in (2.7) is lost. Sometimes (2.7) is also used with an "almost equilibrated" function q which may be a good error indicator. However, guaranteed upper bounds require conforming approximations of Q_f and are efficient not for all equilibrated functions, but *only for those close to ∇u*. In the next section, we will discuss some methods developed for the construction of such functions.

2.4 Estimates of Ostrowski for contractive mappings

A posteriori estimates for iteration methods were derived by A. Ostrowski [221]. Let $\{x_j\}_{j=0}^{\infty}$ be a sequence in a full metric space X with the metric function ρ obtained by the iteration process $x_{j+1} = \mathrm{T}x_j$. It is assumed that a mapping T satisfies the condition

$$\rho(\mathrm{T}x, \mathrm{T}y) \leq q\,\rho(x, y), \qquad q \in (0, 1), \tag{2.9}$$

for any elements x and y of the set $S \subset X$. In view of the Banach fixed point theorem, a contractive mapping T has a fixed point $x_{\odot}$, which is a limit of $\{x_j\}$. The proposition below furnishes upper and lower estimates of the errors $e_j = \rho(x_j, x_{\odot})$, which are easy to compute provided, that the number q (or a sharp upper bound of it) is known and less than 1.

Theorem (A. Ostrowski). *For any x_j, $j \geq 1$, the following estimate holds:*

$$M^j_{\ominus} := \frac{\rho(x_{j+1}, x_j)}{1+q} \leq e_j \leq M^j_{\oplus} := \frac{q}{1-q}\rho(x_j, x_{j-1}). \tag{2.10}$$

Note that

$$\frac{M^j_{\oplus}}{M^j_{\ominus}} = \frac{1+q}{1-q}\frac{\rho(x_j, x_{j-1})}{\rho(x_{j+1}, x_j)}q \geq \frac{1+q}{1-q}.$$

Therefore, the efficiency of the upper and lower bounds in (2.10) depends on the value of $1-q$. Estimates (2.10) and their modifications can be applied to linear and nonlinear algebraic systems, integral equations and other problems solved by iteration methods (see, e.g., A. N. Kolmogorov and S. V. Fomin [175] and E. Zeidler [316]). The major difficulty in the application of the above a posteriori estimates consists of that in practice it may be difficult to find a sharp upper bound of q and establish that is indeed less than 1.

A posteriori methods for various iteration schemes were investigated by many authors. Several references below will help a reader to find more information in the area.

In G. Auchmuty [15], two sided p-norm error bounds for solutions of linear systems are presented. O. Scherzer, H. W. Engl and K. Kunisch [277] studied an a posteriori choice of the regularization parameter in a Tikhonov scheme for ill-posed problems (see also H. W. Engl and O. Scherzer [114]). Further studies of this strategy are presented in Jin Qi-nian and Hou Zong-yi [229]. A posteriori estimates for linear ill-posed problems were investigated in A. Leonov [195]. More information on error estimates for iteration Newton-type methods can be found, e.g., in K. Braune [66], Jin Qi-nian [162], P. Meyer [200], F. Potra [225], K. Tsuruta and K. Ohmori [294], T. Yamamoto [313].

Boundary-value problems are often reformulated in terms of integral equations. A posteriori error estimates for the respective numerical methods (e.g., boundary element methods) are usually derived with the help of approaches that exploit special properties of the iteration operator. Iteration methods and a posteriori estimates for integral type formulations of linear and nonlinear boundary value problems were studied in C. Carstensen [80], Lin Qun and Shi Jan [230], M. Schultz and O. Steinbach [279], S. Shaw and J. R. Whiteman [282], and M. Schulz and W. L. Wendland [280]. A reader will also find there more references related to the topic.

2.5 A posteriori estimates based on monotonicity.

The theory of *monotone operators* gives another way of constructing a posteriori estimates. Monotone operators are defined on the so-called *ordered* (or *partially ordered*) spaces that introduce the relation $x \leq y$ for all (or almost all) elements x, y of the space. We recall that the operator T is called monotone if $x \leq y$ implies $\mathrm{T}x \leq \mathrm{T}y$.

Consider the following fixed point problem

$$x_{\odot} = \mathrm{T}x_{\odot} + f,$$

where T is defined on an ordered (partially ordered) space X. Assume that

$$\mathrm{T} = \mathrm{T}_{\oplus} + \mathrm{T}_{\ominus},$$

$\mathrm{T}_{\oplus}$ is monotone, $\mathrm{T}_{\ominus}$ is antitone (i.e., $x \leq y$ implies $\mathrm{T}x \geq \mathrm{T}y$), and $\mathrm{T}_{\oplus}$, $\mathrm{T}_{\ominus}$ have a common set of images D which is a convex subset of X. Next, let $x_{\ominus\,0}$, $x_{\ominus 1}$, $x_{\oplus\,0}$, and $x_{\oplus 1}$ be such elements that $x_{\ominus 1}, x_{\oplus 1} \in D$, $x_{\ominus\,0} \leq x_{\ominus\,1} \leq x_{\oplus\,1} \leq x_{\oplus\,0}$, and

$$x_{\ominus 1} = \mathrm{T}_{\oplus}x_{\ominus\,0} + \mathrm{T}_{\ominus}x_{\oplus\,0} + f, \qquad x_{\oplus 1} = \mathrm{T}_{\oplus}x_{\oplus\,0} + \mathrm{T}_{\ominus}x_{\ominus\,0} + f.$$

Then, we observe that

$$x_{\ominus 2} = \mathrm{T}_{\oplus}x_{\ominus\,1} + \mathrm{T}_{\ominus}x_{\oplus\,1} + f \geq \mathrm{T}_{\oplus}x_{\ominus\,0} + \mathrm{T}_{\ominus}x_{\oplus\,0} + f = x_{\ominus\,1}$$
$$x_{\oplus 2} = \mathrm{T}_{\oplus}x_{\oplus\,1} + \mathrm{T}_{\ominus}x_{\ominus\,1} + f \leq \mathrm{T}_{\oplus}x_{\oplus\,0} + \mathrm{T}_{\ominus}x_{\ominus\,0} + f = x_{\oplus\,1}.$$

By continuing the iterations we obtain elements such that

$$x_{\ominus\,k} \leq x_{\ominus\,(k+1)} \leq x_{\oplus\,(k+1)} \leq x_{\oplus\,k}.$$

Then $x \to \mathrm{T}x + f$ maps D to itself. If D is compact, then by the Schauder fixed point theorem $x_{\odot} \in D$ exists. Moreover, it is bounded from below and above by the sequences $\{x_{\ominus\,k}\}$ and $\{x_{\oplus\,k}\}$.

Applications of this method are mainly oriented towards systems of linear simultaneous equations and integral equations (a detailed discussion of the monotonicity methods is presented in L. Collatz [102]). For example, consider a system of linear simultaneous equations

$$x = Ax + f$$

that is supposed to have a unique solution $x_{\odot}$. Assume that

$$A = A_{\oplus} - A_{\ominus}, \qquad A_{\ominus} = \{a^{\ominus}_{ij}\} \in \mathbb{M}^{n\times n},$$
$$A_{\oplus} = \{a^{\oplus}_{ij}\} \in \mathbb{M}^{n\times n}, \quad a^{\ominus}_{ij} \geq 0, \quad a^{\oplus}_{ij} \geq 0.$$

We may *partially order* the space $\mathbb{R}^n$ by saying that $x \leq y$ if and only if $x_i \leq y_i$ for $i = 1, 2, \ldots, n$. Compute the vectors

$$x_{\ominus(k+1)} = A_{\oplus} x_{\ominus k} + A_{\ominus} x_{\oplus k} + f, \quad x_{\oplus(k+1)} = A_{\oplus} x_{\oplus k} + A_{\ominus} x_{\ominus k} + f.$$

If $x_{\ominus 0} \leq x_{\ominus 1} \leq x_{\odot} \leq x_{\oplus 1} \leq x_{\oplus 0}$, then for all the components of $x_{\odot}$ we obtain two-sided estimates

$$\boxed{x^{(i)}_{\ominus k} \leq x^{(i)}_{\ominus (k+1)} \leq x^{(i)}_{\odot} \leq x^{(i)}_{\oplus (k+1)} \leq x^{(i)}_{\oplus k}, \quad i = 1, 2, \ldots, n.}$$

It should be remarked that convergence of $x^{(i)}_{\ominus k}$ and $x^{(i)}_{\oplus k}$ to $x_{\odot}$ (and the convergence rate) requires a special investigation, which must use specific features of a particular problem. Two-sided a posteriori estimates for linear simultaneous equations, differential and integral equations within the framework of interval analysis are considered in, e.g., B. S. Dobronets and V. V. Shaĭdurov [106], N. S. Kurpel' and B. A Shuvar [183]. Applications of the above method to ordinary nonlinear differential equations are discussed in E. Geisler, A. Tal, and D. Garg [139].

3 A posteriori error estimates for finite element approximations

It is commonly accepted that the mathematical concept of the finite element method was presented by R. Courant in [104], where the method was introduced as a variant of B. Galerkin [136] and W. Ritz [274] methods using locally supported trial functions. At the end of the 70s, finite element methods have formed one of the main approaches to the numerical analysis of PDE's. The foundations of the finite element method are exposed in the books by Ph. Ciarlét [99], C. Johnson [163], J. T. Oden and J. N. Reddy [218], G. Strang and G. Fix [289], J. L. Segerlind [281], O. C. Zienkiewicz and K. Morgan [319] and many other publications. Advanced versions of the finite element method (such as mixed and hybrid methods) are discussed in, e.g., F. Brezzi and M. Fortin [70], D. Braess [57], V. Girault and P. A. Raviart [141], J. E. Roberts and J.-M. Thomas [275]. Achievements and some unsolved problems of the finite element method are discussed in the papers by I. Babuška [16] and O. C. Zienkiewicz [317] (see also I. Babuška and J. E. Osborn [27]).

Nowadays, mesh-adaptive methods based on a posteriori estimates dominate in the numerical analysis of differential equations (see, e.g., [4, 5, 32, 36, 37, 75, 152, 165, 168, 178, 181, 224, 273, 285, 298, 321] and the papers cited therein). The majority of a posteriori methods for FEM are based either on the analysis of residuals or attract special properties (e.g., additional regularity) of exact solutions.

A posteriori error estimation for finite element approximations is the main subject of the books by M. Ainsworth and J. T. Oden [5], I. Babuška and T. Stroboulis [32], W. Bangerth and R. Rannacher [36], K. Eriksson, D. Estep, P. Hansbo and C. Johnson

[115], and R. Verfürth [298]. They present various approaches, results of numerical experiments and a wide list of references.

3.1 Explicit residual method

From the mathematical point of view, this method is based on deriving an upper bound of the residual functional computed in the image space of the respective differential operator with the help of special interpolation estimates and Galerkin orthogonality property. This way leads to an upper bound of the error presented as a sum of element-wise residuals and interelement jumps with weights given by constants in the so-called Clement's interpolation inequalities [100]. We discuss the main idea of this method on the paradigm of the problem (2.4)–(2.5), whose generalized solution meets the integral identity

$$\int_\Omega \nabla u \cdot \nabla w \, dx = \int_\Omega f w \, dx, \quad w \in V_0 := \overset{\circ}{H}{}^1(\Omega), \tag{3.1}$$

where $\cdot$ denotes the scalar product of vectors. Let $v \in V_0$ be an approximate solution of this problem. Then,

$$\int_\Omega \nabla(u - v) \cdot \nabla w \, dx = \mathcal{F}_v(w), \quad w \in V_0, \tag{3.2}$$

where $\mathcal{F}_v(w) := \int_\Omega (fw - \nabla v \cdot \nabla w) \, dx$ is a linear functional on V_0. It is easy to see that this functional is equal to zero if v coincides with u. In all other cases, the norm of this functional defined by the relation

$$|\!|\!| \mathcal{F}_v |\!|\!| := \sup_{w \in V_0, \, w \neq 0} \frac{| \mathcal{F}_v(w) |}{\|\nabla w\|}, \tag{3.3}$$

is positive. Therefore, it is natural to call $\mathcal{F}_v$ the *error functional*. Let us show that its norm is indeed a measure of the deviation of v from u. By (3.2) and (3.3), we observe that

$$\int_\Omega | \nabla(u - v) |^2 \, dx = \mathcal{F}_v(u - v) \leq |\!|\!| \mathcal{F}_v |\!|\!| \, \|\nabla(u - v)\|.$$

Hence,

$$\|\nabla(u - v)\| \leq |\!|\!| \mathcal{F}_v |\!|\!| . \tag{3.4}$$

However, $| \mathcal{F}_v(w) | \leq \|\nabla(u - v)\| \|\nabla w\|$, so that (3.3) implies the inequality opposite to (3.4). Thus, the norm of the deviation from exact solution coincides with the norm of $\mathcal{F}_v$. The problem consists of that how to practically compute such a norm for a concrete v. In practice, a straightforward computation of the norm by means of (3.3) is hardly possible. A more promising way is to find certain computable upper bounds of $|\!|\!| \mathcal{F}_v |\!|\!|$. In the papers of I. Babuška and W. C. Rheinboldt [28, 29] and some other publications of these authors, it was suggested a way of the derivation of such bounds later called the *explicit residual method*.

The residual method is applicable if v is a *Galerkin approximation* of the exact solution computed on a finite dimensional space $V_h \subset V_0$, i.e., $v = u_h$, where

$$\int_\Omega \nabla u_h \cdot \nabla w_h \, dx = \int_\Omega f w_h \, dx, \qquad w_h \in V_h.$$

In this case, the error $\nabla(u - u_h)$ is orthogonal to ∇w_h for any $w_h \in V_h$ and we find that

$$\mathcal{F}_{u_h}(w) = \int_\Omega \nabla(u - u_h) \cdot \nabla w dx = \int_\Omega \nabla(u - u_h) \cdot \nabla(w - \pi_h w) dx,$$

where $\pi_h : V_0 \to V_h$ is a continuous mapping (typically π_h is defined by Clement's interpolation operator [100]). Let Ω be divided into a collection of subdomains Ω_k, $k = 1, 2, \ldots, M$, and u_h is a smooth function in each subdomain. Then

$$\mathcal{F}_{u_h}(w) := \sum_{k=1}^{M} \int_{\Omega_k} \Delta(u_h - u)(w - \pi_h w) dx + \sum_{k,l=1}^{M} \int_{\partial\Omega_{kl}} \left[\frac{\partial(u - u_h)}{\partial \nu_{kl}} \right] (w - \pi_h w) \, dx,$$

where $\partial\Omega_{kl}$ is a common part of the boundaries of Ω_k and Ω_l, ν_{kl} is the unit normal vector to this boundary, and $[\,]$ denotes the jump of a quantity that arises at the boundary $\partial\Omega_{kl}$. If Ω_k are simplexes, then for π_h Clement's interpolation estimates read as follows:

$$\|v - \pi_h v\|_{\Omega_k} \leq \gamma_{1k} \mathrm{diam}(\Omega_k) \|v\|_{1,2,\omega_{1k}(\Omega_k)}, \tag{3.5}$$

$$\|v - \pi_h v\|_{\partial\Omega_{kl}} \leq \gamma_{2k} |\partial\Omega_{kl}|^{1/2} \|v\|_{1,2,\omega_{2k}(\Omega_k)}, \tag{3.6}$$

where $\| \, \|_{\Omega_k}$ denotes the L^2-norm on Ω_k, $\| \, \|_{1,2,\omega}$ denotes the H^1-norm on ω, and ω_{1k} and ω_{2k} are certain domains (patches of neighbor elements) that contain Ω_k,

$$\mathrm{diam}(\Omega_k) = \sup_{x_1, x_2 \in \Omega_k} |x_1 - x_2|,$$

$|\partial\Omega_{kl}| \sim h^{d-1}$ is the measure of the edge Ω_{kl}, and the interpolation constants γ_{1k} and γ_{2k} depend not only on Ω_k, but also on the form of all the elements in the respective patch. By these estimates we can present the norm of the error functional as the sum of local quantities

$$\eta_k^2 := \mathrm{diam}(\Omega_k)^2 \|\Delta u_h + f\|_{\Omega_k}^2 + \frac{1}{2} \sum_{l=1}^{M} |\partial\Omega_{kl}| \, \| \left[\frac{\partial u_h}{\partial \nu_{kl}} \right] \|_{\partial\Omega_{kl}}^2,$$

that are related to the residual on Ω_k and values of the jumps in the normal component of the gradient on the boundary. The respective estimate of the overall error is as follows:

$$\|\nabla(u - u_h)\|^2 \leq |\!|\!| \mathcal{F}_{u_h} |\!|\!|^2 \leq \sum_{k=1}^{M} c_k \eta_k^2, \tag{3.7}$$

where the constants c_k depend on γ_{1k} and γ_{2k} associated with the sampling considered. The quantities η_k are used as error indicators. By comparing their values on different elements one can suggest an adequate mesh-adaptation procedure. In this case, the estimate is used in the form

$$\boxed{\|\nabla(u-u_h)\|^2 \leq C\sum_{k=1}^{M}\eta_k^2\,,} \tag{3.8}$$

where different c_k are replaced by one common constant C. It should be noted that often the values of two terms in η_k are quite different (see, e.g., C. Carstensen and R. Verfürth [92], where it is shown that usually the second term dominates). For this reason, sometimes error indicators for mesh-adaptation are constructed with the help of only this (dominant) part of η_k.

However, if a guaranteed and sharp upper bound of $\|\nabla(u-u_h)\|$ is required then certain difficulties arise. Indeed, in this method we must find a large amount of local constants γ_{1k} and γ_{2k} that come from interpolation estimates (3.5)–(3.6). In general, finding such a constant requires solving an infinite-dimensional problem on each patch. For example,

$$\gamma_{1k} = \operatorname{diam}(\Omega_k)^{-1} \sup_{v\in H^1(\Omega_k)} \frac{\|v-\pi_h v\|_{\Omega_k}}{\|v\|_{1,2,\omega_{1k}(\Omega_k)}}\,. \tag{3.9}$$

This problem is of the same level as (3.3) (the only difference is that in (3.3) supremum is taken over the $\overset{\circ}{H}{}^1$-functions defined in Ω while in (3.9) over H^1-functions in Ω_k). In the literature, it is sometimes recommended to replace H^1 in (3.9) by a certain set of polynomial functions. This gives a possibility to compute an approximate value of γ_{1k} (which is a lower estimate of γ_{1k}) but since the amount of functions in (3.9) was reduced the quantity computed via such a procedure is not a guaranteed upper bound of the error. Another unpleasant feature of the method consists of that γ_{1k} and γ_{2k} *depend on* $\mathcal{T}_h$ and, consequently, all of them must be recalculated if one sampling is replaced by another one.

Also, it is worth remarking that the estimate (3.7) is derived by formal mathematical transformations that clearly overestimate $|\mathcal{F}_{u_h}|$. Evidently, in (3.8) such an overestimation may be much larger (especially for nonuniform meshes, for which the constants γ_{1k} and γ_{2k} may be very different). Thus, such an estimates may lead to a considerable overestimation of the total error. This fact was observed in some carefully performed tests (see, e.g., C. Carstensen and S. A. Funken [85]). Nevertheless, residual method is widely exploited in practice and the quantities η_k are often used not for the computation of a guaranteed upper bound of the total error, but serve as *error indicators* that give a presentation on the distribution of local errors in Ω.

The amount of publications associated with residual based methods is vast. In addition to the books M. Ainsworth and J. T. Oden [5], I. Babuška and T. Stroboulis [32], K. Eriksson, D. Estep, P. Hansbo and C. Johnson [115], and R. Verfürth [298], discussions of the method are presented in, e.g., M. Ainsworth and J. T. Oden [3, 4], C. Carstensen [78], C. Carstensen and S. A. Funken [84, 85] (where the authors consider ways of computing bounds of the interpolation constants in the residual based error estimator), C. Carstensen and R. Verfürth [92] (the authors show that the "edge component" dominates in the residual error estimator), K. Eriksson and C. Johnson [116], C. Johnson and P. Hansbo [165].

In C. Carstensen and S. Sauter [90], a posteriori error estimates were derived for elliptic PDE's on domains with complicated structures. Estimates for problems with biharmonic operator are analyzed in A. Charbonneau, K. Dossou and R. Pierre [94]. Residual-type estimates for linear first order systems of PDE's were obtained in P. Houston, J. A. Mackenzie, E. Suli and G. Warnecke [159].

A posteriori estimates taking into account the influence of the non-discretized part of the domain on the approximation error are considered in W. Dörfler and M. Rumpf [108]. Estimates in the L^∞-norm can be found, e.g., in S. W. Brady and A. R. Elcrat [56]. A posteriori estimates and adaptive finite elements methods for convection–diffusion problems were considered in C. Johnson [164] and R. Verfürth [300, 301]. In K. Eriksson and C. Johnson [117] and R. Verfürth [298, 299, 302] adaptive methods for evolutionary problems were studied. A posteriori error estimates for anisotropic meshes are presented in K. G. Siebert [283] and G. Kunert [181, 182]. It is hardly possible to overview all publications related to the various aspects of residual type methods. In I. Babuška, R. Duran and R. Rodríguez [22], G. F. Carey and D. L. Humphrey [75], J. T. Oden, L. Demkowicz, W. Rachowicz and T. A. Westermann [216], R. E. Ewing [119], D. W. Kelly, J. R. Gago, O. C. Zienkiewicz and I. Babuška [172], B. I. Wohlmuth and R. H. W. Hoppe [312] readers will find an exposition of some of them and more references.

In an abstract form (for nonlinear mappings), residual type a posteriori estimates were considered in J. Pousin and J. Rappaz [226, 227] and R. Verfürth (see, e.g., [297, 298]). The book [298] contains a systematic consideration of this question.

3.2 Implicit residual methods

In the so-called implicit residual methods, errors are presented as solutions of auxiliary boundary value problems. For example, let u_h be a conforming approximation of the problem (2.4)–(2.5). Then, we have the relation

$$\int_\Omega \nabla(u - u_h) \cdot \nabla w \, dx = \langle r(u_h), w \rangle \quad \forall w \in V_0 \tag{3.10}$$

that shows that $e = u - u_h$ is a solution of the problem

$$\Delta e + r(u_h) = 0 \quad \text{in } \Omega, \tag{3.11}$$

$$e = 0 \quad \text{on } \partial\Omega, \tag{3.12}$$

where $r(u_h) = f + \operatorname{div} \nabla u_h \in H^{-1}(\Omega)$ and (3.11) is understood in the sense of distributions. Formally, this idea can be extended to a wide class of linear problems. Indeed, if $\mathcal{A} : X \to Y$ is a linear operator for which we consider the problem

$$\mathcal{A}u + f = 0 \quad \text{in } \Omega, \tag{3.13}$$

$$u = 0 \quad \text{on } \partial\Omega, \tag{3.14}$$

then, for any $v \in X$,

$$\mathcal{A}e + r(v) = 0 \quad \text{in } \Omega, \qquad r(v) = f + \mathcal{A}v, \tag{3.15}$$

$$e = 0 \quad \text{on } \partial\Omega. \tag{3.16}$$

However, the evaluation of the error by directly using (3.15)–(3.16) may be a difficult task, because the functional $r(v)$ is a distribution. Moreover, the accuracy of an approximate solution of (3.15)–(3.16) obviously affects the accuracy of the error estimation, so that a new error estimation problem arises.

In the literature focused on practical applications of this method, it is often suggested to split the functional $\langle r(u_h), w\rangle$ into a number of functionals defined as solutions of local subproblems.

For example, in R. E. Bank and A. Weiser [38]) error indicators are constructed using local boundary value problems with data defined by residuals and interelement jumps (see also R. Duran and R. Rodríguez [110]). In the *equilibrated residual method* (see, e.g., M. Ainsworth and J. T. Oden [5]) local boundary value problems are constructed on each element using the residuals and suitable Neumann conditions on the element boundary. Local Dirichlét and Neumann problems on patches are also used in R. Verfürth [300, 301]. Applications to nonlinear problems are discussed in, e.g., C. Carstensen, R. Klose and A. Orlando [89].

Finally, we note that implicit type methods are often used for the indication of local errors. Concerning a posteriori methods developed to evaluate local errors of FEM approximations, we refer to, e.g., M. Ainsworth and J. T. Oden [5], I. Babuška and T. Strouboulis [32] and to papers by I. Babuška, F. Ihlenburg, A. Mathur, T. Strouboulis, S. K. Gangaraj, C. S. Upadhyay [24, 33, 34, 35], E. Stein and S. Ohnimus [284], R. Verfürth [300, 301], M. Ainsworth, J. T. Oden and C. Y. Lee [6].

3.3 A posteriori estimates based on post-processing

Post-processing methods exploit certain a priori known properties of exact solutions. For example, it is often known that fluxes (or other parts of the true solution) possess (at least locally) some extra regularity. Another a priori known information is that the flux must be in equilibrium with external forces.

In general terms, for finite element approximations typical situation is as follows. Consider a conforming approximation $u_h \in V_h$. Usually, the function Λu_h (where Λ is a certain linear operator, e.g., ∇) lies in a rather "wide" space U. However, a priori estimates of the solution guarantee that $\Lambda u \in \mathbb{U} \subset U$. Assume that we have a computationally inexpensive continuous mapping G such that

$$\mathrm{G}(\Lambda v_h) \in \mathbb{U} \qquad \forall v_h \in V_h.$$

Then, we may hope that the function $\mathrm{G}(\Lambda v_h)$ (that possesses a priori known regularity properties) would be much closer to Λu than Λv_h. These arguments form the basis of various *post-processing algorithms* that change a computed solution in accordance with some a priori knowledge of properties of the exact solution. Algorithms of such a type are widely used in advanced numerical methods. If the error caused by violations of a priori regularity properties is dominant and the post-processing operator G is properly constructed, then

$$\|\mathrm{G}\Lambda u_h - \Lambda u\| \ll \|\Lambda u_h - \Lambda u\| .$$

In this case, the explicitly computable norm $\|\mathrm{G}\Lambda u_h - \Lambda u_h\|$ can be used to evaluate upper and lower bounds of the error. Indeed, assume that there is a positive number $\alpha < 1$ such that for the mapping Λ the estimate

$$\|\mathrm{G}\Lambda u_h - \Lambda u\| \leq \alpha \|\Lambda u_h - \Lambda u\| \tag{3.17}$$

holds. Then, we have

$$\begin{aligned}(1-\alpha)\|\Lambda u_h - \Lambda u\| &\leq \|\Lambda u_h - \Lambda u\| - \|\mathrm{G}\Lambda u_h - \Lambda u\| \leq \\ &\leq \|\mathrm{G}\Lambda u_h - \Lambda u_h\| \leq \|\mathrm{G}\Lambda u_h - \Lambda u\| + \|\Lambda u_h - \Lambda u\| \leq \\ &\leq (1+\alpha)\|\Lambda u_h - \Lambda u\| .\end{aligned} \tag{3.18}$$

Thus, if $\mathrm{G}\Lambda u_h$ is much closer to Λu than Λu_h (i.e., if $\alpha \ll 1$), then

$$\boxed{\|\Lambda u_h - \Lambda u\| \simeq \|\mathrm{G}\Lambda u_h - \Lambda u_h\| .} \tag{3.19}$$

In this case, the right-hand side of (3.19) can be used as an error indicator. The quality of this indicator is the better the smaller the value of α.

Usually, solutions of elliptic problems with smooth coefficients are regular inside Ω what suggests an idea to project fluxes of approximate solutions to a set of more regular functions (see, e.g., the paper by J. H. Bramble and A. H. Schatz [63] which is one of the earliest publications in this area). Regularization of fluxes (or other solution components) leads to a variety of error indicators. Usually, they are obtained by means of averaging procedures. One of the most well-known methods was suggested in O. C. Zienkiewicz and J. Z. Zhu [320, 321]. Simple and efficient error indicators based upon gradient averaging were in the focus of many publications (see, e.g., M. Ainsworth, J. Z. Zhu, A. W. Craig and O. C. Zienkiewicz [7], I. Babuška and R. Rodríguez [30], B. Boroomand and O. C. Zienkiewicz [53], R. Duran, M. A. Muschietti and R. Rodríguez [109], R. Rodríguez [276], O. C. Zienkiewicz and J. Z. Zhu [322, 323], O. C. Zienkiewicz, B. Boroomand and J. Z. Zhu [318]). In C. Carstensen and S. Bartels [83], it is shown that each averaging technique leads to a certain a posteriori error estimate. Estimates based upon global averaging of gradients were considered in B.-O. Heimsund, X.-C. Tai and J. Wang [154]. In J. Wang [308], a generalization of the patch recovery technique of O. C. Zienkiewicz and J. Z. Zhu is considered. It consists of approximating an approximate solution data by a certain smooth function that is computed by the least-squares method.

Mathematically, these indicators were justified by means of the *superconvergence* phenomenon. Probably earliest results on superconvergence were established in the papers by L. A. Oganesjan and L. A. Ruchovets [220] and M. Zlámal [324, 325]. Also, we refer to J. Brandts [64], S.-S. Chow, G.F. Carey and R.D. Lazarov [98], R.E. Ewing, R.D. Lazarov and J. Wang [120], B.-O. Heimsund, X.-C. Tai and J. Wang [154], I. Hlaváček and M. Křížek [156], M. Křížek and P. Neittaanmäki [180, 208], R. Lazarov [193], R. Verfürth [298], J. Wang [308]. In L. Wahlbin [307], readers can find a detailed exposition of the subject and other references. Surveys on superconvergence are presented, e.g., in M. Křížek and P. Neittaanmäki [179] and J.R. Whiteman and G. Goodsell [311].

A posteriori estimates of the types discussed above exploit specific features of the approximations and the method used. Usually, they are justified only for Galerkin approximations.

Another group of methods is based on post-processing procedures projecting certain components of an approximate solution (e.g., fluxes or stresses) onto the set that contain the respective components of the exact solution. For example, if it is known that the divergence of the exact flux is equal to a given function f, then it is reasonable to post-process our solution in such a way that this relation is exactly (or "almost exactly") satisfied. Thus, post-processing procedures may be sharp or only approximate and lead to a variety of methods. Let us shortly discuss some of them using again our basic problem (2.4)–(2.5). If Π_h is a sharp "equilibration" mapping that transforms ∇u_h into $p_h \in Q_f$, then we apply (2.7) and find that

$$\|\nabla(u - u_h)\| \leq \|\nabla u_h - p_h\|,$$

so that the error is bounded from above by the error in the relation $p = \nabla u$ that must hold for the exact flux. Certainly, if the "equilibration" is only approximate, then the reliability of the upper bound is lost.

Various approaches based on the idea of "equilibration" are presented in, e.g., M. Ainsworth and J.T. Oden [5], D. Braess [57], D.W. Kelly [171], P. Ladevèze and D. Leguillon [186] (the authors show that if the equilibrium equation is exactly satisfied, then the estimate is reduced to the error in constitutive relations; extensions of the method to some other problems can be found in P. Ladevèze, J.-P. Pelle and Ph. Rougeot [187], P. Ladevèze and Ph. Rougeot [188], L. Gallimard, P. Ladevèze and J.-P. Pelle [137], P. Coorevits, J.-P. Dumeau and J.-P. Pelle [103]), E. Stein and S. Ohnimus [284], E. Stein, F.J. Bartold, S. Ohnimus and M. Schmidt [285]).

Post-processing procedures based upon an equilibration of recovered stresses were considered in B. Boroomand and O.C. Zienkiewicz [52]. They result in stress fields that satisfy an equilibrium condition in a weak form. A method of equilibration is presented in P. Destuynder and B. Métivet [105]. This paper also contains numerical tests, in which a posteriori error estimates obtained by equilibration are compared with other estimates.

3.4 A posteriori methods using adjoint problems

In the recent decades, a new approach to a posteriori error control of finite element approximations was developed based on attracting *adjoint boundary value problems*. Probably, first results in the error analysis obtained with the help of adjoint problems are related to the works of J.-P. Aubin [14] and J. Nitsche [213], where a way of deriving a priori rate convergence estimates in weaker norms was suggested.

Among a posteriori approaches using adjoint problems we first of all refer to the so-called *dual-weighted residual method*. One of the advantages of this method is that it allows to avoid difficulties with the evaluation of the interpolation constants.

A reader will find a detailed exposition of the method and applications to various problems in R. Rannacher [231] and W. Bangerth and R. Rannacher [36]. Also, we recommend R. Becker and R. Rannacher [44], C. Johnson and A. Szepessy [168], P. Houston, R. Rannacher and E. Süli [161], and R. Rannacher and F. T. Suttmeier [233, 234]).

Adjoint problems are usually used in the so-called *goal-oriented* a posteriori error estimates that measure errors in terms of special "goal-oriented" quantities instead of global energy norms (see, e.g., W. Bangerth and R. Rannacher [36], J. T. Oden and S. Prudhomme [217], J. Peraire and A. T. Patera [223] and the literature cited in these publications). The idea of this approach can be explained on the paradigm of the problem

$$\int_\Omega (A\nabla u \cdot \nabla w - fw)dx = 0 \quad \forall w \in V_0 := \overset{\circ}{H}{}^1(\Omega),$$

where A is a positive definite real matrix and $f \in L^2(\Omega)$ is a given function. Let $u_h \in V_0$ be an approximate solution computed on the mesh $\mathcal{T}_h$. We need to estimate the quantity $\langle \ell, u - u_h \rangle$, where $\ell \in V_0^*$ is a given vector. Typically, ℓ is an integral functional localized in a certain part of Ω. Define u_ℓ by the relation

$$\int_\Omega A^\star \nabla u_\ell \cdot \nabla w dx = \langle \ell, w \rangle \qquad \forall w \in V_0,$$

in which $A^\star$ is the matrix adjoint to A. Then,

$$\begin{aligned}\langle \ell, u - u_h \rangle &= \\ &= \int_\Omega A^\star \nabla u_\ell \cdot \nabla(u - u_h)dx = \int_\Omega (f \cdot u_\ell - A\nabla u_h \cdot \nabla u_\ell)dx = E(u_\ell, u_h). \quad (3.20)\end{aligned}$$

Hence, $\langle \ell, (u - v) \rangle$ can be easily estimated provided that u_ℓ is defined. Certainly, in practice u_ℓ is replaced by an approximation $u_{\ell\tau}$ computed on an *adjoint mesh* $\mathcal{T}_\tau$ (which may not coincide with $\mathcal{T}_h$). However, if $u_{\ell\tau}$ is a sharp approximation of u_ℓ, then the quantity $E(u_{\ell\tau}, u_h)$ could be a good indicator of the error. To obtain a sharper indicator, we rewrite (3.20) in the form

$$\langle \ell, u - u_h \rangle = E_1(u_h, u_{\ell\tau}) + E_2(u, u_h, u_\ell, u_{\ell\tau}), \tag{3.21}$$

where $E_1(u_h, u_{\ell\tau}) := \int_\Omega (fu_{\ell\tau} - A\nabla u_h \cdot \nabla u_{\ell\tau})dx$ is a directly computable functional and

$$E_2(u, u_h, u_\ell, u_{\ell\tau}) := \int_\Omega A(\nabla u - \nabla u_h) \cdot (\nabla u_\ell - \nabla u_{\ell\tau})dx.$$

In [177, 209], it was suggested to replace unknown functions ∇u and ∇u_ℓ by averaged gradients $G_h \nabla u_h$ and $G_\tau \nabla u_{\ell\tau}$, where G_h and G_τ are the respective averaging operators. It is proved that under the standard assumptions that guarantee superconvergence of the primal and adjoint approximations such a replacement leads to a higher order error. In such a case, the computable quantity

$$\widetilde{E}_2(u, u_h, u_\ell, u_{\ell\tau}) := \int\limits_\Omega A(G_h \nabla u_h - \nabla u_h) \cdot (G_\tau \nabla u_{\ell\tau} - \nabla u_{\ell\tau}) dx$$

can be used instead of E_2. Certainly, the quantity $\langle \ell, u - v \rangle$ cannot completely characterize the error because it vanishes if $u - v$ is orthogonal to ℓ. Therefore, it is desirable to compute the estimates for several different functionals ℓ.

3.5 Literature comments

In this section, we give brief comments on the literature devoted to a posteriori error estimates for some particular classes of problems or approximation methods.

We begin with **a posteriori error control methods in the theory of viscous fluids**. Finite element and other methods for viscous flow problems are discussed in many publications. A systematic exposition can be found in, e.g., M. Feistauer [122], V. Girault and P. A. Raviart [141], R. Rannacher [232], and R. Temam [293]. Various approaches to the numerical solution of Stokes and Navier–Stokes problems are studied in, e.g., R. Glowinski and O. Pironneau [143], L. P. Franca, J. Karam Filho, A. F. D. Loula and R. Stenberg [123], S. Turek [295], A. Bermúdez, R. Durán and R. Rodríguez [48].

Residual type a posteriori methods for this class of problems are considered in the book by R. Verfürth [298]. A posteriori analysis of approximations computed with the help of the backward Euler scheme is presented in C. Bernardi and R. Verfürth [49]. A posteriori error estimators for finite element approximations of the Stokes problem were obtained in R. E. Bank and B. D. Welfert [39] and R. Verfürth [296]. Error indicators for the Navier–Stokes equations in stream function and vorticity formulation are discussed in M. Amara, M. Ben Younes and C. Bernardi [10]. In D. Kay and D. Silvester [170], various a posteriori estimators are investigated for stabilized mixed approximations of the Stokes problem.

Adaptive methods and a posteriori estimates in computational fluid dynamics are exposed in, e.g., C. Johnson, R. Rannacher and M. Boman [167], J. T. Oden, W. Wu and M. Ainsworth [219], R. Rannacher [232], T. Strouboulis and J. T. Oden [290].

A posteriori error estimators for some quasi-newtonian fluids were considered in C. Padra [222] and for combined fluid-solid systems in A. Bermúdez, R. Durán and R. Rodríguez [48]. J. Wang and X. Ye [309] investigated error indicators based on superconvergence of finite element approximations for Stokes and Navier–Stokes equations. In Section 5.4, we will discuss a posteriori error estimates for viscous flow problems that follow from the functional approach.

Approximation methods for **Maxwell's equation** were investigated by many authors (see, e.g., G. Haase, M. Kuhn and U. Langer [147], R. Hiptmair [155], P. Monk [203] and the literature cited in these publications). A posteriori estimates were obtained, e.g., in R. Beck, R. Hiptmair, R. H. W. Hoppe and B. Wohlmuth [41] in the framework of the residual approach and in D. Braess and J. Schöberl [59] with the help of equilibrated approach. Various estimators for the Maxwell equation are compared in S. Cochez-Dhondt and S. Nicaise [101].

In the recent decades **mixed and hybrid finite element methods** gain high popularity among numerical analysts. A posteriori error estimates for mixed finite element methods were considered in A. Alonso [9], I. Babuska and G. N. Gatica [23], C. Carstensen [77], C. Carstensen and G. Dolzmann [82], and C. Carstensen and R. H. W. Hoppe [86]. In G. N. Gatica [138], the efficiency of residual-based a posteriori error estimators for mixed approximations is studied. A posteriori estimates for the approximations using Raviart–Thomas elements are presented in D. Braess and R. Verfürth [60]. A posteriori estimates based on superconvergence were analyzed by J. H. Brandts [64]. A posteriori error estimators for mixed finite element methods in linear elasticity are investigated in M. Lonsing and R. Verfürth [196] and for mixed approximations of the Stokes problem in X. Cheng, W. Han and H. Huang [96] and M. Farhloul, S. Nicaise, L. Paquet [121]. Dual-mixed adaptive finite element error control for Lamé and Stokes equations were studied in C. Carstensen, P. Causin and R. Sacco [81]. In B. Engelmann, R. H. W. Hoppe, B. Wolmuth, Yu. Kuznetsov, Yu. Iliash and Yu. Vasilevskii [113], adaptive methods for hybrid finite element approximations were considered. New mixed approximations of diffusion problems on highly distorted meshes were suggested in [184, 185]. These papers also contain an analysis of the respective errors. Further discussion of a posteriori estimates for mixed approximations is presented in Section 5.8.

Another rapidly developing group of methods is related to **nonconforming approximations** of boundary value problems. For this group, a posteriori error estimation methods started receiving serious attention later than for "classical" FEM and, therefore, they are relatively less developed. However, in recent years interesting researches appeared in the area (see, e.g., C. Carstensen and R. H. W. Hoppe [87], C. Carstensen, J. Hu, and A. Orlando [88]). A posteriori estimates for *Discontinuous Galerkin* approximations of elliptic type equations were studied in R. Becker, P. Hansbo and M. G. Larsson [42] and R. Bustinza, G. N. Gatica and B. Cockburn [72], where a modification of the residual based estimate for the energy norm of the error was suggested. In P. Castillo [93], a posteriori estimates in L^2-norm were derived for the so-called "local DG method" applied to an elliptic boundary value problem. A posteriori error estimates for DG approximations were also obtained for other classes of problems. In particular, in S. Sun and M. F. Wheeler [291] time-dependent (transport) equations were considered and in P. Houston, I. Perugia and D. Schotzau [160] the authors investigated elliptic problems of the Maxwell type. The paper by J. Ma and H. Brunner [197] deals with a posteriori error estimates for DG approximations of integral equations. In A. Ern and J. Proft [118], a posteriori estimates were obtained for DG approximations of the convection–diffusion equation. In Section 5.9, we discuss an approach to a posteriori error control of DG approximations based on functional a posteriori estimates.

Concerning a posteriori estimates for *finite volume methods* we refer to, e.g., Y. Achdou, C. Bernardi and F. Coquel [1], A. Bergam, Z. Mghazli and R. Verfürth [47], C. Carstensen, R. Lazarov and S. Tomov [91], V. Jovanovic and C. Rohde [169], S. Nicaise [211].

Adaptive methods and a posteriori estimates for **optimal control problems** are considered in R. Becker, H. Kapp and R. Rannacher [43], R. H. W. Hoppe, Y. Iliash, C. Iyyunni and N. H. Sweilam [157], and R. Becker and B. Vexler [45]. In these publications, a reader will find overviews of the results and other references. An approach to a posteriori error estimation of optimization problems based on functional error majorants was considered in [132, 133].

Finally, we conclude this overview by comments on a posteriori estimates for **nonlinear boundary value problems**. General residual type approach is exposed in the already cited book by R. Verfürth [298]. Certain particular classes of nonlinear problems were analyzed by many authors. For example, residual type estimates for nonlinear diffusion–convection problems were obtained in J. Medina, M. Picasso and J. Rappaz [199] and for nonlinear parabolic problems in R. Verfürth [299]. Adaptive methods and a posteriori estimates for problems in plasticity theory were investigated in C. Carstensen, R. Klose, and A. Orlando [89], W. Han [149, 150], W. Han and D. C. Reddy [151], R. Rannacher and F. T. Suttmeier [234], S. Repin and L. S. Xanthis [271], M. Schulz and W. L. Wendland [280], and E. Stein et al. [285]. A posteriori estimates for variational inequalities related to problems with obstacles were analyzed in M. Ainsworth, J. T. Oden and C. Y. Lee [6], C. Bartels and C. Carstensen [40], V. Bostan and W. Han [54], D. Braess [58], Z. Chen and R. H. Nochetto [95], D. French, S. Larsson and R. H. Nochetto [125], R. H. W. Hoppe and R. Kornhuber [158], R. Kornhuber [176], A. Vesser [303]. An overview of a posteriori error estimation methods for free boundary problems is presented in R. Nochetto [214]. In V. Bostan, W. Han and B. D. Reddy [55], a posteriori error estimates for a class of variational inequalities were derived by methods of the convex analysis and duality theory. These estimates differ from the functional a posteriori estimates derived for variational inequalities and other nonlinear problems that we discuss in Section 6.4.

4 Functional approach to a posteriori error estimation

Functional a posteriori estimates were obtained as a tool for solving the following mathematical problem:

> **Given the data (coefficients, domain, boundary conditions) of a boundary value problem and a function v from the respective energy space V compute radius ρ of a ball $\mathcal{B}(v,\rho)$ centered at v that contains the exact solution u.**

More precisely, the task is to find a method able to verify that for $\rho_1 \leq \rho_2$

$$u \notin \mathcal{B}(v,\rho_1) \quad \text{and} \quad u \in \mathcal{B}(v,\rho_2).$$

We say that the method is *sharp* if it is possible to find ρ_1 and ρ_2 such that $\rho_2 - \rho_1 \leq \epsilon$ for any given ϵ. Above relations are verifiable provided that for a class of problems considered we have established the estimate

$$M_\ominus(\mathcal{D}, v) \leq \|u - v\|_V \leq M_\oplus(\mathcal{D}, v) \qquad \forall v \in V, \tag{4.1}$$

where $\mathcal{D}$ denotes the set of known data and the functionals $M_{\oplus}$ (error majorant) and $M_{\ominus}$ (error minorant) are *directly computable*. If the estimate (4.1) is obtained, then the above *Main Error Control Problem* is solved for the respective class of boundary value problems. This estimate establishes two-sided functional a posteriori estimates that give *guaranteed error bounds for conforming approximations of all the types*. Since the estimates show deviations from exact solutions we also call them *deviation estimates*.

It should be outlined that if a boundary value problem is considered in the context of *reliable computer simulation*, then such estimates are indeed required. Therefore, for a class of boundary value problems such estimates are not less important than asymptotic rate convergence or regularity estimates.

In many cases, it is also required to find guaranteed upper bounds of local norms or other nonnegative quantities of the deviation $v - u$. Estimates of such a type have the form

$$\Phi(u - v) \leq M_{\Phi}(\mathcal{D}, v) \qquad \forall v \in V, \tag{4.2}$$

where $\Phi : V \to \mathbb{R}_{+}$ is a given functional (we consider such estimates in Section 5.7). It is clear that the majorant M (and M_{Φ}) must be not only computable but also continuous in the sense that

$$M(\mathcal{D}, v) \to 0 \qquad \text{if } v \to u \quad \text{in } V.$$

At the end of the 20th century estimates of the type (4.1) have been derived by purely functional methods that are usually applied in the theory of partial differential equations (see [236–241, 271, 272]). A systematic exposition of these results is given in [209, 242, 243]). These estimates can be derived by two different methods: variational (which is applicable for problems having a variational formulation) and nonvariational (which is based on transformations of integral identities). Below we discuss both methods developed in the above cited papers for the derivation of functional a posteriori estimates.

4.1 Variational method

The variational method consists of the following main steps. First, it is required to obtain an estimate of the difference $v - u$ in terms of the difference $J(v) - J(u)$, where J is the respective variational functional. Then, $J(u)$ is estimated from below by $I^*(y^*)$, where I^* is the functional of the dual problem.[1] As a result, an upper bound is presented as a certain nonnegative functional defined on the pair (v, y^*). Regrettably, in many important cases, the dual variable y^* must lie in an affine set defined by differential relations. Therefore, on the last step it is required to extend the estimate to a much wider set of y^* that admits simple approximations.

Let us first consider the simplest variant of the above scheme using again the problem (2.4)–(2.5) as an example. In this case, the first step results in (2.8). By attracting the dual variational problem we obtain (2.7), which we now rewrite in a slightly different form

$$\|\nabla(v - u)\| \leq \|\nabla v - q^*\|, \qquad q^* \in Q_f.$$

[1] Traditionally, the elements of a dual space are marked by stars. Hereafter, we follow this rule. It also helps to make the structure of the estimates clearer.

Take arbitrary $y^* \in L^2(\Omega)$. Then,

$$\|\nabla(v-u)\| \leq \|\nabla v - y^*\| + \inf_{q^* \in Q_f} \|y^* - q^*\|.$$

Let us show that

$$\inf_{q^* \in Q_f} \|y^* - q^*\| \leq |\!|\!|\operatorname{div} y^* + f|\!|\!| \qquad y^* \in L^2(\Omega), \tag{4.3}$$

$$\inf_{q^* \in Q_f} \|y^* - q^*\| \leq C_\Omega \|\operatorname{div} y^* + f\| \qquad y^* \in H(\Omega, \operatorname{div}). \tag{4.4}$$

To prove these estimates we consider an auxiliary problem

$$\Delta\eta + f + \operatorname{div} y^* = 0 \qquad \text{in } \Omega, \qquad \eta = 0 \text{ on } \partial\Omega,$$

whose solution satisfies the relation

$$\int_\Omega \nabla\eta \cdot \nabla w \, dx = \int_\Omega (f + \operatorname{div} y^*) w dx = \int_\Omega (fw - y^* \cdot \nabla w) dx \qquad \forall w \in V_0.$$

Hence

$$\int_\Omega (\nabla\eta + y^*) \cdot \nabla w \, dx = \int_\Omega fw \, dx \qquad \forall w \in V_0$$

and we observe that $\bar{q}^* := \nabla\eta + y^* \in Q_f$. Also, we have

$$\|\nabla\eta\| \leq |\!|\!|\operatorname{div} y^* + f|\!|\!| := \sup_{w \in V_0} \frac{\int_\Omega (y^* \cdot \nabla w - fw) dx}{\|\nabla w\|}.$$

Therefore,

$$\inf_{q^* \in Q_f} \|y^* - q^*\| \leq \|y^* - \bar{q}^*\| = \|\nabla\eta\| \leq |\!|\!|\operatorname{div} y^* + f|\!|\!|.$$

If y^* has a square summable divergence, then

$$|\!|\!|\operatorname{div} y^* + f|\!|\!| = \sup_{w \in V_0} \frac{\int_\Omega (\operatorname{div} y^* + f) w dx}{\|\nabla w\|} \leq C_\Omega \|\operatorname{div} y^* + f\|,$$

where C_Ω is the constant in the Friederichs–Steklov inequality for the domain Ω. Thus, for any $y^* \in H(\Omega, \operatorname{div})$ we obtain

$$\boxed{\|\nabla(v-u)\| \leq \|\nabla v - y^*\| + C_\Omega \|\operatorname{div} y^* + f\|.} \tag{4.5}$$

This estimate is one of the simplest a posteriori estimates of the functional type, which, however, preserves all the main features (we discuss them in Section 4.3).

4.2 Method of integral identities

Modern theory of partial differential equations considers integral relations as one of the major mathematical objects. Integral identities define generalized solutions of differential equations and provide a basis for their a priori analysis (see, e.g., D. Gilbarg and N. S. Trudinger [140], O. Ladyzhenskaya and N. Uraltseva [191]). Let us show that estimates of the type (4.1) and (4.2) can be also derived from integral identities. Originally, this *modus operandi* was suggested in [243], where functional a posteriori estimates for linear elliptic problems were derived by means of transformations of the respective integral identities. It should come as no surprise that they coincide with those earlier derived by the variational method. Later, the method was applied to parabolic problems (see [134, 244]) and mixed approximations of elliptic problems [256].

As in the previous cases, we explain the method on the paradigm of the problem (2.4)–(2.5). By (3.1), we have

$$\int_\Omega \nabla(u-v)\cdot\nabla w dx = \int_\Omega (fw - \nabla v\cdot\nabla w)\, dx.$$

Now, we transform this identity by means of the functional relation

$$\int_\Omega (w \operatorname{div} y^* + \nabla w\cdot y^*)\, dx = 0 \qquad \forall w\in V_0,$$

where $y^*\in H(\Omega,\operatorname{div})$. Set $w=u-v$, then we obtain

$$\|\nabla(u-v)\|^2 \le \|\nabla v - y^*\|\,\|\nabla(u-v)\| + \|f+\operatorname{div} y^*\|\,\|u-v\|.$$

Hence,

$$\|\nabla(u-v)\| \le \|\nabla v - y^*\| + C_\Omega\|f+\operatorname{div} y^*\|$$

and, therefore, the estimate (4.5) is derived by another method.

4.3 Properties of functional a posteriori estimates

First of all, we note that both terms in the right-hand side of (4.5) have a clear sense: they present measures of the errors in the basic relations

$$p^* = \nabla u, \qquad \text{and} \qquad \operatorname{div} p^* + f = 0 \quad \text{in}\,\Omega$$

that jointly form the equation.

The multipliers of the terms $\|\nabla v - y^*\|$ and $\|f+\operatorname{div} y^*\|$ are *sharp* in the sense that the estimate is not true if the first term has a factor less than 1 and the second one less than C_Ω. Indeed, set in (4.5) $v=0$ and $y^*=0$. Then, we obtain the energy estimate for the generalized solution

$$\|\nabla u\| \le C_\Omega\|f\|.$$

Therefore, no constant less than C_Ω can be a multiplier of the second term. Next, if set $y^* = \nabla u$, then (4.5) holds as the equality, so that a constant less than 1 cannot be a multiplier of the first term. Also, we observe that (4.5) is *sharp* in the sense that the set of admissible y^* contains a function such that it holds as the equality.

By means of the algebraic Young's inequality $2ab \leq \beta a^2 + \frac{1}{\beta} b^2$ valid for $\beta > 0$ we rewrite this estimate in the form of a quadratic functional

$$\|\nabla(u - v)\|^2 \leq (1+\beta)\|\nabla v - y^*\|^2 + \frac{1+\beta}{\beta} C_\Omega^2 \|\operatorname{div} y^* + f\|^2. \tag{4.6}$$

This form is more convenient because it makes it possible to apply well-known minimization methods for finding an "optimal" function y^* that makes the majorant close to the true value of the error.

Denote the right-hand side of (4.6) by $M_\oplus(v, y^*, \beta, C_\Omega, f)$. This functional deserves a special consideration. For any $v \in V_0$, it provides an upper bound for the norm of the deviation of v from u. Therefore, it is natural to call it the *deviation majorant*. It is worth noting, that $M_\oplus$ gives *another variational formulation* for the original boundary value problem (2.4)–(2.5): for a given $\beta > 0$ find $u \in V_0$ and $p^* \in H(\Omega, \operatorname{div})$ such that

$$M_\oplus(u, p^*, \beta, C_\Omega, f) = \inf_{v \in V_0,\; y \in H(\Omega, \operatorname{div})} M_\oplus(v, y^*, \beta, C_\Omega, f).$$

It is not difficult to prove that the minimal value of $M_\oplus(v, y^*, \beta, C_\Omega, f)$ is zero for any positive β, and that it is attained if and only if $v = u$ and $y^* = A\nabla u$.

4.4 How to use functional a posteriori estimates

There are several ways to use functional a posteriori estimates in practice.

1. Direct substitution of recovered gradients on the mesh $\mathcal{T}_h$. Let $u_h \in V_h$, then

$$p_h^* := \nabla u_h \in L^2(\Omega, \mathbb{R}^d), \quad p_h^* \notin H(\Omega, \operatorname{div}).$$

Use an averaging operator $G_h : L^2(\Omega, \mathbb{R}^d) \to H(\Omega, \operatorname{div})$ and have a directly computable estimate

$$\|\nabla(u - u_h)\| \leq \|\nabla u_h - G_h p_h^*\| + C_\Omega \|\operatorname{div} G_h p_h^* + f\|.$$

2. Quantitative form of Runge's rule. Let $u_1, u_2, \ldots, u_k, \ldots$ be a sequence of approximations on meshes $\mathcal{T}_{h_k}$. Compute $p_k^* := \nabla u_k$ and average it by G_k. Then the accuracy of the approximation u_{k-1} can be measured by the estimate

$$\|u - u_{k-1}\| \leq \|\nabla u_{k-1} - G_k p_k^*\| + C_\Omega \|\operatorname{div} G_k p_k^* + f\|.$$

This estimate involves approximate solutions computed on two consequent meshes and, in fact, gives *a quantitative form of Runge's rule.*

3. Minimization of $M_\oplus$ with respect to y^*. Select a certain subspace Y_τ in $H(\Omega, \mathrm{div})$. Generally, Y_τ may be constructed on another mesh $\mathcal{T}_\tau$ and with the help of different trial functions. Then

$$\|\nabla(u-u_h)\| \leq \inf_{y_\tau^* \in Y_\tau} \left\{ \|\nabla u_h - y_h^*\| + C_\Omega \,\|\mathrm{div}\, y_h^* + f\| \right\}.$$

The wider $Y_\tau \subset H(\Omega, \mathrm{div})$ the sharper is the upper bound.

4. Using partially equilibrated fluxes. Assume that by post-processing of ∇u_h we have constructed a function $y_{\bar f}^*$ such that $\mathrm{div}\, y_{\bar f}^* + \bar f = 0$ and $\bar f$ is sufficiently close to f in L^2-norm (so that the value of $C_\Omega \|f - \bar f\|$ is smaller than the desired accuracy level). Set $y = y_{\bar f}^* - \mathrm{curl}\tau$ in (4.5), where τ is a vector-valued function in $H(\Omega, \mathrm{curl})$. Then, we obtain the estimate

$$\|\nabla(u-v)\| \leq \|\nabla v + \mathrm{curl}\tau - y_{\bar f}^*\| + C_\Omega \|f - \bar f\|,$$

where τ can be used to reduce the value of the first term in the right-hand side.

5. Minimization of $M_\oplus$ with respect to both variables v and y^*. In principle, we can select certain sequences of subspaces for both primal and dual variables, i.e., $\{V_{hk}\} \in V_0$ and $\{Y_{hk}\} \in H(\Omega, \mathrm{div})$. Solve finite dimensional problems

$$\inf_{\substack{v \in V_{hk},\, \beta > 0, \\ y^* \in Y_{hk}}} M_\oplus(v, y^*, \beta, C_\Omega, f) = M_\oplus(v_k, y_k^*, \beta_k, C_\Omega, f) := \gamma_k.$$

If the subspaces are limit dense in the respective functional spaces, then it is possible to prove that approximate solutions (v_k, y_k^*) tend to (u, p^*) and the sequence of numbers γ_k tends to zero. A more detailed discussion of the questions related to numerical implementation of functional a posteriori estimates is contained in [129, 209, 239, 257, 261].

5 Functional a posteriori estimates for linear elliptic and parabolic problems

5.1 Abstract elliptic problem

In this section, we consider problems that are represented in the following general form: to find $u \in V_0 + u_0$ such that

$$(\mathcal{A}\Lambda u, \Lambda w) + \langle l, w \rangle = 0 \quad \forall w \in V_0. \tag{5.1}$$

Here V_0 is a subspace of a reflexive Banach space V, Λ is a linear bounded operator acting from V to a Hilbert space U with scalar product $(\cdot\,,\cdot)$, $l \in V_0^*$, and $\mathcal{A} \in \mathcal{L}(U, U)$

is a self-adjoint operator. In this section, $\| \, \|$ stands for the norm in U. We assume that V is compactly embedded in U and the operators Λ and $\mathcal{A}$ satisfy the relations

$$c_1\|y^*\|^2 \leq (\mathcal{A}y^*, y^*) \leq c_2\|y^*\|^2 \qquad \forall y^* \in U, \tag{5.2}$$

and

$$\|\Lambda w\| \geq c_3\|w\|_V \quad \forall w \in V_0, \tag{5.3}$$

with positive constants c_1, c_2, and c_3. We take l in the form $\langle l, w\rangle = (f, w)_{\mathcal{V}} + (g, \Lambda w)$, where $g \in U$ and $f \in \mathcal{V}$; $\mathcal{V}$ is a Hilbert space such that $V \subset \mathcal{V} \subset V_0^*$.

For our analysis, it is convenient to introduce two more spaces. The quantity $(\mathcal{A}y^*, y^*)^{1/2}$ determines a new norm $\| y^* \|$, which is equivalent to the original norm $\|y^*\| = (y^*, y^*)^{1/2}$. Another equivalent norm is

$$\| y^* \|_* = (\mathcal{A}^{-1}y^*, y^*)^{1/2},$$

where $\mathcal{A}^{-1}$ is the operator inverse to $\mathcal{A}$. The spaces Y and Y^* contain elements of U equipped with the norms $\| \cdot \|$ and $\| \cdot \|_*$, respectively.

Note that the solution u is the minimizer of the variational problem

$$J(u) = \inf_{e \in V_0 + u_0} J(w),$$

where $J(w) := \frac{1}{2} \| \Lambda w \|^2 + \langle l, w\rangle$.

First, we present the result (see [238, 242] for the proofs and discussions) that forms a basis of the variational approach to the derivation of functional a posteriori estimates for the problem (5.1).

Theorem. *For any $v \in V_0 + u_0$ and $q^* \in Q_l^*$, the following relations hold:*[1]

$$\| \Lambda(v - u) \|^2 + \| q^* - p^* \|_*^2 = \; 2\,(J(v) - I^*(q^*)), \tag{5.4}$$

$$\| \Lambda(v - u) \|^2 + \| q^* - p^* \|_*^2 = \; 2\,D(\Lambda v, q^*), \tag{5.5}$$

where $D(\Lambda v, q^*) := \frac{1}{2} \| \Lambda v \|^2 + \frac{1}{2} \| q^* \|_*^2 - (q^*, \Lambda v)$ is the so-called compound functional,

$$Q_l^* := \left\{ y^* \in Y^* \middle| (y^*, \Lambda w) + \langle l, w\rangle = 0 \quad \forall w \in V_0 \right\},$$

and

$$I^*(q^*) := (q^*, \Lambda u_0) - \frac{1}{2} \| q^* \|_*^2 + \langle l, u_0\rangle \qquad \forall q^* \in Q_l^*.$$

In [238, 242, 243] the following *general form* of a posteriori estimates for the above class of elliptic problems were derived.

$$\frac{1}{2} \| \Lambda(v - u) \|^2 \leq (1 + \beta)D(\Lambda v, y^*) + \left(1 + \frac{1}{\beta}\right)\frac{1}{2} \, | \, l + \Lambda^* y^* \, |^2, \tag{5.6}$$

[1] Note that (2.6) is a particular form of (5.5). Generalizations of (5.4), (5.5), and (5.8) to a wide class of convex variational problems are presented in [243], some nonconvex variational problems are discussed in [237].

where $\Lambda^* : U \to V^*$ is the operator adjoint to Λ, i.e.,

$$(y, \Lambda w) = \langle \Lambda^* y, w \rangle, \quad w \in V_0 \tag{5.7}$$

and

$$| l + \Lambda^* y^* | := \sup_{w \in V_0} \frac{\langle l + \Lambda^* y^*, w \rangle}{\| \Lambda w \|}.$$

Denote the right-hand side of (5.6) by $\mathcal{M}_\oplus(v, \beta, y^*)$. From the computational point of view, the majorant $\mathcal{M}_\oplus(v, \beta, y^*)$ has a drawback: it contains the norm $|\ \ |$, which is defined as a supremum over a functional space. In practice, it is impossible to directly compute such a quantity. However, in the majority of practically interesting cases we can find a computable upper bound of it. Assume that $l \in \mathcal{V}$ and $\langle l, w \rangle = (l, w)_\mathcal{V}$. Additionally, let

$$y^* \in Q^* := \{y^* \in Y^* | \ \Lambda^* y^* \in \mathcal{V}\},$$

so that for any $w \in V_0$, we have $\langle l + \Lambda^* y^*, w \rangle = (l + \Lambda^* y^*, w)_\mathcal{V}$. In this case,

$$| l + \Lambda^* y^* | = \sup_{w \in V_0} \frac{\langle l + \Lambda^* y^*, w \rangle}{\| \Lambda w \|} \leq \sup_{w \in V_0} \frac{\|l + \Lambda^* y^*\|_\mathcal{V} \|w\|_\mathcal{V}}{\| \Lambda w \|} \leq$$
$$\leq \|l + \Lambda^* y^*\|_\mathcal{V} c_1^{-1} \sup_{w \in V_0} \frac{\|w\|_\mathcal{V}}{\|\Lambda w\|} \leq \frac{c_4}{c_1} \|l + \Lambda^* y^*\|_\mathcal{V},$$

where c_4 is a constant in the estimate $\|w\|_\mathcal{V} \leq c\|\Lambda w\|$ that holds for any $w \in V_0$.

Now, the majorant is represented in the form

$$\| \Lambda(u - v) \|^2 \leq M_\oplus(v, \beta, y^*) = (1 + \beta) 2D(\Lambda v, y^*) + \left(1 + \tfrac{1}{\beta}\right) c^2 \|l + \Lambda^* y^*\|^2_\mathcal{V}, \tag{5.8}$$

where $c = c_4/c_1$ and the second term is expressed by the norm of the Hilbert space $\mathcal{V}$ (e.g., by L^2-norm). Note that if $l \in \mathcal{V}$, then $p^* \in Q^*$. Therefore, we set $y^* = p^*$ in (5.8) and observe that

$$\inf_{\substack{y^* \in Q^* \\ \beta > 0}} M_\oplus(v, \beta, y^*) \leq M_\oplus(v, \varepsilon, p^*) = (1 + \varepsilon) \| \Lambda(u - v) \|^2,$$

where $\varepsilon > 0$ may be taken arbitrarily small. Hence,

$$\inf_{\substack{y^* \in Q^* \\ \beta > 0}} M_\oplus(v, \beta, y^*) = \| \Lambda(u - v) \|^2 \leq \mathcal{M}_\oplus(v, \beta, y^*) \leq M_\oplus(v, \beta, y^*). \tag{5.9}$$

In other words, the majorant $M_\oplus$ is sharp and using this functional it is possible to find an upper bound of the error as accurate as required. Functional a posteriori estimate

(5.8) can be also obtained by transformations of the integral identity (5.1). By (5.7) we observe that for $y^* \in Q^*$ and $l \in \mathcal{V}$

$$\begin{aligned}(\mathcal{A}\Lambda(u-v),\Lambda(u-v)) &= \langle -l, u-v\rangle - (\mathcal{A}\Lambda v, \Lambda(u-v)) = \\ &= (y^* - \mathcal{A}\Lambda v, \Lambda(u-v)) - \langle l + \Lambda^* y^*, u-v\rangle \leq \\ &\leq (c\,\|l+\Lambda^* y^*\|_{\mathcal{V}} + \| (y^* - \mathcal{A}\Lambda v \|_*)\, \| \Lambda(u-v) \| \,. \quad (5.10)\end{aligned}$$

From (5.10) we conclude that

$$\| \Lambda(u-v) \| \leq c\,\|l + \Lambda^* y^*\|_{\mathcal{V}} + \| y^* - \mathcal{A}\Lambda v \|_* \,. \tag{5.11}$$

It is easy that (5.8) coincides with (5.11) if the parameter β is selected optimally.

We conclude this section with a comment concerning the *error reduction property.* In practice, it is often interesting to predict how essentially an approximation error would decrease as a result of mesh refinement. Assume that a coarse mesh $\mathcal{T}_h$ is replaced by a refined one $\mathcal{T}_{h\mathrm{REF}}$. For problems associated with quadratic functionals (as, e.g., (5.1)), the value of error reduction is easy to compute by means of the equality (5.4). Indeed, we observe that for any $v \in V_0 + u_0$

$$\frac{1}{2} \| \Lambda(u-v) \|^2 = J(v) - I^*(p^*) = J(v) - J(u).$$

Therefore,

$$\frac{1}{2} \| \Lambda(u-u_h) \|^2 = J(u_h) - J(u), \quad \frac{1}{2} \| \Lambda(u-u_{h\mathrm{REF}}) \|^2 = J(u_{h\mathrm{REF}}) - J(u).$$

From these relations, it follows a simple *error reduction relation*

$$\| \Lambda(u - u_{h\mathrm{REF}}) \|^2 = \| \Lambda(u - u_h) \|^2 - 2\big(J(u_h) - J(u_{h\mathrm{REF}})\big). \tag{5.12}$$

We observe that the reduction of approximation error is equal to the difference of the values of the respective functionals. If $\| \Lambda(u-u_h) \|^2 \leq M_\oplus$, then for the refined mesh we have the estimate

$$\| \Lambda(u - u_{h\mathrm{REF}}) \|^2 \leq M_\oplus - 2\big(J(u_h) - J(u_{h\mathrm{REF}})\big).$$

We note that the right-hand side of this estimate contains only explicitly computable quantities. From this estimate, it also follows that an "optimal" mesh-adaptation strategy (oriented to the energy error norm) should refine meshes in such a way that *to minimize the energy functional as much as possible.*

5.2 Diffusion problem

Linear diffusion equation gives a class of boundary value problems that are particular forms of the general problem (5.1). Consider the stationary diffusion problem

$$\operatorname{div} A\nabla u + f = 0 \quad \text{in } \Omega; \tag{5.13}$$

$$u = u_0 \quad \text{on } \partial_1\Omega; \tag{5.14}$$

$$n \cdot A\nabla u = F \quad \text{on } \partial_2\Omega, \tag{5.15}$$

where $\Omega \in \mathbb{R}^n$ is a bounded connected domain with Lipschitz continuous boundary that consists of two measurable disjoint parts $\partial_1\Omega$ and $\partial_2\Omega$. Assume that

$$c_1^2|\xi|^2 \leq A\xi\cdot\xi \leq c_2^2|\xi|^2 \qquad \forall\xi \in \mathbb{R}^n. \tag{5.16}$$

In the case considered,

$$\| y^* \|^2 := \int_\Omega Ay^*\cdot y^*\,dx, \qquad \| y^* \|_*^2 := \int_\Omega A^{-1}y^*\cdot y^*\,dx,$$

$$U := L^2(\Omega,\mathbb{R}^n), \quad V_0 := \{w \in H^1(\Omega) \mid w = 0 \text{ on } \partial_1\Omega\},$$

and (5.1) has the form

$$\int_\Omega A\nabla u\cdot\nabla w\,dx = \int_\Omega fwdx + \int_{\partial_2\Omega} Fw\,ds \qquad \forall w \in V_0(\Omega). \tag{5.17}$$

We apply (5.8) and observe that for any

$$y^* \in Q^* := \{y^* \in Y^* \mid \operatorname{div} y^* \in L^2(\Omega),\ y^*\cdot n \in L^2(\partial_2\Omega)\}$$

we have the estimate

$$\| \nabla(u-v) \|^2 \leq (1+\beta) \| A\nabla v - y^* \|_*^2 + \\ + C^2\frac{1+\beta}{\beta}\left(\|f + \operatorname{div} y^*\|^2 + \|F - y^*\cdot n\|_{\partial_2\Omega}^2\right). \tag{5.18}$$

Here, β is an arbitrary positive number, and C is any constant greater than the quantity $\lambda^{-1}(\Omega,\partial_2\Omega)$ defined by the relation

$$\lambda^2(\Omega,\partial_2\Omega) = \inf_{w\in V_0} \frac{\| \nabla w \|^2}{\|w\|^2 + \|w\|_{\partial_2\Omega}^2}. \tag{5.19}$$

Note that (5.18) also holds for the Neumann problem. In this case, $\partial\Omega = \partial_2\Omega$ and the constant C must be greater than $\lambda^{-1}(\Omega)$, where

$$\lambda^2(\Omega) = \inf_{w\in\widetilde{V}_0} \frac{\| \nabla w \|^2}{\|w\|^2 + \|w\|_{2,\partial\Omega}^2} \tag{5.20}$$

and

$$\widetilde{V}_0 := \left\{w \in H^1(\Omega) \,\middle|\, \int_\Omega w\,dx = 0\right\}.$$

Functional a posteriori estimates for diffusion problem were derived and analyzed in [236, 240, 241, 242, 243] and their performance in practical computations was tested in [128, 129, 241, 258]. In [259], a functional a posteriori estimate was derived for the *reaction–diffusion* problem (with the help of the integral identities method). For the *convection–diffusion* problem such an estimate was obtained in [252]. The same method can be applied for the *convection–reaction–diffusion* problem

$$-\operatorname{div} A\nabla u + a\cdot\nabla u + \lambda u = f \quad \text{in } \Omega, \tag{5.21}$$

$$u = u_0 \quad \text{on } \partial\Omega. \tag{5.22}$$

We assume that

$$a \in L_\infty(\Omega,\mathbb{R}^d),\ \lambda > 0, \quad \operatorname{div} a \in L_\infty, \quad -\frac{1}{2}\operatorname{div} a + \lambda := \lambda_a \geq \lambda_0 \geq 0. \tag{5.23}$$

Generalized solution u is defined by the integral identity

$$\int\limits_\Omega (A\nabla u\cdot\nabla w + (a\cdot\nabla u)w + \lambda w)\,dx = \int\limits_\Omega fw dx, \quad w \in \overset{\circ}{H}{}^1(\Omega). \tag{5.24}$$

Let $v \in \overset{\circ}{H}{}^1 + u_0$ be an approximate solution. Then,

$$\int\limits_\Omega (A\nabla(u-v)\cdot\nabla w + (a\cdot\nabla(u-v))w + \lambda(u-v)w)\,dx =$$

$$= \int\limits_\Omega (fw - A\nabla v\cdot\nabla w - (a\cdot\nabla v)w - \lambda vw)dx.$$

Note that

$$\frac{1}{2}\int\limits_\Omega (\operatorname{div} a)(u-v)^2\,dx = -\int\limits_\Omega (u-v)a\cdot\nabla(u-v)dx.$$

Set $w = u - v$. For the natural (energy) norm we have

$$|[u-v]|^2 := \| \nabla(u-v) \|^2 + \int\limits_\Omega \lambda_a(u-v)^2\,dx =$$

$$= \int\limits_\Omega (f - a\cdot\nabla v - \lambda v + \operatorname{div} y^*)(u-v) + (y^* - A\nabla v)\cdot\nabla(u-v)dx, \tag{5.25}$$

where $y^*(x)$ is an arbitrary function in $H(\Omega, \operatorname{div})$. This identity implies two estimates

$$|[u-v]| \leq \| y^* - A\nabla v \|_* + C_\Omega \| f - a\cdot\nabla v - \lambda v + \operatorname{div} y^* \|, \tag{5.26}$$

and (for the case $\lambda_0 > 0$)

$$|[u-v]|^2 \leq \| \frac{1}{\sqrt{\lambda_a}}(f - a\cdot\nabla v - \lambda v + \operatorname{div} y^*) \|^2 + \| y^* - A\nabla v \|_*^2 . \tag{5.27}$$

These and other a posteriori estimates for the reaction–convection–diffusion problem are presented in S. Nicaise and S. Repin [212].

5.3 Linear elasticity

Assume that Ω satisfies the same conditions as in the previous section and $\partial\Omega$ consists of two disjoint parts $\partial_1\Omega$ and $\partial_2\Omega$ and $\operatorname{meas}\partial_1\Omega > 0$. The classical formulation of the linear elasticity problem is to find a tensor-valued function σ^* (stress) and a vector-valued function u (displacement) satisfying the system of equations

$$\sigma^* = \mathbb{L}\varepsilon \quad \text{in } \Omega, \tag{5.28}$$

$$\operatorname{div}\sigma^* = f \quad \text{in } \Omega, \tag{5.29}$$

$$u = u_0 \quad \text{on } \partial_1\Omega, \tag{5.30}$$

$$\sigma^* n + F = 0 \quad \text{on } \partial_2\Omega. \tag{5.31}$$

Here f and F are given forces and $\mathbb{L} = \{L_{ijkm}\}$ is the tensor of elasticity constants, which is subject to the conditions

$$C_1|\varepsilon|^2 \leq \mathbb{L}\varepsilon : \varepsilon \leq C_2|\varepsilon|^2 \;\forall \varepsilon \in \mathbb{M}_s^{n\times n}, \tag{5.32}$$

where $\mathbb{M}_s^{n\times n}$ denotes the space of symmetric $n \times n$ matrices (tensors), whose scalar product is denoted by two dots. Henceforth, we assume that

$$\mathbb{L}_{ijkm} = \mathbb{L}_{jikm} = \mathbb{L}_{kmij}, \; \mathbb{L}_{ijkm} \in L^\infty(\Omega), \tag{5.33}$$

and

$$f \in L^2(\Omega, \mathbb{R}^n), \quad F \in L^2(\partial_2\Omega, \mathbb{R}^n). \tag{5.34}$$

Then, a generalized solution $u \in V_0 + u_0$ is defined by the identity

$$\int_\Omega \mathbb{L}\,\varepsilon(u) : \varepsilon(w)\,dx + \langle l, w\rangle \;=\; 0 \;\forall w \in V_0, \tag{5.35}$$

where

$$\langle l, w\rangle = \int_\Omega f \cdot w\,dx + \int_{\partial_2\Omega} F \cdot w\,ds.$$

The existence and uniqueness of u follows from Korn's inequality.

Let v and y^* be some approximations of u and σ^*. Estimates of $v - u$ and $y^* - \sigma^*$ follow from the general scheme if we set

$$U = L^2(\Omega, \mathbb{M}_s^{n\times n}), \;\; V = H^1(\Omega, \mathbb{R}^n), \;\; V_0 = \{w \in V \mid w = 0 \text{ on } \partial_1\Omega\},$$

$$\|\, y \,\|^2 = \int_\Omega \mathbb{L}y : y\,dx, \qquad \|\, y^* \,\|_*^2 = \int_\Omega \mathbb{L}^{-1}y^* : y^*\,dx,$$

and identify Λv with $\varepsilon(v) = \frac{1}{2}(\nabla v + (\nabla v)^T)$. In this case,

$$\langle \Lambda^* y^*, w\rangle = \int_\Omega y^* : \varepsilon(w)\,dx \;\forall w \in V_0. \tag{5.36}$$

By applying the general estimate, we find the upper bound for the error

$$\frac{1}{2} \parallel \varepsilon(v-u) \parallel^2 \leq M_{\oplus}(v,\beta,y^*), \tag{5.37}$$

where $v \in V_0 + u_0$, $\beta > 0$, and

$$y^* \in Q^* := \{y^* \in Y^* \mid \operatorname{div} y^* \in L^2(\Omega, \mathbb{M}^{n\times n}),\ y^*n \in L^2(\partial_2\Omega, \mathbb{R}^n)\},$$

and

$$M_{\oplus}(\varepsilon(v),y^*) = \frac{1+\beta}{2} \int\limits_{\Omega} (\varepsilon(v) - \mathbb{L}^{-1}y^*) : (\mathbb{L}\,\varepsilon(v) - y^*)\,dx +$$

$$+ \frac{1+\beta}{2\beta c_1} C_\Omega^2 \left(\|\operatorname{div} y^* - f\|_\Omega^2 + \|F + y^*n\|_{\partial_2\Omega}^2 \right).$$

Here, y^* can be viewed as an image of the true stress tensor σ^*, β is a parameter used to properly balance the two terms of $M_\oplus$, and C_Ω is a constant in the inequality

$$\int\limits_{\Omega} |w|^2\,dx + \int_{\partial_2\Omega} |w|^2\,ds \leq C_\Omega^2 \|\varepsilon(w)\|_\Omega^2 \ \forall w \in V_0. \tag{5.38}$$

Note that the existence of such a constant follows from Korn's inequality.

Estimate (5.37) has a clear physical meaning. The first term of $M_\oplus$ is nonnegative and vanishes if and only if

$$y^* = \mathbb{L}\varepsilon(v).$$

It penalizes violations of the *constitutive relation* (Hooke's law) (5.28). The meaning of the second term is obvious: it contains L^2-norms of two other relations (5.29) and (5.31), which show errors in the *equilibrium equation* and natural *boundary condition*. Thus, the majorant does not only give a bound of the overall value of the error, but also shows its physically sensible parts. The latter information may be of utmost interest, because it suggests a correct way of finding a better approximation. A more detailed discussion of functional a posteriori estimates for linear elasticity problems is contained in [241, 242, 209] and some tests in [206]. Extensions of the above theory to thermoelastic problems were investigated in [207].

5.4 Stokes problem

The Stokes model describes slow stationary flow of an incompressible fluid whose viscosity is characterized by a positive constant ν. It consists of finding u (velocity), σ (stress), and p (pressure) that satisfy the system

$$-\operatorname{div}\sigma + \nabla p = f \quad \text{in } \Omega, \tag{5.39}$$

$$\sigma = \nu\varepsilon(u) \quad \text{in } \Omega, \tag{5.40}$$

$$\operatorname{div} u = 0 \quad \text{in } \Omega, \tag{5.41}$$

$$u = u_0 \quad \text{on } \partial\Omega. \tag{5.42}$$

Let us denote by $\overset{\circ}{J}{}^1_2(\Omega)$ the closure of the set of smooth solenoidal functions with compact supports in Ω with respect to the norm $\|\nabla v\|$. As in the previous section, $V := H^1(\Omega, \mathbb{R}^d)$, and $Y^* := L^2(\Omega, \mathbb{M}^{d\times d})$. V_0 is a subspace of V that consists of functions with zero traces on $\partial\Omega$. The set $\overset{\circ}{J}{}^1_2(\Omega) + u_0$ consists of the functions $w + u_0$, where $w \in \overset{\circ}{J}{}^1_2(\Omega)$. The tensor of small deformations $\varepsilon(v)$ is considered as an operator from V to Y^*. Also, we define a Hilbert space $Y^*_{\text{div}}(\Omega)$, which is a subspace of Y^* containing tensor-valued functions τ such that $\operatorname{div}\tau \in L^2$. The scalar product in this space is defined by the relation

$$(\tau, \eta) := \int_\Omega (\tau : \eta + \operatorname{div}\tau \cdot \operatorname{div}\eta)\, dx\,.$$

The space of square summable functions with zero mean is defined by $\widetilde{L}_2(\Omega)$. Assume that $f \in L^2(\Omega, \mathbb{R}^d)$, $u_0 \in V$, and $\operatorname{div} u_0 = 0$. The generalized solution u of the Stokes problem is defined as the function in $\overset{\circ}{J}{}^1_2(\Omega) + u_0$ that meets the integral identity

$$\nu \int_\Omega \varepsilon(u) : \varepsilon(v)\, dx = \int_\Omega f \cdot v\, dx \quad \forall v \in \overset{\circ}{J}{}^1_2(\Omega). \tag{5.43}$$

A unique solution to this problem exists. It minimizes the functional

$$I(v) = \int_\Omega \left(\frac{\nu}{2}|\varepsilon(v)|^2 - f \cdot v\right) dx \tag{5.44}$$

on the set $\overset{\circ}{J}{}^1_2(\Omega) + u_0$. One of the major difficulties of the Stokes problem is that admissible fields must be subject to the incompressibility condition $\operatorname{div} u = 0$.

Functional a posteriori estimates for the Stokes problem can be derived by the variational method (which is applied to the functional (5.44)) (see [246]) and by transformations of the integral identity (5.43) (see [252]). Both methods give the same estimate:

$$\boxed{\nu\|\varepsilon(u - v)\| \le \|y^* - \nu\varepsilon(v)\| + C_\Omega\|f - \nabla q + \operatorname{div} y^*\|.} \tag{5.45}$$

Here v is *any* conforming approximation of u (i.e., $v \in \overset{\circ}{J}{}^1_2(\Omega) + u_0$), y^* is *any* tensor-function in $Y^*_{\text{div}}(\Omega)$ (it is an "image" of the deviatoric part of the stress field), and $q \in H^1(\Omega) \cap \widetilde{L}_2(\Omega)$ is an "image" of the pressure function. The structure of (5.45) is analogous to (4.5). It is easy to see that if $y^* = \nu\nabla u$ and $q = p$, then (5.45) holds as the equality. We may also rewrite this estimate in a different form, which is valid for a wider set of pressure functions q. For this purpose, we set $y^* = \eta + q\mathbb{I}$, where $\mathbb{I}$ is the unit tensor, $q \in \widetilde{L}_2(\Omega)$, and $\eta \in Y^*_{\text{div}}(\Omega)$. Then, the estimate(5.45) has a somewhat different form

$$\nu\,\|\varepsilon(u - v)\| \le \|\nu\varepsilon(v) - \eta - q\mathbb{I}\| + C_\Omega\|\operatorname{div}\eta + f\|. \tag{5.46}$$

Set $q = \frac{1}{d}\mathrm{tr}\,\eta$, where $\mathrm{tr}\,\eta := \sum\limits_i \eta_i i$ and let the mean value of $\mathrm{tr}\,\eta$ be equal to zero. Then, we have another form of the estimate

$$\nu\|\nabla(u-v)\| \leq \|\nu\varepsilon^D(v) - \eta^D\| + C_\Omega\|\mathrm{div}\eta + f\|, \tag{5.47}$$

which contains only one "free" variable η.

Estimates (5.45)–(5.45) are valid only for solenoidal approximations. Extensions to a wider class of vector-valued functions are possible. However, prior to exposing the respective results, it is necessary to recall one important result in the theory of functions.

Lemma. *For any* $g \in \overset{\circ}{L}_2(\Omega)$ *there exists a function* $\bar{u} \in V_0$ *satisfying the relation* $\mathrm{div}\,\bar{u} = g$ *and the condition*

$$\|\nabla\bar{u}\| \leq \kappa_\Omega\|g\|, \tag{5.48}$$

where κ_Ω *is a positive constant that depends on* Ω*.*

A proof is given in I. Babuška and A. K. Aziz [19] and O. Ladyzhenskaya and V. Solonnikov [190]. Also, the lemma can be considered as a special case of the *closed range lemma* (see, e.g., F. Brezzi and M. Fortin [70], K. Yosida [314]).

The lemma implies several important corollaries. The first one is related to the so-called Inf-Sup or Ladyzhenskaya–Babuška–Brezzi (LBB)-condition [17, 68] that reads: there exists a positive constant $\mathbb{C}_{\mathrm{LBB}}$ such that

$$\inf_{g\in\overset{\circ}{L}_2(\Omega),\, g\neq 0}\ \sup_{w\in V_0,\, w\neq 0} \frac{\displaystyle\int_\Omega g\,\mathrm{div}\,w\,dx}{\|\phi\|\,\|\nabla w\|} \geq \mathbb{C}_{\mathrm{LBB}}\,. \tag{5.49}$$

It is easy to show that the lemma implies (5.49). Indeed, any $g \in \overset{\circ}{L}_2(\Omega)$ has a counterpart function $v_g \in V_0$ that meets the conditions

$$\mathrm{div}\,v_g = g, \qquad \|\nabla v_\phi\| \leq \kappa_\Omega\|g\|.$$

Therefore,

$$\sup_{v\in V_0, w\neq 0} \frac{\int_\Omega g\,\mathrm{div}\,v\,dx}{\|\nabla v\|\,\|g\|} \geq \frac{\int_\Omega g\,\mathrm{div}\,v_g\,dx}{\|\nabla v_g\|\,\|g\|} = \frac{\|g\|}{\|\nabla v_g\|} \geq \frac{1}{\kappa_\Omega}$$

and, consequently, (5.49) holds with $\mathbb{C}_{\mathrm{LBB}} = \kappa_\Omega^{-1}$. Note that (5.49) can be presented in the form

$$\sup_{w\in V_0,\, w\neq 0} \frac{\displaystyle\int_\Omega g\,\mathrm{div}\,w\,dx}{\|\nabla w\|} \geq \mathbb{C}_{\mathrm{LBB}}\,\|\,g\| \qquad \text{for all } g \in \overset{\circ}{L}_2(\Omega).$$

The expression in the left-hand side of the above inequality is the norm of ∇g in the space topologically dual to V_0, namely

$$|\nabla g| := \sup_{w\in V_0} \frac{<\nabla g, w>}{\|\nabla w\|}.$$

Then, we arrive to the Nečas inequality

$$\|p\| \leq \kappa_\Omega \, \|\nabla p\| \qquad \forall\; p \in \overset{\circ}{L}_2(\Omega). \tag{5.50}$$

A simple proof of the Nečas inequality for domains with Lipschitz boundaries is given in the paper by J. Bramble [61], where it is also shown that the well-known Korn's inequality follows from the inf-sup condition.

The constants $\mathbb{C}_{\rm LBB}$ and κ_Ω play an important role in the numerical analysis of the Stokes problem as well as in the theoretical one. They affect the stability of mixed-type formulations and the efficiency of iteration methods (see, e.g., G. Kobelkov and M. Olshanskii [174], U. Langer and W. Queck [192]). Therefore, it is very desirable to have a unified numerical technology able to compute values of $\mathbb{C}_{\rm LBB}$ for arbitrary Lipschitz domains. To the best of our knowledge, at present such a unified technology does not exist. An upper bound of $\mathbb{C}_{\rm LBB}$ can be expressed throughout the constants in the Friederichs and Poincaré inequalities for the domain Ω (see [247, 252]). However, in practice we are more interested in the lower bound. For rectangular domains, two-sided bounds for the constant were derived in E. Chizhonkov and M. Olshanskii [97], and G. Stoyan [288]. In L. Halpern [148], it was shown that for a unitary disc the constant equals $1/\sqrt{2}$. Some ideas numerical evaluation of the LBB constant are contained in the above cited publications and in the paper by M. Dobrowolski [107], where the case of stretched domains is considered.

In the context of our analysis, we also mention two important corollaries of the lemma (a reader can find the proofs in, e.g., [251, 252]). The first one is that by the lemma we easily establish the existence of a saddle point to the Lagrangian

$$L(v,q) = \int\limits_\Omega \Big(\frac{\nu}{2}|\nabla v|^2 - f\cdot v - q\,{\rm div}\, v\Big)\, dx.$$

Another corollary is the projection type inequality

$$\inf_{v\in \overset{\circ}{J}{}^1_2(\Omega)} \|\nabla(\widehat{v}-v)\| \leq \kappa_\Omega \|{\rm div}\,\widehat{v}\|, \tag{5.51}$$

which shows that the distance to $\overset{\circ}{J}{}^1_2(\Omega)$ is measured by the value of $\|{\rm div}\,\widehat{v}\|$. By (5.51), the estimate (5.45) is extended to the class on non-solenoidal approximations. Let $\widehat{v} \in V_0 + u_0$ be such an approximation. In [251, 252], it is proved that

$$\boxed{\nu\|\nabla(u-\widehat{v})\| \leq \|\nu\nabla\widehat{v}-y^*\| + C_\Omega\|{\rm div}\, y^* + f - \nabla q\| + \kappa(\widehat{v}),} \tag{5.52}$$

where $\kappa(\widehat{v}) := \frac{2\nu}{\mathbb{C}_{\rm LBB}}\|{\rm div}\,\widehat{v}\|$. Three terms in the right-hand side of the estimate present three natural parts of the error, namely *errors in the constitutive law, differential equation* and *incompressibility condition.*

Similarly to (5.46) and (5.47) we derive two other forms of the above estimate:

$$\nu\,\|\nabla(u-\widehat{v})\| \;\leq\; \|\nu\nabla\widehat{v}-\eta-q\mathbb{I}\| + C_\Omega\|{\rm div}\,\eta + f\| + \kappa(\widehat{v}), \tag{5.53}$$

$$\nu\,\|\nabla(u-\widehat{v})\| \;\leq\; \sqrt{\|(\nu\nabla v-\eta)^D\|^2 + \kappa\{{\rm tr}\,\eta\}^2} + C_\Omega\|{\rm div}\,\eta + f\| + \kappa(\widehat{v}). \tag{5.54}$$

Thus, if the constants c_Ω and $\mathbb{C}_{\rm LBB}$ are known (or we know suitable upper bounds for them), then (5.52)–(5.54) are directly computable estimates of $\|\nabla(u-\widehat{v})\|$.

Numerical testing of functional a posteriori estimates for the Stokes problem was performed in [145]. An extended discussion of the above estimates can be also found in [209, 251].

Variational approach was also used to derive functional a posteriori estimates in some viscous flow problems ([242, 243, 244]). In [50, 131] such estimates were obtained for stationary viscous flow problems with non-quadratic dissipative potentials.

In [146], functional a posteriori estimates were derived for problems with rotation (in these models the equation of motion includes an additional term $\omega \times u$, where $\omega = |\omega|\mathbf{i}$ and $\mathbf{i}$ is the unit rotation vector).

A posteriori estimates for the generalized Stokes problem

$$\begin{aligned} -\operatorname{div}\sigma + \varrho^2 u + \nabla p = f \quad & \text{in } \Omega, \\ \sigma = \nu\varepsilon(u) \quad & \text{in } \Omega, \\ \operatorname{div} u = 0 \quad & \text{in } \Omega, \\ u = u_0 \quad & \text{on } \partial\Omega \end{aligned}$$

were obtained in [268]. Another generalized Stokes formulation was analyzed in the context of a posteriori error estimation in [269].

5.5 Biharmonic problem

Consider the problem

$$\operatorname{div}\operatorname{div}(B\nabla\nabla u) = f \quad \text{in } \Omega, \tag{5.55}$$

$$u = \frac{\partial u}{\partial \nu} = 0 \quad \text{on } \partial\Omega. \tag{5.56}$$

Here $f \in L^2(\Omega)$ and $B = \{b_{ijkl}\} \in \mathcal{L}(\mathbb{M}_s^{n\times n}, \mathbb{M}_s^{n\times n})$ satisfies the conditions

$$\begin{aligned} & b_{ijkl} = b_{jikl} = b_{klij} \in L^\infty(\Omega), \\ & \alpha_1|\eta|^2 \le B\eta : \eta \le \alpha_2|\eta|^2 \quad \forall \eta \in \mathbb{M}_s^{n\times n}. \end{aligned}$$

To apply the general scheme, we set

$$\begin{aligned} & U = L^2(\Omega, \mathbb{M}_s^{n\times n}), \qquad V = H^2(\Omega), \\ & V_0 = \{w \in V \mid w = \frac{\partial w}{\partial \nu} = 0 \ \text{ on } \partial\Omega\}, \end{aligned}$$

and define Λ as the Hessian operator. Generalized solution is defined as the function in V_0 satisfying the integral identity

$$\int_\Omega B\nabla\nabla u : \nabla\nabla w \, dx = \int_\Omega f w \, dx \quad \forall w \in V_0. \tag{5.57}$$

Define the norms

$$\| y \|^2 := \int_\Omega By : y\, dx, \qquad \| y^* \|_*^2 := \int_\Omega B^{-1} y^* : y^* dx.$$

Let $v \in V_0$ be an approximation of u. In [209], a variational method was used to derive functional a posteriori estimates of $\| \nabla\nabla(v-u) \|$. The first estimate is a particular form of (5.8). It reads

$$\frac{1}{2} \| \nabla\nabla(v-u) \|^2 \leq (1+\beta) D(\nabla\nabla v, y^*) + \left(1+\frac{1}{\beta}\right) \frac{C_{1\Omega}^2}{2\alpha_1^2} \|\mathrm{div}\,\mathrm{div}\, y^* - f\|_\Omega^2. \quad (5.58)$$

Here, y^* is an arbitrary tensor-valued function from

$$H(\mathrm{div}\,\mathrm{div}, \Omega) := \left\{ \eta \in L^2(\Omega, \mathbb{M}_s^{n\times n}) \mid \mathrm{div}\,\mathrm{div}\, \eta \in L^2(\Omega) \right\},$$

β is a positive real number, and $C_{1\Omega}$ is a constant in the inequality

$$\|w\|_\Omega \leq C_{1\Omega} \|\nabla\nabla w\|_\Omega \qquad \forall w \in V_0. \quad (5.59)$$

However, the requirement that y^* must have square summable div div is rather demanding. To avoid technical difficulties that rise from this restriction, another estimate was obtained:

$$\frac{1}{2} \| \nabla\nabla(v-u) \|^2 \leq (1+\beta) D(\nabla\nabla v, y^*) +$$
$$+ \left(1+\frac{1}{\beta}\right) \frac{1}{2\alpha_1^2} \left(C_{2\Omega} \|\mathrm{div}\, y^* - \eta\|_\Omega + C_{1\Omega} \|\mathrm{div}\, \eta - f\|_\Omega \right)^2. \quad (5.60)$$

It has two "free" functions: $\eta \in H(\mathrm{div}, \Omega)$ and y^* (which must be such that $\mathrm{div}\, y^* \in L^2(\Omega, \mathbb{R}^d)$). They can be viewed as images of the double gradient and gradient, respectively. In (5.60) $C_{2\Omega}$ is a constant in the inequality

$$\|\nabla w\|_\Omega \leq C_{2\Omega} \|\nabla\nabla w\|_\Omega \qquad \forall w \in V_0. \quad (5.61)$$

It is worth noting that functional a posteriori estimates also follow from the integral identities in the framework of the general scheme suggested in [243]. Let us show this on the paradigm of the biharmonic problem

$$\int_\Omega \Delta u \Delta w\, dx = \int_\Omega f w\, dx \quad \forall w \in V_0.$$

For $y^* \in H^2(\Omega)$ we have

$$\int_\Omega \Delta(u-v)\Delta w\, dx = \int_\Omega (fw - \Delta v \Delta w)\, dx = \int_\Omega ((f - \Delta y^*) w + (y^* - \Delta v)\Delta w\, dx.$$

From here, we find that

$$\|\Delta(u-v)\| \leq C_\Omega \|f - \Delta y^*\| + \|y^* - \Delta v\|, \quad (5.62)$$

where C_Ω is a constant in the inequality $\|w\| \leq C_\Omega \|\Delta w\|$ for $w \in V_0$. Numerical testing of (5.62) was performed in [127], where this estimate was derived by means of the variational method.

Finally, we mention several results related to the theory of plates. In the context of a posteriori estimates, classical Kirchhoff–Love plate model was considered in [210]. Estimates for the Reissner–Mindlin model were obtained by the variational and non-variational methods in [130, 257].

5.6 Evolutionary problems

Functional a posteriori error estimates can be obtained by transformations of integral relations that define weak solutions to evolutionary problems. We consider their basic structure on the paradigm of the linear parabolic problem: find a function $u(x,t)$ such that

$$u_t - \operatorname{div}(A\nabla u) = f \quad \text{in } Q_T := \Omega \times (0,T), \tag{5.63}$$

$$u(x,0) = \phi(x), \qquad x \in \Omega, \quad T > 0, \tag{5.64}$$

$$u(x,t) = 0, \qquad (x,t) \in S_T := \partial\Omega \times [0,T]. \tag{5.65}$$

Here, $A(x,t)$ is a symmetric matrix satisfying the condition

$$\nu_1 |\xi|^2 \leq A(x,t)\xi \cdot \xi \leq \nu_2 |\xi|^2 \ \forall \xi \in \mathbb{R}^n, \quad 0 < \nu_1 \leq \nu_2 < \infty.$$

Weak formulation of the problem (5.63)–(5.65) is defined in accordance with the general theory of parabolic partial differential equations (see, e.g., [189]). It is to find $u \in \overset{\circ}{W}{}_2^{1,0}(Q_T) := L^2((0,T); \overset{\circ}{H}{}^1(\Omega))$, such that

$$\int\limits_{Q_T} A\nabla u \cdot \nabla \eta \, dxdt - \int\limits_{Q_T} u\eta_t \, dxdt + \\ + \int\limits_{\Omega} (u(x,T)\eta(x,T) - \phi(x)\eta(x,0))dx = \int\limits_{Q_T} f\eta \, dxdt \tag{5.66}$$

for any trial function η in

$$W^1_{2,0}(Q_T) := \{ w \in H^1(Q_T) \mid w(x,t) = 0 \text{ on } S_T \}.$$

Let $v \in W^1_{2,0}(Q_T)$ be a given approximation of u. We derive an upper bound of the deviation $e := u - v$ in terms of the quantity

$$[e]^2_{(\lambda,\mu)} := \lambda \|e(x,T)\|^2 + \mu \, \|\nabla e\|^2_{Q_T}$$

where

$$\|\nabla e\|^2_{Q_T} := \int\limits_{Q_T} A\nabla e \cdot \nabla e \, dx\, dt,$$

γ_1 and γ_2 are some positive numbers. In [244], a way of deriving computable upper bounds of $[e]^2_{(\gamma_1,\gamma_2)}$ was suggested based on transformations of (5.66). In the simplest case $A = \mathbb{I}$, the respective estimate is as follows:

$$[e]^2_{(2-\delta,1)} \leq M_{\oplus}(v,y,\beta) :=\| v(x,0) - \phi(x) \|^2 + \\ + \frac{1}{\delta} \int_0^T \left[(1+\beta)\|\nabla v - y\|^2 + C_\Omega^2 \left(1 + \frac{1}{\beta}\right) \|f - v_t + \operatorname{div} y\|^2 \right] dt, \quad (5.67)$$

where $\delta \in (0, 2]$, C_Ω is the constant in the Friedrichs inequality, $y \in H(Q_T, \operatorname{div})$, and $\beta = \beta(t)$ is a positive valued function. It is easy to observe that $M_\oplus$ has a clearly motivated structure: the terms of $M_\oplus$ are certain penalties for possible violations of the relations

$$u_t - \operatorname{div} y = f, \quad y = \nabla u, \\ u(x,0) = \phi(x), \quad u(x,t) = 0 \text{ for } (x,t) \in S_T.$$

Note that $M_\oplus(v,y,\beta)$ vanishes if and only if $v = u$ and $y = \nabla u$. Therefore, $M_\oplus$ also provides a *variational formulation* of the heat transfer problem.

For the problem (5.63)–(5.65) the majorant $M^2_\oplus(v,y,\beta)$ differs from (5.67) in one term (see [134]), namely $\|\nabla v - y\|^2$ is replaced by

$$\int_\Omega \left(A\nabla v \cdot \nabla v + A^{-1} y \cdot y - 2\nabla v \cdot y\right) dx.$$

In [244], a sharper upper bound of the error was also suggested. For the case $A = \mathbb{I}$ it has the form

$$[e]^2_{(2-\delta,\mu)} \leq \widehat{M}_\oplus(v,w,y,\beta) := \gamma\|w(x,T)\|^2 + \\ + \frac{1}{\delta} \int_0^T \left[(1+\beta)\,\|y - \nabla v + \nabla w\|^2 + \left(1 + \frac{1}{\beta}\right) C_\Omega^2 \|f - v_t - w_t + \operatorname{div} y\|^2 \right] dt + \\ + 2 \int_{Q_T} (\mathrm{A}\nabla v \cdot \nabla w + v_t w - f w)\, dx\, dt + \\ + \int_\Omega \left(|\phi(x) - v(x,0)|^2 - 2w(x,0)(\phi(x) - v(x,0))\right) dx, \quad (5.68)$$

where $\mu = 1 - \frac{1}{\gamma}$, $y \in Y_{\text{div}}(Q_T)$, $\delta \in (0,2]$, $\gamma \geq 1$, $\beta = \beta(t)$ is a positive valued function and $w \in \overset{\circ}{W}{}^1_{2,0}(Q_T)$ is *an additional free variable*. Extended version of this estimate valid for any conforming approximation of (5.63)–(5.65) is presented in [134].

Estimates (5.67) and (5.68) imply directly computable functionals defined on approximate solutions computed on the coarse and refined mesh. Let U_{TH} be an approximate solution of the problem computed on the mesh with mesh size T for time variable and H for spatial variables and let $u_{\tau,h}$ be another approximate solution computed on a finer mesh (τ, h) (e.g., $\tau = \mathcal{T}/2$, $h = H/2$). Then,

$$[u - U_{\mathcal{T}H}]^2_{(2-\delta,1)} \leq M_{\oplus}(U_{\mathcal{T}H}, G_{\mathcal{T}H}(\nabla u_{\tau h}), \beta), \tag{5.69}$$

$$[u - U_{\mathcal{T}H}]^2_{(2-\delta,\mu)} \leq \widehat{M}_{\oplus}(U_{\mathcal{T}H}, u_{\tau h} - U_{\mathcal{T}H}, G_{\mathcal{T}H}(\nabla u_{\tau h}), \beta). \tag{5.70}$$

Estimates (5.69) and (5.70) give a justified form of Runge's rule. Numerical tests performed in [134] have shown the efficiency of these estimates.

5.7 Estimates in terms of non-energy quantities

It is often required to measure the accuracy of approximate solutions in terms of functionals other than the energy norm. Let the basic *error control problem* formulated at the beginning of Section 4 be solved and we know $M_{\ominus}$ and $M_{\oplus}$ satisfying (4.1). Then it is not difficult to construct computable and sharp error estimates for any error functional governed by the energy norm.

Assume that we are interested in the value of $\Phi(v-u)$, where $\Phi : V \to \mathbb{R}$ is a given functional such that

$$|\Phi(w_1 + w_2)| \leq \mu_1 (|\Phi(w_1)| + |\Phi(w_2)|) \qquad \forall w_1, w_2 \in V, \tag{5.71}$$

$$|\Phi(w)| \leq \mu_2 \|w\|_V \qquad \forall w \in V, \tag{5.72}$$

where $\mu_1 \geq 1$ and $\mu_2 > 0$. From (5.71) it follows that

$$|\Phi(w_1)| - \mu_1 |\Phi(w_2)| \leq \mu_1 |\Phi(w_1 - w_2)| \qquad \forall w_1, w_2 \in V. \tag{5.73}$$

For any admissible function $\phi \in V_0$, we have

$$|\Phi(u - v)| \leq \mu_1 |\Phi(\phi)| + \mu_1 \mu_2 M_{\oplus}(v + \phi, \mathcal{D}), \tag{5.74}$$

$$|\Phi(u - v)| \geq \frac{1}{\mu_1} \Phi(\phi) - \mu_2 M_{\oplus}(v + \phi, \mathcal{D}). \tag{5.75}$$

Since the energy norm majorant possesses the property $M_{\oplus}(u, \mathcal{D}) = 0$, we observe that if $\mu_1 = 1$ than by choosing the function ϕ the above upper and lower error bounds can be found as close to $\Phi(u-v)$ as it is required.

Typically, the functional Φ is either a linear functional or a local norm, i.e.,

$$\Phi(u - v) = \ell(u - v) \qquad \text{or} \qquad \Phi(u - v) = \| u - v \|_{\omega},$$

where ω is a subdomain of Ω. In these cases, $\mu_1 = 1$ and we have a two-sided estimate

$$\Phi(\phi) - M_{\oplus}(v + \phi, \mathcal{D}) \leq \Phi(u - v) \leq \Phi(\phi) + M_{\oplus}(v + \phi, \mathcal{D}). \tag{5.76}$$

In [250, 254], a more detailed consideration of the topic is presented. In particular, for the diffusion problem guaranteed upper bounds of local errors are given by the following estimate:

$$\|\!|\nabla(u-v)\|\!|_\omega^2 \leq \frac{1+\beta}{1-\gamma}\Big\{\|\!| A\nabla(v+\phi)-y^*\|\!|_*^2 + \\ +\frac{1}{\beta}C^2\left(\|f+\operatorname{div} y^*\|^2+\|F-y^*\cdot\nu\|^2_{\partial_2\Omega}\right)\Big\}+\frac{1}{\gamma}\|\!|\nabla\phi\|\!|_\omega^2, \quad (5.77)$$

where $\beta>0$ and $\gamma\in(0,1)$.

5.8 Estimates for mixed approximations

Mixed approximations are used if a boundary value problem is presented in the so-called *minimax form* (see, e.g., F. Brezzi and M. Fortin [70], P.-A. Raviart and J.-M. Thomas [235]). In such a formulation, "solution" means a pair of functions. For example, the problem (2.4)–(2.5) can be formulated as a minimax problem for the Lagrangian

$$L(v,y^*)=\int\limits_\Omega\left(y^*\cdot\nabla v-\frac{1}{2}|\nabla v|^2-fv\right)dx,$$

where $v\in\overset{\circ}{H}{}^1(\Omega)$ and $y^*\in L^2(\Omega,\mathbb{R}^n)$. Another version of the minimax formulation (dual-mixed method) operates with a different pair of basic spaces, namely $(u,p^*)\in L^2(\Omega)\times H(\Omega,\operatorname{div})$. Approximations in $\overset{\circ}{H}{}^1(\Omega)\times H(\Omega,\operatorname{div})$ are used in the least square mixed formulations (see, e.g., T. Belytschko and T. Blacker [46], J. H. Bramble, R. D. Lazarov and J. E. Pasciak [62], J. H. Brandts and Y. Chen [65], Z. Cai, R. Lazarov, T. A. Manteuffel and S. F. McCormick [73], Z. Cai, T. A. Manteuffel, S. F. McCormick and S. V. Parter [74], G. F. Carey and A. I. Pehlivanov [76], L. P. Franca and R. Stenberg [124]).

Functional a posteriori estimates for the primal, dual and least square mixed formulations of elliptic boundary value problems were investigated in [260] and [267] where a detailed discussion of the topic is contained. Here, we mention only one result having a principal value. We formulate it in terms of the problem (2.4)–(2.5) (generalizations are presented in [267]). Let (v,y^*) be a pair of approximations in $\overset{\circ}{H}{}^1(\Omega)\times H(\Omega,\operatorname{div})$. To measure the deviation from (u,p^*) it is natural to use *full primal-dual norm*:

$$\|(u-v,p^*-y^*)\|^2_{1\times\operatorname{div}}:=\|\nabla(u-v)\|^2+\|p^*-y^*\|^2_{\operatorname{div}}.$$

Then,

$$\|(u-v,p^*-y^*)\|_{1\times\operatorname{div}} \leq \|\nabla v-y^*\|+\left(1+2C^2_\Omega\right)^{1/2}\|\operatorname{div} y^*+f\|, \quad (5.78)$$

$$\|(u-v,p^*-y^*)\|_{1\times\operatorname{div}} \geq \frac{1}{\sqrt{3}}\left(\|\nabla v-y^*\|+\|\operatorname{div} y^*+f\|\right). \quad (5.79)$$

Estimates (5.78) and (5.79) show that the majorant $M_\oplus$ is *equivalent to the full error norm* with the factors independent of the mesh and approximations used. This important fact confirms that the majorant is an adequate error control tool having the right structure.

5.9 Estimates for nonconforming approximations

Nonconforming methods (e.g., finite difference (FD) or Discontinuous Galerkin (DG)) are widely used in computer simulation. Functional a posteriori estimates exposed above are applicable for conforming approximations. However, they can be also used for the nonconforming ones provided that a suitable projection operator mapping the approximations to the energy space is constructed. In such a case, the functional a posteriori estimate is extended to a wider set of functions. It contains a new term, which is a *penalty for the nonconformity*. For DG approximations of elliptic problems this idea has been analyzed and numerically tested in [194].

Numerical analysts often use approximations which are nonconforming in the sense that the Dirichlét conditions are not exactly satisfied. The necessity of them arises if the Dirichlét boundary has a complicated structure. For example, in computational fluid dynamics, the so-called fictitious domain method is rather popular (see, e.g., Ph. Angot [11], R. Glowinski, T.-W. Pan and J. Periaux [144]). A posteriori error estimates for nonconforming approximations of such a type were obtained in [261]. For the problem (2.4)–(2.5) the respective estimate differs from (4.5) by the additional term that can be viewed as a penalty for possible violation of the prescribed boundary condition.

6 Functional a posteriori estimates for convex variational problems

6.1 General form of the a posteriori estimate

A wide class of practically interesting variational problems has one common general form. Let $\mathcal{V}$ be a reflexive Banach space and

$$J(v) := G(\Lambda v) + F(v) \qquad \forall v \in \mathcal{V} \tag{6.1}$$

where G and F are convex continuous functionals and Λ is a linear continuous operator that maps $\mathcal{V}$ to another Banach space Y. It is assumed that

$$\|\Lambda w\| \geq c_1 \|w\|_{\mathcal{V}} \quad \forall v \in \mathcal{V}. \tag{6.2}$$

Also, we assume that J is coercive on $\mathcal{V}$, i.e.,

$$J(v) \to +\infty \quad \text{as } \|v\|_{\mathcal{V}} \to +\infty. \tag{6.3}$$

In order to explain the form of functional a posteriori estimates this class of problems (see ([236, 238, 242])) we need to introduce an additional notation. The space $\mathcal{V}$ has a topologically dual counterpart $\mathcal{V}^*$. The product of $v \in \mathcal{V}$ and $v^* \in \mathcal{V}^*$ is denoted by $\langle v^*, v\rangle$. The operator $\Lambda^* : \mathcal{V}^* \to Y^*$ satisfying the relation

$$(y^*, \Lambda w) = \langle \Lambda^* y^*, w\rangle \quad \forall w \in \mathcal{V} \tag{6.4}$$

is *conjugate* to Λ. Here, (y^*, y) denotes the value of a functional $y^* \in Y^*$ on $y \in Y$. By G^* we denote the Fenchel conjugate of G, i.e.,

$$G^*(y^*) = \sup_{y \in Y} \left((y^*, y) - G(y) \right).$$

We say that G and G^* are *uniformly convex in the balls* $B_\delta := B(0, \delta) \in Y$ and $B^*_{\delta^*} := B(0, \delta^*) \in Y^*$, respectively, if they satisfy the relations

$$G\Big(\frac{y_1 + y_2}{2}\Big) + \Upsilon_\delta\Big(\frac{y_1 - y_2}{2}\Big) \le \frac{1}{2}\left(G(y_1) + G(y_2)\right), \tag{6.5}$$

$$G^*\Big(\frac{y_1^* + y_2^*}{2}\Big) + \Upsilon^*_{\delta^*}\Big(\frac{y_1^* - y_2^*}{2}\Big) \le \frac{1}{2}\left(G^*(y_1) + G^*(y_2)\right), \tag{6.6}$$

where $\Upsilon_\delta : \mathcal{V} \to \mathbb{R}_+$ and $\Upsilon^*_{\delta^*} : Y^* \to \mathbb{R}_+$ are certain nonnegative functionals vanishing at zero.

The general form of the functional type a posteriori error majorant is given by the theorem below. It gives an upper bound of the differences $v - u$ (for the primal problem) and $y^* - p^*$ (for the dual one) evaluated in terms of the functionals Υ and Υ^*.

Theorem ([236, 242]). *Let u be a minimizer of the problem $\inf\limits_{v \in \mathcal{V}} J(v)$, the functionals F and G satisfy the above conditions, and $\Lambda u \in B_\delta$ and $p^* \in B^*_{\delta^*}$. Then for any $v \in \mathcal{V}$ and $y^* \in Y^*$ such that $\Lambda v \in B_\delta$ and $y^* \in B^*_{\delta^*}$, the estimate*

$$\Upsilon_\delta\Big(\frac{\Lambda(v - u)}{2}\Big) + \Upsilon^*_{\delta^*}\Big(\frac{y^* - p^*}{2}\Big) \le \mathcal{M}_\oplus(v, y^*), \tag{6.7}$$

holds, where

$$2\mathcal{M}_\oplus(v, y^*) := \Big[F(v) + F^*(-\Lambda^* y^*) + \langle \Lambda^* y^*, v \rangle\Big] + \Big[G(\Lambda v) + G^*(y^*) - (y^*, y)\Big].$$

A consequent exposition of the variational approach can be found in [238, 242, 251] and in the book [209].

6.2 Diffusion problem with Robin boundary condition

Let us apply the above general estimate to diffusion problem with Robin type boundary condition. Consider the variational problems for the functional

$$J(v, \nabla v) = \int\limits_\Omega \Big(\frac{1}{2}|\nabla v|^2 + \frac{\delta}{2}|v|^2\Big)dx + \int\limits_{\partial_2 \Omega} \Big(\frac{\alpha}{2}|v|^2 - gv\Big)\, ds.$$

Our problem is to minimize J on the set of functions $V_0 \subset H^1(\Omega)$ vanishing at $\partial_1 \Omega$.

Minimizer u of this variational problem is related to the system

$$-\Delta u + \delta u = 0 \quad \text{in } \Omega,$$

$$\frac{\partial u}{\partial n} + \alpha u - g = 0 \quad \text{on } \partial_2 \Omega.$$

On $\partial_2\Omega$ the solution satisfies the so-called *Robin* boundary condition. Functional a posteriori estimate for this problem is a special case of (6.7) if Λ is associated with the operator ∇v,

$$G(\Lambda w) = \int_\Omega \frac{1}{2}|\nabla v|^2 dx, \quad F(v) = \int_\Omega \frac{\delta}{2}|v|^2 dx + \int_{\partial_2\Omega} (\frac{\alpha}{2}|v|^2 - gv)\, ds\,.$$

Since

$$\int_\Omega y^* \cdot \nabla v dx = \int_\Omega -\operatorname{div} y^* v dx + \int_{\partial_2\Omega} (y^* n) v ds \quad \forall v \in V_0,$$

we observe that $\Lambda^* y^* = \{-\operatorname{div} y^* \mid_\Omega, y^* \cdot n \mid_{\partial_2\Omega}\}$. In the considered case,

$$\langle y^*, y\rangle := \int_\Omega y^* \cdot y\, dx, \quad G^*(-y^*) = \int_\Omega \frac{1}{2}|y^*|^2 dx.$$

Therefore,

$$G(\Lambda v) + G^*(-y^*) + \langle y^*, \Lambda v\rangle = \int_\Omega (\frac{1}{2}|\nabla v|^2 + \frac{1}{2}|y^*|^2 + \nabla v \cdot y^*) dx.$$

If y^* is sufficiently regular, then $\langle \Lambda^* y^*, v\rangle = \int_\Omega -\operatorname{div} y^* v dx + \int_{\partial_2\Omega} y^* \cdot n v ds$ and

$$F^*(\Lambda^* y^*) = \sup_{v\in V_0} \Big\{ \int_\Omega -\operatorname{div} y^* v dx + \int_{\partial_2\Omega} y^* \cdot n v ds - F(v) \Big\} \leq$$

$$\leq \sup_{v\in L_2(\Omega)} \int_\Omega (-\operatorname{div} y^* v - \frac{\delta}{2}|v|^2) dx + \sup_{\varrho\in L_2(\partial_2\Omega)} \int_{\partial_2\Omega} ((y^* \cdot n)\varrho - \frac{\alpha}{2}|\varrho|^2 + g\varrho) ds =$$

$$= \int_\Omega \frac{1}{2\delta}|\operatorname{div} y^*|^2 dx + \int_{\partial_2\Omega} \frac{1}{2\alpha}|y^* \cdot n + g|^2 ds.$$

We obtain

$$F(v) + F^*(\Lambda^* y^*) - \langle \Lambda^* y^*, v\rangle \leq \int_\Omega \frac{1}{2\delta}(\operatorname{div} y^* + \delta v)^2 dx + \int_{\partial_2\Omega} \frac{1}{2\alpha}|y^* \cdot n + g - \alpha v|^2 ds,$$

$$G(\Lambda v) + G^*(-y^*) + \langle y^*, \Lambda v\rangle = \int_\Omega \frac{1}{2}|\nabla v + y^*|^2 dx.$$

The two terms above form the error majorant, which vanishes if and only if

$$\begin{aligned} \operatorname{div} y^* + \delta v = 0 \quad & \text{in } \Omega, \\ y^* \cdot n + g - \alpha v = 0 \quad & \text{on } \partial_2\Omega, \\ y^* = -\nabla v \quad & \text{in } \Omega. \end{aligned}$$

Since v vanishes at $\partial_1\Omega$ these relations implicate that v coincides with the exact solution.

A general approach to the derivation of functional a posteriori estimates for boundary-value problems with nonlinear boundary conditions is presented in [270].

6.3 Nonlinear problems in the theory of solids and fluids

Functional a posteriori estimates for the p-Laplacian problem $p \geq 2$ were suggested in [239]. A complete investigation of the question and better estimates valid for $p \in (1, 2]$ and $p \geq 2$ are presented in [51]. By the variational method, functional a posteriori estimates were also obtained for some nonlinear problems in solid mechanics (see [245, 249, 252, 271]).

Generalized Newtonian fluids present another practically interesting class of nonlinear boundary value problems. Bingham and Powell–Euring models of viscous fluids expose strongly nonlinear behavior. In the context of functional a posteriori estimates, these problems were investigated in [50, 131, 247].

6.4 Variational inequalities

Variational inequalities form an important class of nonlinear problems with wide applications in mechanics (see, e.g., H. Brezis [67], R. Glowinski [142], G. Duvaut and G.-L. Lions [111], A. Friedman [126], D. Kinderlehrer and G. Stampacchia [173]). Functional a posteriori estimates were derived for some classes of variational inequalities in [71, 209, 240]. We briefly discuss their properties on the paradigm of a problem with an obstacle. In this problem, it is required to find

$$u \in K_\phi := \{v \in V_0 \mid v(x) \geq \phi(x) \text{ a.e. in } \Omega\}$$

such that

$$\int_\Omega A\nabla u \cdot \nabla(w-u)\,dx \geq \int_\Omega f(w-u)dx \quad \forall w \in K_\phi, \tag{6.8}$$

where $f \in L^2(\Omega)$, $\phi \in H^2(\Omega)$, and $\phi(x) \leq 0$ on $\partial\Omega$. In general, Ω is divided into two sets:

$$\Omega_c^u := \{x \in \Omega \mid u(x) = \phi(x)\} \quad \text{and} \quad \Omega_0^u := \{x \in \Omega \mid \phi(x) < u(x)\}.$$

Here, Ω_c^u is the so-called *coincidence set* and Ω_0^u is an open set where u satisfies the differential equation. Thus, the problem (6.8) involves free boundaries, which are *unknown a priori*.

Let $v \in K_\phi$ be an approximate solution and

$$\Omega_c^v := \{x \in \Omega \mid v(x) = \phi(x)\} \quad \text{and} \quad \Omega_0^v := \{x \in \Omega \mid \phi(x) < v(x)\}$$

be two sets defined by means of v. Then, a computable upper bound of the deviation norm $\|\!| \nabla(v-u) \|\!|$ can be expressed in several forms. Below we present the simplest one:

$$\| \nabla(v-u) \|^2 \leq (1+\beta) \| \nabla v - A^{-1}y^* \|^2 + \\ + C^2_{\Omega,A}\left(1+\frac{1}{\beta}\right)\left(\int\limits_{\Omega^u_0} |r(y^*)|^2\,dx + \int\limits_{\Omega^u_c} [r(y^*)]^2_+\,dx\right), \quad (6.9)$$

where $r(y^*) = \operatorname{div} y^* + f$, $y^* \in H(\Omega, \operatorname{div})$, $\beta > 0$, $C_{\Omega,A}$ is a constant in the inequality

$$\|w\| \leq C_{\Omega,A} \| w \| \quad \forall w \in V_0,$$

and $[r(y^*)]_+$ denotes the positive part of $r(y^*)$.

In [240], functional a posteriori estimates were derived for elliptic problems with two obstacles, elasto-plastic bending problem, and for a model problem with friction type boundary conditions. Results of numerical tests and a discussion can be found in [209].

6.5 Principal structure of functional a posteriori estimates

As we have seen, functional a posteriori majorants are constructed by the relations that jointly define the exact solution. Typically, they contain global constants that come from the functional inequalities (e.g., Friederichs, Poincaré, LBB) or from inequalities for boundary traces. In other words, a majorant of $\| u-v \|$, where u is the exact solution of a boundary value problem and v is an arbitrary approximation from the respective energy class, is the sum of terms that can be thought of penalties for unconformity in each of the basic relation. The respective multipliers are defined by constants in the *embedding inequalities* for the spaces assocoiated with the mathematical formulation of the problem.

7 Evaluation of errors arising due to uncertainty in the data

In the fully reliable modeling, the errors caused by uncertainties in the problem data must be measured as well as the approximation errors. However, this very important topic is much less investigated than, e.g., error control of approximation errors. From the general scientific positions, such a situation cannot be accepted because any effort to reduce the approximation error has no sense if it does not take into account the accuracy limit imposed by uncertainty in the data. Also, it is clear that practically meaningful results in the area can be obtained only in the framework of a posteriori analysis. This subject started receiving serious attention only recently. Among the publications, we refer to the works of I. Babuška and J. Chleboun [20, 21] (where effects caused by uncertainty in the boundary conditions were considered), I. Babuška, F. Nobile and R. Tempone [26], and J. T. Oden, I. Babuška, F. Nobile, Y. Feng and R. Tempone [215].

In [253], on the paradigm of a linear elliptic problem a method of evaluating errors caused by uncertainties in the coefficients with the help of functional a posteriori estimates was suggested. Below we briefly discuss the idea of this approach.

Consider the problem (5.1) where the operator $\mathcal{A}$ and the functional ℓ are *defined with an indeterminacy.* In such a case, instead of particular $\mathcal{A}$ and ℓ it is only known that

$$\mathcal{A} \in \mathcal{U}_{\mathcal{A}} \subset \mathcal{L}(U,U) \qquad \text{and} \quad \ell \in \mathcal{U}_{\ell} \subset V_0^*,$$

where $\mathcal{U}_{\mathcal{A}}$ and $\mathcal{U}_{\ell}$ are certain bounded sets of "admissible" data. In order to guarantee solvability of all the problems, we assume that (5.3) holds and

$$c_1\|y\|^2 \leq (\mathcal{A}y, y) \leq c_2\|y\|^2 \qquad \forall y \in U \; \forall \mathcal{A} \in \mathcal{U}_{\mathcal{A}}.$$

All possible solutions of the problem with such a data form the set

$$\Upsilon(\mathcal{U}_{\mathcal{A}}, \mathcal{U}_{\ell}) := \left\{ \tilde{u} \in V_0 + u_0 \,\middle|\, \Lambda^* \mathcal{A} \Lambda \tilde{u} = \ell \;\text{ for some } \mathcal{A} \in \mathcal{U}_{\mathcal{A}},\; \ell \in \mathcal{U}_{\ell} \right\}.$$

Let $v \in V_0 + u_0$ be an approximation of an unknown exact solution. Since the data are indeterminate, the error estimation problem comes in two different forms. The first task is to find the quantity

$$e_{\min}(v, \Upsilon) = \inf_{\tilde{u} \in \Upsilon} \|\Lambda(v - \tilde{u})\| . \tag{7.1}$$

The quantity $e_{\min}$ measures the distance between v and the set Υ. It is equal to zero if $\Lambda^* \mathcal{A} \Lambda v = \ell$ for some pair $(\mathcal{A}, \ell) \in \mathcal{U}_{\mathcal{A}} \times \mathcal{U}_{\ell}$. This quantity provides the lowest possible bound of the true error or the error in the *best-case scenario*. The second task is to find the quantity

$$e_{\max}(v, \Upsilon) = \sup_{\tilde{u} \in \Upsilon} \|\Lambda(v - \tilde{u})\| , \tag{7.2}$$

which add computational and indeterminacy errors. It gives the error *in the worst-case scenario*. Evidently, $e_{\max}(v, \Upsilon)$ is always positive.

Theoretically, it is also important to obtain estimates of

$$\mathbb{D}(\Upsilon) := \sup_{u_1, u_2 \in \Upsilon} \|\Lambda(u_1 - u_2)\| \tag{7.3}$$

that shows the diameter of the set $\Upsilon(\mathcal{U}_{\mathcal{A}}, \mathcal{U}_{\ell})$. The quantity $\mathbb{D}(\Upsilon)$ gives a presentation on the accuracy limit reasoned by the indeterminacy in the data.

In general, the exact values of $e_{\min}$, $e_{\max}$, and $\mathbb{D}(\Upsilon)$ cannot be found. However, functional a posteriori estimates open a way of finding their computable bounds. For the first two quantities, the respective method is exposed in [253] (see also [209]).

Note that

$$\|\Lambda(u_1 - u_2)\| \geq \|\Lambda(v_1 - v_2)\| - \|\Lambda(u_1 - v_1)\| - \|\Lambda(u_2 - v_2)\| , \tag{7.4}$$

where v_1 and v_2 are conforming approximations of u_1 and u_2, respectively. Therefore, a lower bound of the quantity $\mathbb{D}(\Upsilon)$ can be computed as follows. Assume that v_i are approximate solutions of the boundary value problems with $\mathcal{A}_i \in \mathcal{U}_\mathcal{A}$ and $\ell_i \in \mathcal{U}_\ell$ and ϵ_i $(i = 1, 2, \ldots, m)$ are the respective error bounds computed with the help of the majorant (5.8), i.e.,

$$\epsilon_i \leq \frac{1}{\sqrt{c_1}} \left(\sqrt{2D_i(\Lambda v, y^*)} + c\|\ell_i + \Lambda^* y^*\| \right). \tag{7.5}$$

Then

$$\mathbb{D}(\Upsilon) \geq \mathbb{D}_\ominus(\Upsilon, m) := \sup_{i,j=1,2,\ldots,m} \{\|\Lambda(v_i - v_j)\| - \epsilon_i - \epsilon_j\}. \tag{7.6}$$

It is worth outlining that $\mathbb{D}_\ominus(\Upsilon, m)$ is indeed computable because all the computations are reduced to solving finite dimensional problems. The more approximate solutions v_i are considered, the sharper lower bound of $\mathbb{D}(\Upsilon)$ is given by (7.6).

It should be remarked that the influence of various indeterminacies (e.g., roundoff errors) affecting computations can be also investigated in terms of the so-called *interval analysis*. In it, the operations are performed for intervals instead of numbers. This theory deserves a special consideration, which is beyond the subject of the present paper. The respective discussion is contained in, e.g., G. Alefeld and J. Herzberger [8], B. S. Dobronets and V. V. Shaidurov [106], Y. V. Matijasevich [198], R. E. Moore [204] and in the references cited.

8 Evaluation of modeling errors

Simplified models are often used in computer simulation if it is a priori clear that the major part of a solution can be computed with the help of simpler mathematical problem. Typical examples give the models constructed by *dimension reduction* and *homogenization*.

Dimension reduction models lead to certain errors caused by replacing a certain n-dimensional boundary value problem by a simplified $\widehat{n}$-dimensional one (where $\widehat{n} = n - k$ and $k \in (1, n-1)$). Let u and $\widehat{u}$ be the solutions of the original and simplified problems, respectively. In general, they belong to different functional spaces V and $\widehat{V}$, so that a projection operator $\Pi : \widehat{V} \to V$ is required in order to compare u and $\widehat{u}$. Then, the quantity

$$e_{\text{mod}} := \|u - \Pi\widehat{u}\|_V$$

is the *dimension reduction error*. Estimation of modeling errors is an important task that was investigated by several author (see, e.g., C. Schwab [278], I. Babuška, I. Lee and C. Schwab [25, 31], M. Ainsworth [2], M. Vogelius and I. Babuška [305, 306]).

Functional a posteriori estimates are valid for all conforming approximations, so that the function $\Pi\widehat{u}$ can be compared with u in the framework of the unified approach discussed above (see [248, 249, 263, 264, 266] where a posteriori estimates for modeling errors were obtained and tested). Let us briefly discuss main features of such estimates on the paradigm of *plain stress problem* in linear elasticity. Let $\Omega = \widehat{\Omega} \times (-d, +d)$,

$\widehat{\Omega} \in R^2$ with boundary γ. Assume that

$$d \ll \operatorname{diam}(\widehat{\Omega}) := \sup_{(x_1,x_2)\in\widehat{\Omega}} |x_1 - x_2| \,.$$

In this model, the solution (u, σ) is approximated by the pair of functions $(\widehat{u}, \widehat{\sigma})$, where $\widehat{u} = (\widehat{u}_1, \widehat{u}_2)$ and $\widehat{\sigma}$ is a 2×2 tensor. Let

$$\widetilde{u} = (\widehat{u}_1, \widehat{u}_2, \phi(x_1, x_2, x_3)); \; \widetilde{\sigma}_{\alpha\beta} = \widehat{\sigma}_{\alpha\beta}, \quad \widetilde{\sigma}_{3\alpha} = 0,$$

where $\phi \in H^1(\Omega)$ and satisfies the boundary condition stated for u_3 on the Dirichlét part of $\partial\widehat{\Omega} \times (-d, d)$. A posteriori estimate for the difference between exact solutions of the original (3d) model and the simplified one is as follows:

$$C_\varepsilon \, \| \, \varepsilon(\widetilde{u} - u)\|_\Omega^2 + C_\tau \, \| \, \widetilde{\sigma} - \sigma\|_\Omega^2 \leq$$
$$\leq \left(\frac{K_0}{2} + \frac{2\mu}{3}\right) \int\limits_\Omega \left(\rho\left(\widehat{u}_{1,1} + \widehat{u}_{2,2}\right) + \phi_{,3}\right)^2 dx + \frac{\mu}{2} \int\limits_\Omega \left(\phi_{,1}^2 + \phi_{,2}^2\right) dx.$$

Here μ and K_0 are the elasticity coefficients, $\rho = \frac{3K_0 - 2\mu}{3K_0 + 4\mu}$, $C_\varepsilon = \min\{\frac{1}{\mu}, \frac{2}{3\mu}\}$, and $C_\tau = \min\{4\mu, 6K_0\}$.

Computable estimates for dimension reduction models related to the problem

$$\operatorname{div}(A\nabla u) + f = 0, \quad u = u_0 \text{ on } \partial\Omega$$

were derived and investigated in [263]. In this paper, more complicated "thin" domains were considered (namely, $\Omega = \Gamma \times [-d, d]$, where Γ is a certain surface). In a particular case of a plate with plane parallel faces, the general estimate obtained in [263] converts to

$$\| \, u - \widehat{u} \, \| \leq \sqrt{\frac{d_0}{3}} \left(\int_{\widehat{\Omega}} \frac{F_\oplus^2 + F_\ominus^2 - F_\oplus F_\ominus}{a_{33}} d\widehat{x} \right)^{1/2} + C_\Omega \|f - \widetilde{f}\|_{L^2(\Omega)} \, ,$$

where d_0 is the thickness parameter, $F_\oplus(\widehat{x})$ and $F_\ominus(\widehat{x})$ are the forces acting on the upper and lower faces, and $\widetilde{f}$ is the function obtained by averaging of f with respect to the x_3-coordinate. If, in addition, we assume that $f = 0$, $a_{33} = 1$ and $F_\oplus = F_\ominus = F(\widehat{x})$, then

$$\| \, u - \widehat{u} \, \| \leq \sqrt{\frac{d_0}{3}} \, \|\widehat{F}\|_{L^2(\widehat{\Omega})},$$

which is exactly the estimate obtained in [25, 31] for zero-order models.

Bibliography

[1] Y. Achdou, C. Bernardi, and F. Coquel, *A priori and a posteriori analysis of finite volume discretizations of Darcy's equations*, Numer. Math. 96 (2003), pp. 17–42.

[2] M. Ainsworth, *A posteriori error estimation for fully discrete hierarchic models of elliptic boundary value problems in thin domains,* Numer. Math. 80 (1998), pp. 325–362.

[3] M. Ainsworth and J.T. Oden, *A posteriori error estimation in the finite element method,* Numer. Math. 60 (1992), pp. 429–463.

[4] ______, *A unified approach to a posteriori error estimation using element residual methods,* Numer. Math. 65 (1993), pp. 23–50.

[5] ______, A posteriori error estimation in finite element analysis, Wiley and Sons, New York, 2000.

[6] M. Ainsworth, J.T. Oden, and C.Y. Lee, *Local a posteriori error estimators for variational inequalities*, Numer. Methods for PDE 9 (1993), pp. 23–33.

[7] M. Ainsworth, J.Z. Zhu, A.W. Craig, and O.C. Zienkiewicz, *Analysis of the Zienkiewicz-Zhu a posteriori error estimator in the finite element method*, Int. J. Numer. Methods Engrg. 28 (1989), pp. 2161–2174.

[8] G. Alefeld and J. Herzberger, *Introduction to interval computations,* Academic Press, New York, 1983.

[9] A. Alonso, *Error estimators for a mixed method*, Numer. Math. 74 (1996), pp. 385–395.

[10] M. Amara, M. Ben Younes, and C. Bernardi, *Error indicators for the Navier-Stokes equations in stream function and vorticity formulation*, Numer. Math. 80 (1998), pp. 181–206.

[11] Ph. Angot, *Analysis of singular perturbations on the Binkman problem for ficticious domain models of viscous flows,* Math. Meth. Appl. Sci. 22 (1999), pp. 1395–1412.

[12] A.M. Arthurth, *Complementary variational principles,* Clarendon Press, Oxford, 1980.

[13] S.N. Atluri (ed.), *Computational Methods in the mechanics of Fracture*, North-Holland, New York, 1986.

[14] J.P. Aubin, *Behavior of the error of the approximate solutions of boundary value problems for linear elliptic operators by Galerkin's and finite difference methods*, Ann. Scuola Norm. Sup. Pisa 21 (1967), pp. 599–637.

[15] G. Auchmuty, *A posteriori error estimates for linear equations,* Numer. Math. 61 (1992), pp. 1–6.

[16] I. Babuška, *Courant element: before and after*, Fifty Years of Courant Element, M. Křížek, P. Neittaanmäki and R. Stenberg, (eds.), Marcel Dekker, 1994, pp. 37–51.

[17] ______, *The finite element method with Lagrangian multipliers,* Numer. Math. 20 (1973), pp. 179–192.

[18] ______, *Fifty years of applications of mathematics,* Applications of mathematics 51 (2006), pp. 1–4.

[19] I. Babuška and A.K. Aziz, *Survey lectures on the mathematical foundations of the finite element method*. The mathematical formulations of the finite element method with applications to partial differential equations, Academic Press, New York, 1972, pp. 1–359.

[20] I. Babuška and J. Chleboun, *Effects of uncertainties in the domain on the solution of Neumann boundary value problems in two spatial dimensions,* Math. Comput. 71 (2001), pp. 1339–1370.

[21] ———, *Effects of uncertainties in the domain on the solution of Drichlet boundary value problems.* Numer. Math. 93 (2003), pp. 583–610.

[22] I. Babuška, R. Duran, and R. Rodríguez, *Analysis of the efficiency of a posteriori error estimator for linear triangular elements,* SIAM J. Numer. Anal. 29 (1992), pp. 947–964.

[23] I. Babuska and G.N. Gatica, *On the mixed finite element method with Lagrange multipliers,* Numer. Meth. Partial Diff. Eq. 19 (2003), pp. 192–210.

[24] I. Babuška, F. Ihlenburg, T. Strouboulis, and S.K. Gangaraj, *A posteriori error estimation for finite element solutions on Helmholtz' equation – Part II: estimation of the pollution error,* Int. J. Numer. Meth. Engrg. 40 (1997), pp. 3883–3900.

[25] I. Babuška, I. Lee, and C. Schwab, *On the a posteriori estimation of the modeling error for the heat conduction in a plate and its use for adaptive hierarchical modeling,* Appl. Numer. Math. 14 (1994), pp. 5–21.

[26] I. Babuška, F. Nobile, and R. Tempone, *Worst case scenario analysis for elliptic problems with uncertainty,* Numer. Math. 101 (2005), pp. 185–219.

[27] I. Babuška and J.E. Osborn, *Can a finite element method perform arbitrarily badly?,* Math. Comput. 69 (2000), pp. 443–462.

[28] I. Babuška and W.C. Rheinboldt, *A-posteriori error estimates for the finite element method.* Internat. J. Numer. Meth. Engrg. 12 (1978), pp. 1597–1615.

[29] ———, *Error estimates for adaptive finite element computations.* SIAM J. Numer. Anal. 15 (1978), pp. 736–754.

[30] I. Babuška and R. Rodríguez, *The problem of the selection of an a posteriori error indicator based on smoothing techniques,* Internat. J. Numer. Meth. Engrg. 36 (1993), pp. 539–567.

[31] I. Babuška and C. Schwab, *A posteriori error estimation for hierarchic models of elliptic boundary value problems on thin domains,* SIAM J. Numer. Anal. 33 (1996), pp. 221–246.

[32] I. Babuška and T. Strouboulis, *The finite element method and its reliability*, Oxford University Press, New York, 2001.

[33] I. Babuška, T. Strouboulis, C.S. Upadhyay, and S.K. Gangaraj, *A posteriori estimation and adaptive control of the pollution-error in the h-version of the FEM,* Int. J. Numer. Meth. Engrg. 38 (1995), pp. 4207–4235.

[34] I. Babuška, T. Strouboulis, S.K. Gangaraj, and C.S. Upadhyay, *Pollution-error in the h-version of the FEM and the local quality of recovered derivatives,* Comput. Methods Appl. Mech. Engrg. 140 (1997), pp. 1–37.

[35] I. Babuška, T. Strouboulis, and C.S. Upadhyay, *A model study of the quality of a posteriori error estimators for linear elliptic problems. Error estimation in the interior of patchwise uniform grids of triangles,* Comp. Meth. Appl. Mech. Engrg. 114 (1994), pp. 307–378.

[36] W. Bangerth and R. Rannacher, *Adaptive finite element methods for differential equations,* Birkhäuser, Berlin, 2003.

[37] R.E. Bank and R.K. Smith, *Mesh smoothing using a posteriori error estimates,* SIAM J. Numer. Anal. 34 (1997), pp. 979–997.

[38] R.E. Bank and A. Weiser, *Some a posteriori error estimators for elliptic partial differential equations,* Math. Comput. 44 (1985), pp. 283–301.

[39] R.E. Bank and B.D. Welfert, *A posteriori error estimates for the Stokes problem,* SIAM J. Numer. Anal. 28 (1991), pp. 591–623.

[40] C. Bartels and C. Carstensen, *Averaging techniques yield reliable a posteriori finite element error control for obstacle problems,* Numer. Math. 99 (2004), pp. 225–249.

[41] R. Beck, R. Hiptmair, R.H.W. Hoppe, and B. Wohlmuth. Residual based a posteriori error estimators for eddy current computation. RAIRO – Math. Model. Numer. Anal. 34 (2000), 159–182.

[42] R. Becker, P. Hansbo, and M.G. Larsson, *Energy norm a posteriori error estimation for discontinuous Galerkin methods,* Comput. Methods Appl. Mech. Engrg. 192 (2003), pp. 723–733.

[43] R. Becker, H. Kapp, and R. Rannacher, *Adaptive finite element methods for optimal control of partial differential equations: Basic concept,* SIAM J. Control Optim. 39 (2000), pp. 113–132.

[44] R. Becker and R. Rannacher, *A feed-back approach to error control in finite element methods: Basic approach and examples,* East-West J. Numer. Math. 4 (1996), pp. 237–264.

[45] R. Becker and B. Vexler, *A posteriori error estimation for finite element discretization of parameter identification problems,* Numer. Math. 96 (2004), pp. 435–459.

[46] T. Belytschko and T. Blacker, *Enhanced derivative recovery through least square residual penalty,* Appl. Num. Math. 14 (1994), pp. 55–68.

[47] A. Bergam, Z. Mghazli, and R. Verfürth, *A posteriori estimators for the finite volume discretization of elliptic problems,* Numer. Math. 95 (2003), pp. 599–624.

[48] A. Bermúdez, R. Durán, and R. Rodríguez, *Finite element analysis of compressible and incompressible fluid-solid systems,* Math. Comp. 67 (1998), pp. 111–136.

[49] C. Bernardi and R. Verfürth, *A posteriori error analysis of the fully discretized time-dependent Stokes equations,* ECIAM:M2NL 38 (2004), pp. 437–455.

[50] M. Bildhauer, M. Fuchs and S. Repin, *A posteriori error estimates for stationary slow flows of power-law fluids,* International Journal of Non-Newtonian Fluids (in press).

[51] M. Bildhauer and S. Repin, *Estimates for the deviation from exact solutions of variational problems with power growth functionals,* 336 (2006), pp. 5–24.

[52] B. Boroomand and O.C. Zienkiewicz, *Recovery by equilibrium in patches (REP),* Int. J. Numer. Meth. Engrg. 40 (1997), pp. 137–164.

[53] ———, *An improved REP recovery and the effectivity robustness test,* Int. J. Numer. Meth. Engrg. 40 (1997), pp. 3247–3277.

[54] V. Bostan and W. Han, *A posteriori error analysis for finite element solutions of a frictional contact pronlem,* Comput. Meth. Appl. Math. Engrg. 195 (2006), pp. 1252–1274.

[55] V. Bostan, W. Han, and B.D. Reddy, *A posteriori error estimation and adaptive solution of elliptic variational inequalities of the second kind,* Appl. Num. Math. 52 (2005), pp. 13–38.

[56] S.W. Brady and A.R. Elcrat, *Some results on a posteriori error estimation for approximate solution of second order elliptic problems,* SIAM J. Numer. Anal. 16 (1979), pp. 877–889.

[57] D. Braess, *Finite elements*, Cambridge University Press, Cambridge, 3rd edition, 2007.

[58] ———, *A posteriori error estimators for obstacle problems – another look,* Numer. Math. 101 (2005), pp. 415–421.

[59] D. Braess and J. Schöberl, *Equilibrated residual error estimator for Maxwell's equation,* Preprint, 2006.

[60] D. Braess and R. Verfürth, *A posteriori error estimators for the Raviart-Thomas element,* SIAM J. Numer. Anal. 33 (1996), pp. 2431–2444.

[61] J.H. Bramble, *A proof of the inf-sup condition for the Stokes equations on Lipschitz domains,* Math. Models Methods Appl. Sci. 13 (2003), pp. 361–371.

[62] J.H. Bramble, R.D. Lazarov, and J.E. Pasciak, *Least-squares for second-order elliptic problems,* Comput. Methods Appl. Mech. Engrg. 152 (1998), pp. 195–210.

[63] J.H. Bramble and A.H. Schatz, *Higher order local accuracy by averaging in the finite element method,* Math. Comp. 31 (1977), pp. 94–111.

[64] J.H. Brandts, *Superconvergence and a posteriori error estimation for triangular mixed finite elements,* Numer. Math. 68 (1994), pp. 311–324.

[65] J.H. Brandts and Y. Chen, *An alternative to the least-squares mixed finite element method for elliptic problems,* Numerical Mathematics and Advanced Applications, M. Feistauer, V. Dolejší, P. Knobloch, and K. Najzar (eds.), Springer, Berlin, 2004, pp. 169–175.

[66] K. Braune, *A posteriori error bounds in the computation of inverse standard functions with the aid of the Newton method,* Z. Angew. Math. Mech., 70 (1990), pp. T579–T581 (German).

[67] H. Brezis, *Problémes unilatéraux,* J. Math. Pures Appl. 9 (1971), pp. 1–168.

[68] F. Brezzi, *On the existence, uniqueness and approximation of saddle-point problems arising from Lagrange multipliers,* R.A.I.R.O., Annal. Numer. 8 (1974), pp. 129–151.

[69] F. Brezzi, J.J. Douglas, R. Duran, and M. Fortin, *Mixed finite elements for second order elliptic problems in three variables,* Numer. Math. 51 (1987), pp. 237–250.

[70] F. Brezzi and M. Fortin, *Mixed and hybrid finite element methods*, Springer Series in Computational Mathematics 15, New York, 1991.

[71] H. Buss and S. Repin, *A posteriori error estimates for boundary value problems with obstacles,* Numerical mathematics and advanced applications (Jyvaskyla, 1999), World Sci. Publishing, River Edge, NJ, 2000, pp. 162–170.

[72] R. Bustinza, G.N. Gatica, and B. Cockburn, *An a posteriori error estimate for the local Discontinuous Galerkin method applied to linear and nonlinear diffusion problems,* Journal of Scientific Computing, 22/23 (2005), pp. 147–185.

[73] Z. Cai, R. Lazarov, T.A. Manteuffel, and S.F. McCormick, *First-order system least squares for second-order partial differential equations: Part I,* SIAM J. Num. Anal. 31 (1994), pp. 1785–1799.

[74] Z. Cai, T.A. Manteuffel, S.F. McCormick, and S.V. Parter, *First-order system least squares (FOSLS) for planar linear elasticity: Pure traction problem,* SIAM J. Num. Anal. 35 (1998), pp. 320–335.

[75] G.F. Carey and D.L. Humphrey, *Mesh refinement and iterative solution methods for finite element computations,* Int. J. Numer. Meth. Engrg. 17 (1981), pp. 1717–1734.

[76] G.F. Carey and A.I. Pehlivanov, *Local error estimation and adaptive remeshing scheme for least-squares mixed finite elements,* Comput. Methods Appl. Mech. Engrg. 150 (1997), pp. 125–131.

[77] C. Carstensen, *A posteriori estimate for the mixed finite element method,* Math. Comp. 66 (1997), pp. 465–476.

[78] ———, *Quasi-interpolation and a posteriori error analysis of finite element methods,* Math. Model. Numer. Anal. 33 (1999), pp. 1187–1202.

[79] ———, *A posteriori error estimate for the mixed finite element method,* Math. Comp. 66 (1997), pp. 465–476.

[80] ———, *An a posteriori error estimate for first-kind integral equation,* Math. Comp. 66 (1997), pp. 139–155.

[81] C. Carstensen, P. Causin, and R. Sacco, *A posteriori dual-mixed adaptive finite element error control for Lamé and Stokes equations,* Numer. Math. 101 (2005), pp. 309–332.

[82] C. Carstensen and G. Dolzmann, *A posteriori error estimates for mixed FEM in elasticity,* Numer. Math. 81 (1998), pp. 187–209.

[83] C. Carstensen and S. Bartels, *Each averaging technique yields reliable a posteriori error control in FEM on unstructured grids. I: Low order conforming, nonconforming, and mixed FEM,* Math. Comp. 71 (2002), pp. 945–969.

[84] C. Carstensen and S.A. Funken, *Fully reliable localized error control in the FEM,* SIAM J. Sci. Comput. 21 (2000), pp. 1465–1484.

[85] ———, *Constants in Clément-interpolation error and residual based a posteriori error estimates in Finite Element Methods,* East-West J. Numer. Math. 8 (2000), pp. 153–175.

[86] C. Carstensen and R.H.W. Hoppe, *Error reduction and convergence for an adaptive mixed finite element method,* Math. Comp. 75 (2006), pp. 1033–1042.

[87] ———, *Convergence analysis of an adaptive nonconforming finite element method,* Numer. Math. 103 (2006), pp. 251–266.

[88] C. Carstensen, J. Hu, and A. Orlando, *Framework for the A Posteriori Error Analysis of Nonconforming Finite Elements*, SIAM J. Numer. Anal. 45 (2007), pp. 68–82.

[89] C. Carstensen, R. Klose, and A. Orlando, *Reliable and efficient equilibrated a posteriori finite element error control in elastoplasticity and elastoviscoplasticity with hardening,* Comput. Methods Appl. Mech. Engrg. 195 (2006), pp. 2574-2598.

[90] C. Carstensen and S. Sauter, *A posteriori error analysis for elliptic PDEs on domains with complicated structures,* Numer. Math. 96 (2004), pp. 691–721.

[91] C. Carstensen, R. Lazarov, and S. Tomov, *Explicit and averaging a posteriori error estimates for adaptive finite volume methods,* SIAM J. Numer. Anal. 42 (2005), pp. 2496–2521.

[92] C. Carstensen and R. Verfürth, *Edge residuals dominate a posteriori error estimates for low order finite element methods,* SIAM J. Numer. Anal. 36 (1999), pp. 1571–1587.

[93] P. Castillo, *An a posteriori error estimate for the local Discontinuous Galerkin method,* Journal of Scientific Computing, 22/23 (2005), pp. 187–2004.

[94] A. Charbonneau, K. Dossou, and R. Pierre, *A residual-based a posteriori error estimator for Ciarlet–Raviart formulation of the first biharmonic problem,* Numer. Meth. PDE 13 (1997), pp. 93–111.

[95] Z. Chen and R.H. Nochetto, *Rezidual type a posteriori error estimates for elliptic obstacle problems,* Numer. Math. 84 (2000), pp. 527–548.

[96] X. Cheng, W. Han, and H. Huang, *Analysis of some mixed elements for the Stokes problem,* Journal of Computational and Applied Mathematics 85 (1997), pp. 19–35.

[97] E. Chizhonkov and M. Olshanskii, *On the domain geometry dependence of the LBB condition,* Math. Model. Numer. Anal. 34 (2000), pp. 935–951.

[98] S.-S. Chow, G.F. Carey, and R.D. Lazarov, *Natural and post-processed superconvergence in semilinear problems,* Numer. Meth. PDE 7 (1991), pp. 245–259.

[99] P.G. Ciarlet, *The finite element method for elliptic problems.* North Holland, New York, Oxford, 1978.

[100] Ph. Clément, *Approximations by finite element functions using local regularization,* RAIRO Anal. Numér. 9 (1975), pp. 77–84.

[101] S. Cochez-Dhondt and S. Nicaise, *Robust a posteriori error estimation for the Maxwell equations*, Comput. Methods Appl. Mech. Engrg. 196 (2007), pp. 2583–2595.

[102] L. Collatz, *Funktionanalysis und Numerische Mathematik*, Springer, Berlin, 1964.

[103] P. Coorevits, J.-P. Dumeau, and J.-P. Pelle, *Error estimator and adaptivity for three-dimensional finite element analysis,* Advances in Adaptive Computational Methods in Mechanics, P. Ladevèze and J.T. Oden (eds.), Elsevier, New York, 1998, pp. 443–457.

[104] R. Courant, *Variational methods for some problems of equilibrium and vibration,* Bulletin of AMS 49 (1943), pp. 1–23.

[105] P. Destuynder and B. Métivet, *Explicit error bounds of a conforming finite element method,* Math. Comp. 68 (1999), pp. 1379–1396.

[106] B.S. Dobronets and V.V. Shaĭdurov, *Two-sided numerical methods*, Nauka Sibirsk. Otdel., Novosibirsk, 1990 (Russian).

[107] M. Dobrowolski, *On the LBB constant on stretched domains,* Math. Nachr. 254–255 (2003), pp. 64–67.

[108] W. Dörfler and M. Rumpf, *An adaptive strategy for elliptic problems including a posteriori controlled boundary approximation,* Math. Comp. 67 (1998), pp. 1361–1362.

[109] R. Duran, M.A. Muschietti, and R. Rodríguez, *On the asymptotic exactness of error estimators for linear triangle elements,* Numer. Math. 59 (1991), pp. 107–127.

[110] R. Duran and R. Rodríguez, *On the asymptotic exactness of Bank–Weiser's estimator,* Numer. Math. 62 (1992), pp. 297–303.

[111] G. Duvaut and J.-L. Lions, *Les inequations en mecanique et en physique*, Dunod, Paris, 1972.

[112] I. Ekeland and R. Temam, *Convex analysis and variational problems*, North-Holland, Amsterdam, 1976.

[113] B. Engelmann, R.H.W. Hoppe, B. Wolmuth, Yu. Kuznetsov, Yu. Iliash, and Yu. Vasilevskii, *Adaptive macro-hybrid finite element methods,* Proceedings of the 2nd European Conference on Numerical Mathematics and Advanced Applications, Heidelberg, 1997, World Sci, Publishing, River Edge, New York, 1998, pp. 294–302.

[114] H.W. Engl and O. Scherzer, *Convergence rates results for iterative methods for solving nonlinear ill-posed problems,* Surveys on solution methods for inverse problems, Springer, Vienna, 2000, pp. 7–34.

[115] K. Eriksson, D. Estep, P. Hansbo, and C. Johnson, *Computational differential equations*, Cambridge University Press, Cambridge, 1996.

[116] K. Eriksson and C. Johnson, *An adaptive finite element method for linear elliptic problems,* Math. Comp. 50 (1988), pp. 361–383.

[117] ———, *Adaptive finite element methods for parabolic problems. I. A linear model problem,* SIAM J. Numer. Anal. 28 (1991), pp. 43–77.

[118] A. Ern and J. Proft, *A posteriori Discontinuous Galerkin error estimates for transient convection-diffusion equations,* Appl. Math. Lett. 18 (2005), pp. 833–841.

[119] R.E. Ewing, *A posteriori error estimation,* Comput. Meth. Appl. Mech. Engrg. 82 (1990), pp. 59–72.

[120] R.E. Ewing, R.D. Lazarov, and J. Wang, *Superconvergence of the velocity along the Gauss lines in mixed finite element methods,* SIAM J. Numer. Anal. 18 (1991), pp. 1015–1029.

[121] M. Farhloul, S. Nicaise, L. Paquet, *A posteriori error estimation for the dual mixed finite element method of the Stokes problem*, Numer. Funct. Anal. Optim. 27 (2006), pp. 831–846.

[122] M. Feistauer, *Mathematical methods in fluid dynamics.* Longman, Harlow, 1993.

[123] L.P. Franca, J. Karam Filho, A.F.D. Loula, and R. Stenberg, *A convergence analysis of a stabilized method for the Stokes flow,* Mat. Apl. Comput. 10 (1991), pp. 19–26.

[124] L.P. Franca and R. Stenberg, *Error analysis of Galerkin least squares methods for the elasticity equations,* SIAM J. Numer. Anal. 28 (1991), pp. 1680–1697.

[125] D. French, S. Larsson, and R.H. Nochetto, *Pointwisw a posteriori error analysis for an adaptive penalty finite element method for the obstacle problem,* Comput. Methods Appl. Math. 1 (2001), pp. 18–38.

[126] A. Friedman, *Variational principles and free-boundary problems*, Wiley and Sons, New York, 1982.

[127] M. Frolov, *On the efficiency of the dual majorant method for the quality estimation of approximate solutions of fourth-order elliptic boundary value problems,* Russian J. Numer. Anal. Math. Modelling 19 (2004), pp. 407–418.

[128] M. Frolov, P. Neittaanmäki, and S. Repin, *On the reliability, effectivity and robustness of a posteriori error estimation methods*, Numerical Methods for Scientific Computing, Variational problems and applications, pp. 153–175, CIMNE, Barcelona, 2003.

[129] ______, *On computational properties of a posteriori error estimates based upon the method of duality error majorants*, Numerical mathematics and advanced applications, Springer, Berlin, 2004, pp. 346–357.

[130] ______, *Guaranteed functional error estimates for the Reissner-Mindlin plate problem,* J. Math. Sci., New York 132 (2006), pp. 553–561.

[131] M. Fuchs and S. Repin, *Estimates for the deviation from the exact solutions of variational problems modeling certain classes of generalized Newtonian fluids,* Math. Meth. Appl. Sci. 29 (2006), pp. 2225–2244.

[132] A. Gaevskaya, R.H.W. Hoppe, and S. Repin, *A posteriori estimates for cost functionals of optimal control problems,* Proceedings of the 6th European Conference on Numerical Mathematics and Advanced Applications, Santiago de Compostela, 2005, Springer, Berlin, 2006, pp. 291–298.

[133] A. Gaevaskaya, R.H.W. Hoppe, and S. Repin, *Functional approach to a posteriori error estimation for elliptic optimal control problems with distributed control*, J. Math. Sci., New York 144 (2007), pp. 4535–4547.

[134] A. Gaevskaya and S. Repin, *A posteriori error estimates for approximate solutions of linear parabolic problems,* Differential Equations 41 (2005), pp. 970–983 (Russian); Translation in Differ. Equ. 41 (2005), pp. 970–983.

[135] H. Gajewski, K. Gröger, and K. Zacharias, *Nichtlineare Operatorgleichungen and Operatordifferentialgleichungen,* Akademie-Verlag, Berlin, 1974.

[136] B.G. Galerkin, *Beams and plates. Series in some questions of elastic equilibrium of beams and plates.* Vestnik Ingenerov 19 (1915), pp. 897–908 (Russian).

[137] L. Gallimard, P. Ladevèze, and J.-P. Pelle, *Error estimation and adaptivity in elastoplasticity,* Int. J. Numer. Meth. Engrg. 39 (1996), pp. 189–217.

[138] G.N. Gatica, *A note on the efficiency of residual-based a-posteriori error estimators for some mixed finite element methods,* Electron. Trans. Numer. Anal. 17 (2004), pp. 218-233.

[139] E. Geisler, A. Tal, and D. Garg, *On a posteriori error bounds for the solution of ordinary nonlinear differential equations,* Computers and Mathematics with Applications, Pergamon, Oxford, 1976, pp. 407–416.

[140] D. Gilbarg and N.S. Trudinger, *Elliptic partial differential equations of second order*, Springer, Berlin, 1977.

[141] V. Girault and P.A. Raviart, *Finite element approximation of the Navier–Stokes equations*, Springer, Berlin, 1986.

[142] R. Glowinski, *Numerical methods for nonlinear variational problems*, Springer, New York, 1982.

[143] R. Glowinski and O. Pironneau, *Numerical methods for the first biharmonic equation and for the two-dimensional Stokes problem,* SIAM Review 21 (1999), pp. 167–212.

[144] R. Glowinski, T.-W. Pan, and J. Periaux, *A fictitious domain method for Dirichlet problem and applications,* Comput. Methods Appl. Mech. Engrg. 111 (1994), pp. 283–303.

[145] E. Gorshkova, P. Neittaanmäki, and S. Repin, *Comparative study of a posteriori error estimates for the Stokes problem,* Proc. of the 6th European Conference on Numerical Mathematics and Advanced Applications, Santiago de Compostela, 2005, Springer, Berlin, 2006, pp. 255–262.

[146] E. Gorshkova, A. Mahalov, P. Neittaanmäki, and S. Repin, *Functional a posteriori estimates for viscous flow problems with rotation,* University of Jyväskyl]"a, Preprint 2006 (to appear in J. Math. Sci., New York, 2007).

[147] G. Haase, M. Kuhn, and U. Langer, *Parallel multigrid 3D Maxwell solvers*, Parallel Computing 27 (2001), pp. 761–775.

[148] L. Halpern, *Spectral methods in polar coordinates for the Stokes problem, Application to computation in unbounded domains,* Math. Comp. 65 (1996), pp. 507–531.

[149] W. Han, *Finite element analysis of a holonomic elasto-plastic problem,* Numer. Math. 60 (1992), pp. 493–508.

[150] ______, *A posteriori error analysis for linearization of nonlinear elliptic problems and their discretization,* Math. Meth. Appl. Sci. 17 (1994), pp. 487–508.

[151] W. Han and D.C. Reddy, *On the finite element method for mixed variational inequalities arising in elastoplasticity,* SIAM J. Numer. Anal. 32 (1995), pp. 1776–1807.

[152] P. Hansbo, *Generalized Laplacian smoothing of unstructured grids,* Commun. Numer. Methods Eng. 11 (1995), pp. 455–464.

[153] Y. Hayashi, *On a posteriori error estimation in the numerical solution of systems of ordinary differential equations,* Hiroshima Math. J. 9 (1979), pp. 201–243.

[154] B.-O. Heimsund, X.-C. Tai, and J. Wang, *Superconvergence for the gradient of finite element approximations by L^2 projections,* SIAM J. Numer. Anal. 40 (2002), pp. 1263–1280.

[155] R. Hiptmair, *Multigrid method for Maxwell's equations,* SIAM J. Numer. Anal. 36 (1999), pp. 204–225.

[156] I. Hlaváŭek and M. Křížek, *On a superconvergence finite element scheme for elliptic systems. I. Dirichlet boundary conditions,* Aplikace Matematiky 32 (1987), pp. 131–154.

[157] R.H.W. Hoppe, Y. Iliash, C. Iyyunni, and N.H. Sweilam, *A posteriori error estimates for adaptive finite element discretizations of boundary control problems,* J. Numer. Math. 14 (2006), pp. 57–82.

[158] R.H.W. Hoppe and R. Kornhuber, *Adaptive multilevel methods for obstacle problems,* SIAM J. Numer. Anal. 31 (1994), pp. 301–323.

[159] P. Houston, J.A. Mackenzie, E. Suli, and G. Warnecke, *A posteriori analysis for numerical approximations of Friederichs systems,* Numer. Math. 82 (1999), pp. 433–470.

[160] P. Houston, I. Perugia, and D. Schotzau, *A posteriori error estimation for discontinuous Galerkin discretizations of $H(curl)$elliptic partial differential equations,* IMA Journal of Numerical Analysis 25 (2005), pp. 1–25.

[161] P. Houston, R. Rannacher, and E. Süli, *A posteriori error analysis for stabilised finite element approximations of transport problems,* Comput. Methods Appl. Mech. Engrg. 190 (2000), pp. 1483–1508.

[162] Jin Qi-nian, *Error estimates of some Newton-type methods for solving nonlinear inverse problems in Hilbert scales,* Inverse Problems 16 (2000), pp. 187–197.

[163] C. Johnson, *Numerical Solution of Partial Differential Equations by the Finite Element Methods*, Cambridge University Press, Cambridge, 1987.

[164] ———, *Adaptive finite element methods for diffusion and convection problems,* Comput. Methods Appl. Mech. Engrg. 82 (1990), pp. 301–322.

[165] C. Johnson and P. Hansbo, *Adaptive finite elements in computational mechanics,* Comput. Methods Appl. Mech. Engrg. 101 (1992), pp. 143–181.

[166] C. Johnson and B. Mercier, *Some equilibrium finite element methods two-dimensional elasticity problems,* Numer. Math. 30 (1978), pp. 101–116.

[167] C. Johnson, R. Rannacher, and M. Boman, *Numerics in hydrodynamic stability. Towards error control in CFD,* SIAM J. Numer. Anal. 32 (1995), pp. 1058–1079.

[168] C. Johnson and A. Szepessy, *Adaptive finite element methods for conservation laws based on a posteriori error estimates,* Commun, Pure and Appl. Math. XLVIII (1995), pp. 199–234.

[169] V. Jovanovic and C. Rohde, *Finite-volume schemes for Friedrichs systems in multiple space dimensions: a priori and a posteriori error estimates,* Numer. Methods PDE 21 (2005), pp. 104–131.

[170] D. Kay and D. Silvester, *A posteriori error estimation for stabilised mixed approximations of the Stokes equations,* SIAM J. Sci. Comput. 21 (2000), pp. 1321–1336.

[171] D.W. Kelly, *The self equilibration of residuals and complementary error estimates in the finite element method,* Internat. J. Numer. Meth. Engrg. 20 (1984) pp. 1491–1506.

[172] D.W. Kelly, J.R. Gago, O.C. Zienkiewicz, and I. Babuška, *A posteriori error analysis and adaptive processes in the finite element method. Part I - error analysis,* Internat. J. Numer. Meth. Engrg. 19 (1983), pp. 1593–1619.

[173] D. Kinderlehrer and G. Stampacchia, *An introduction to variational inequalities and their applications*, Academic Press, New York, 1980.

[174] G. Kobelkov and M. Olshanskii, *Effective preconditioning of Uzawa type schemes for a generalized Stokes problem,* Numer. Math. 86 (2000), pp. 443–470.

[175] A.N. Kolmogorov and S.V. Fomin, *Introductory real analysis*, Dover Publications, Inc., New York, 1975.

[176] R. Kornhuber, *A posteriori error estimates for elliptic variational inequalities,* Comput. Math. Appl. 31 (1996), pp. 49–60.

[177] S. Korotov, P. Neittaanmäki, and S. Repin, *A posteriori error estimation of goal-oriented quantities by the superconvergence patch recovery,* J. Numer. Math. 11 (2003), pp. 33–59.

[178] M. Křížek, P. Neittaanmäki, and R. Stenberg (eds.), *Finite Element Methods. Superconvergence, Post-Processing and A Posteriori Error Estimates*, Lecture Notes in Pure and Applied Mathematics 196, Marcel Dekker, New York, 1998.

[179] M. Křížek and P. Neittaanmäki, *Finite Element Approximations of Variational Problems and Applications*. Wiley and Sons, New York, 1990.

[180] ———, *Superconvergence phenomenon in the finite element method arising from averaging of gradients,* Numer. Math. 45 (1984), pp. 105–116.

[181] G. Kunert, *An a posteriori residual error estimator for the finite element method on anisotropic tetrahedral meshes,* Numer. Math. 86 (2000), pp. 471–490.

[182] ———, *Towards anisotropic mesh construction and error estimation in the finite element method,* Numer. Meth. PDE 18 (2002), pp. 625–648.

[183] N.S. Kurpel' and B.A Shuvar, *Two-sided operator inequalities and their applications*, Naukova Dumka, Kiev, 1980.

[184] Yu. Kuznetsov and S. Repin, *Convergence analysis and error estimates for mixed finite element method on distorted meshes,* Journal of Numerical Mathematics 13 (2005), pp. 22–51.

[185] ———, *New mixed finite element method on polygonal and polyhedral meshes,* Russian J. Numer. Anal. Math. Modelling 18 (2003), pp. 261–278.

[186] P. Ladevèze and D. Leguillon, *Error estimate procedure in the finite element method and applications,* SIAM J. Numer. Anal. 20 (1983), pp. 485–509.

[187] P. Ladevèze, J.-P. Pelle, and Ph. Rougeot, *Error estimation and mesh optimization for classical finite elements,* Engineering Computations 8 (1991), pp. 69–80.

[188] P. Ladevèze and Ph. Rougeot, *New advances on a posteriori error on constitutive relation in f.e. analysis,* Comput. Methods Appl. Mech. Engrg. 150 (1997), pp. 239–249.

[189] O.A. Ladyzhenskaya, *The boundary value problems of mathematical physics*, Springer, New York, 1985.

[190] O.A. Ladyzenskaja and V.A. Solonnikov, *Some problems of vector analysis, and generalized formulations of boundary value problems for the Navier-Stokes equation,* Zap. Nauchn, Sem. Leningrad. Otdel. Mat. Inst. Steklov. (LOMI) 59 (1976), pp. 81–116 (Russian).

[191] O.A. Ladyzhenskaya and N.N. Uraltseva, *Linear and Quasilinear Elliptic equations*, Academic Press, New York, 1968.

[192] U. Langer and W. Queck, *On the convergence factor of Uzawa's algorithm,* J. Comput. Appl. Math. 15 (1986), pp. 191–202.

[193] R. Lazarov, *Superconvergence of the gradient for triangular and tetrahedral finite elements of a solution of linear problems in elasticity theory,* Computational Processes and Systems 6 (1988), pp. 180–191 (Russian).

[194] R. Lazarov, S. Repin, and S. Tomar, *Functional a posteriori error estimates for Discontinuous Galerkin method,* Radon Institute of Computational and Applied Mathematics, Preprint 2006–40, Linz.

[195] A.S. Leonov, *Some a posteriori stopping rules for iterative methods for solving linear ill-posed problems,* Comput. Math. Math. Phys. 34 (1994), pp. 121–126.

[196] M. Lonsing and R. Verfürth, *A posteriori error estimators for mixed finite element methods in linear elasticity,* Numer. Math. 97 (2004), pp. 757–778.

[197] J. Ma and H. Brunner, *A posteriori error estimates of Discontinuous Galerkin methods for non-standard Volterra integro-differential equations,* IMA J. Numer. Anal. 26 (2006), pp. 78–95.

[198] Y.V. Matijasevich, *A posteriori interval analysis,* Lecture Notes in Computer Sciences 204 (1985), pp. 328–334.

[199] J. Medina, M. Picasso, and J. Rappaz, *Error estimates and adaptive finite elements for nonlinear diffusion-convection problems*, Ecole Polytechnique Federale de Lausanne, Preprint CH-1015, 1995.

[200] P. Meyer, *A unifying theorem on Newton's method,* Numer. Funct. Anal. Optim. 13 (1992), pp. 463–473.

[201] S.G. Mikhlin, *Variational methods in mathematical physics*, Pergamon, Oxford, 1964.

[202] ———, *Error Analysis in Numerical Processes,* Wiley and Sons, Chiester, New York, 1991.

[203] P. Monk, *Finite Element Methods for Maxwell's Equations,* Clarendon Press, Oxford, 2003.

[204] R.E. Moore, *Interval analysis,* Prentice-Hall, Englewood Cliffs, New Jersey, 1966.

[205] P. Mosolov and V. Myasnikov, *Mechanics of rigid plastic bodies*, Nauka, Moscow, 1981 (Russian).

[206] A. Muzalevsky and S. Repin, *On two-sided error estimates for approximate solutions of problems in the linear theory of elasticity,* Russian J. Numer. Anal. Math. Modelling 18 (2003), pp. 65–85.

[207] ———, *On error estimates for approximate solutions in problems of the linear theory of thermoelasticity,* Russian Math. (Iz. VUZ) 49 (2005), pp. 60–68.

[208] P. Neittaanmäki and M. Křížek, *On* $0(h^4)$ *superconvergence of piecewise bilinear FE-approximations,* Teubner-Texte Math. 107, Teubner, 1988, pp. 250–255.

[209] P. Neittaanmäki and S. Repin, *Reliable methods for computer simulation, Error control and a posteriori estimates*, Elsevier, New York, 2004.

[210] ———, *A posteriori error estimates for boundary value problems related to the biharmonic operator,* East-West J. Numer. Math. 9 (2001), pp. 157–178.

[211] S. Nicaise, *A posteriori error estimation of some cell centered finite volume methods for diffusion-convection-reaction problems*, SIAM J. Numer. Anal. 44 (2006), pp. 949–978.

[212] S. Nicaise and S. Repin, *Functional a posteriori error estimates for the reaction-convection-diffusion problem*, Zapiski Nauchnich Seminarov POMI (to appear).

[213] J.A. Nitsche, *Ein Kriterium für die Quasi-Optimalität des Ritzchen Verfahrens,* Numer. Math. 11 (1968), pp. 346–348.

[214] R. Nochetto, *A posteriori error control of free boundary problems*, Free boundary problems (Trento, 2002), Internat. Ser. Numer. Math. 147, Birkhäuser, Basel, 2004, pp. 207–223.

[215] J.T. Oden, I. Babuška, F. Nobile, Y. Feng, and R. Tempone, *Theory and methodology for estimation and control of errors due to modeling, approximation, and uncertainty,* Comput. Methods Appl. Mech. Engrg. 194 (2005), pp. 195–204.

[216] J.T. Oden, L. Demkowicz, W. Rachowicz, and T.A. Westermann, *Towards a universal* $h-p$ *adaptive finite element strategy. Part 2. A posteriori error estimation,* Comput. Methods Appl. Mech. Engrg. 77 (1989), pp. 113–180.

[217] J.T. Oden and S. Prudhomme, *Goal-oriented error estimation and adaptivity for the finite element method,* Comput. Math. Appl. 41 (2001), pp. 735–756.

[218] J.T. Oden and J.N. Reddy, *Mathematical theory of finite elements*, Wiley and Sons, 1976.

[219] J.T. Oden, W. Wu, and M. Ainsworth, *An a posteriori error estimate for finite element approximations of the Navier–Stokes equations,* Comput. Methods. Appl. Mech. Engrg. 111 (1994), pp. 185–202.

[220] L.A. Oganesjan and L.A. Ruchovec, *Study of the rate of convergence of variational difference scheme for second order elliptic equations in two-dimensional region with a smooth boundary,* U.S.S.R. Comput. Math. and Math. Phy. 9 (1969), pp. 158–183.

[221] A. Ostrowski, *Les estimations des erreurs a posteriori dans les procédés itératifs,* C.R. Acad.Sci, Paris Sér. A–B 275 (1972), pp. A275–A278.

[222] C. Padra, *A posteriori error estimators for nonconforming approximation of some quasi-newtonian flows,* SIAM J. Numer. Anal. 34 (1997), pp. 1600–1615.

[223] J. Peraire and A.T. Patera, *Bounds for linear-functional outputs of coercive partial differential equations: Local indicators and adaptive refinement,* Advances in Adaptive Computational Methods in Mechanics, P. Ladevèze and J.T. Oden (eds.), Elsevier, New York, 1998, pp. 199–228.

[224] T. Plewa, T. Linde, and V.G. Weirs (eds.), *Adaptive mesh refinement – theory and applications.* Proceedings of the Chicago workshop on adaptive mesh refinement methods, Chicago, IL, USA, September 3–5, 2003, Springer, Berlin, 2005.

[225] F. Potra, *Sharp error bounds for a class of Newton-like methods,* Libertas Math. 5 (1985), pp. 71–84.

[226] J. Pousin and J. Rappaz, *Consistance, stabilité, erreurs a priori et a posteriori pour des problèmes non linéaries,* C. R. Acad. Sci. Paris 312 (1991), pp. 699–703.

[227] ———, *Consistency, stability, a priori and a posteriori errors for Petrov-Galerkin methods applied to nonlinear problems,* Numer. Math. 69 (1994), pp. 213–231.

[228] W. Prager and J.L. Synge, *Approximation in elasticity based on the concept of function space,* Quart. Appl. Math. 5 (1947), pp. 241–269.

[229] Jin Qi-nian and Hou Zong-yi, *On a posteriori parameter choice strategy for Tikhonov regularization of nonlinear ill–posed problems,* Numer. Math. 83 (1999), pp 139–159.

[230] Lin Qun and Sci Jan, *Iterative correctins and a posteriori error estimate for integral equations,* Journal of Computational Math. 11 (1993), pp. 297–300.

[231] R. Rannacher, *The dual-weighted-residual method for error control and mesh adaptation in finite element methods,* The mathematics of finite elements and applications X, MAFELAP 1999, Proceedings of the 10th conference, Brunel Univ., Uxbridge, Middlesex, GB, June 22–25, 1999, J. Whiteman (ed.), Elsevier, Amsterdam, 2000, pp 97–116.

[232] ———, *Finite element methods for the incompressible Navier–Stokes equations,* Fundamental directions in mathematical fluid mechanics, G.P. Galdi (ed.), Birkhäuser, Basel, 2000, pp. 191–293.

[233] R. Rannacher and F.T. Suttmeier, *A feed-back approach to error control in finite element methods: application to linear elasticity,* IWR, Preprint 96–42 (SFB 359), Heidelberg, 1996.

[234] ———, *A posteriori error control and mesh adaptation for F.E. models in elasticity and elasto-plasticity,* Advances in Adaptive Computational Methods in Mechanics, P. Ladevèze and J.T. Oden (eds.), Elsevier, New York, 1998, pp. 275–292.

[235] P.-A. Raviart and J.-M. Thomas, *A mixed finite element for second order elliptic problems,* Mathematical Aspects of Finite Element Methods, I. Galligani and E. Magenes (eds.), Springer, Berlin, 1977, pp. 292–315.

[236] S. Repin, *A posteriori estimates for approximate solutions of variational problems with strongly convex functionals,* Problems of Mathematical Analysis 17 (1997), pp. 199–226 (Russian); Translation in J. Math. Sci., New York 97 (1999), pp. 4311–4328.

[237] ———, *A posteriori estimates of the accuracy of variational methods for problems with nonconvex functionals,* Algebra i Analiz 11 (1999), pp. 151–182 (Russian); Translation in St.-Petersburg Mathematical Journal 11 (2000), pp. 651–672.

[238] ———, *A posteriori error estimation for nonlinear variational problems by duality theory,* Zapiski Nauchnych Seminarov POMI 243 (1997), pp. 201–214.

[239] ———, *A posteriori error estimation for variational problems with power growth functionals based on duality theory,* Zapiski Nauchnych Seminarov POMI 249 (1997), pp. 244–255.

[240] ———, *A posteriori error estimates for approximate solutions of variational problems,* Proc. 2nd European Conference on Numerical Mathematics and Advanced Applications, (Heidelberg) 1997, World Sci, Publishing, River Edge, New York, 1998, pp. 524–531.

[241] ———, *A unified approach to a posteriori error estimation based on duality error majorants,* Mathematics and Computers in Simulation 50 (1999), pp. 313–329.

[242] ———, *A posteriori error estimation for variational problems with uniformly convex functionals,* Math. Comp. 69 (2000), pp. 481–500.

[243] ———, *Two-sided estimates of deviation from exact solutions of uniformly elliptic equations,* Proc. St. Petersburg Math. Society IX (2001), pp. 143–171 (Russian); Translation in Amer. Math. Soc. Transl. Ser. 2, 209, AMS, Providence, RI, 2003.

[244] ———, *Estimates of deviation from exact solutions of initial-boundary value problems for the heat equation,* Rend. Mat. Acc. Lincei 13 (2002), pp. 121–133.

[245] ———, *Estimates of deviations from exact solutions of elliptic variational inequalities,* Zapiski Nauchnych Seminarov POMI 271 (2000), pp. 188–203.

[246] ———, *Aposteriori estimates for the Stokes problem,* J. Math. Sci., New York 109 (2002), pp. 1950–1964.

[247] ———, *Estimates of deviations for generalized Newtonian fluids,* Zapiski Nauchnych Seminarov POMI 288 (2002), pp. 178–203.

[248] ———, *Estimates for errors in two-dimensional models of elasticity theory,* J. Math. Sci., New York 106 (2001), pp. 3027–3041.

[249] ———, *A posteriori estimates of the accuracy of dimensional reduction models in 3-D elasticity theory,* Finite element methods (Jyvaskyla), 2000, GAKUTO Internat. Ser. Math. Sci, Appl. 15, Gakkotosho, Tokyo, 2001, pp. 240–253.

[250] ———, *A posteriori estimates in local norms. Problems in mathematical analysis,* J. Math. Sci., New York 124 (2004), pp. 5026–5035.

[251] ———, *Functional type a posteriori error estimates for approximate solutions of problems with incompressibility condition,* European Congress on Computational Methods in Applied Sciences and Engineering, ECCOMAS 2004 P. Neittaanmäki, T. Rossi, K. Majava, and O. Pironeau (eds.), O. Nevanlinna and R. Rannacher (assoc. eds.), Jyväskylä, 24–28 July, 2004 (electronic).

[252] ———, *Estimates of deviations from exact solutions for some boundary value problems with incompressibility condition,* Algebra and Analiz 16 (2004), pp. 124–161.

[253] ———, *A posteriori error estimates taking into account indeterminacy of the problem data,* Russian J. Numer. Anal. Math. Modelling 18 (2003), pp. 507–519.

[254] ———, *Functional Approach to Locally Based A Posteriori Error Estimates for Elliptic and Parabolic Problems,* Proc. 6th European Conference on Numerical Mathematics and Advanced Applications (Santiago de Compostela) 2005, Springer, Berlin, 2006, pp. 133–148.

[255] ______, *Local a posteriori estimates for the Stokes problem,* Zapiski Nauchnych Seminarov POMI 318 (2004), pp. 233–245.

[256] ______, *On the derivation of functional a posteriori estimates from integral identities,* Advances in numerical mathematics, Moscow, 2006, pp. 219–244.

[257] S. Repin and M. Frolov, *An estimate for deviations from the exact solution of the Reissner-Mindlin plate problem,* J. Math. Sci., New York 132 (2006), pp. 331–338.

[258] ______, *A posteriori error estimates for approximate solutions of elliptic boundary value problems,* Zh. Vychisl. Mat. Mat. Fiz. 42 (2002), pp. 1774–1787 (Russian).

[259] S. Repin and S. Sauter, *Functional A Posteriori Estimates for the Reaction-Diffusion Problem,* C. R. Acad. Sci. Paris, Ser. 1, 343 (2006), pp. 349–354.

[260] S. Repin and A. Smolianski, *Functional-type a posteriori error estimates for mixed finite element methods,* Russian J. Numer. Anal. Math. Modelling 20 (2005), pp. 365–382.

[261] S. Repin, S. Sauter, and A. Smolianski, *A posteriori error estimation for the Dirichlet problem with account of the error in the approximation of boundary conditions,* Computing 70 (2003), pp. 205–233.

[262] ______, *Duality based a posteriori error estimator for the Dirichlet problem,* Proc. Appl. Math. Mech. 2 (2003), pp. 513–514.

[263] ______, *A posteriori estimation of dimension reduction errors for elliptic problems in thin domains,* SIAM J. Numer. Anal. 42 (2004), pp. 1435–1451.

[264] ______, *A Posteriori Control of Dimension Reduction Errors on Long Domains,* Proceedings in Applied Mathematics and Mechanics 4 (2004), pp. 714–715.

[265] ______, *A posteriori error estimation for the Poisson equation with mixed Dirichlet/Neumann boundary conditions,* J. Comput. Appl. Math. 164/165 (2004), pp. 601–612.

[266] ______, *A posteriori estimation of dimension reduction errors,* Proc. 5th European Conference on Numerical Mathematics and applications (Prague) 2004, Springer, pp. 717–725.

[267] ______, *Two-Sided A Posteriori Error Estimates for Mixed Formulations of Elliptic Problems,* SIAM J. Numer. Anal. 45 (2007), pp. 928–945.

[268] S. Repin and R. Stenberg, *A posteriori error estimates for the generalized Stokes problem,* J. Math. Sci., New York (to appear).

[269] ______, *Two-sided a posteriori estimates for the generalized Stokes problem,* Institute of Mathematics, Helsinki University of Technology, Preprint A-516, Espoo 2006.

[270] S. Repin and J. Valdman, *A posteriori error estimates for problems with nonlinear boundary conditions,* Radon Institute of Computational and Applied Mathematics, Preprint 2006–25, Linz.

[271] S. Repin and L.S. Xanthis, *A posteriori error estimation for elasto-plastic problems based on duality theory,* Comput. Methods Appl. Mech. Engrg. 138 (1996), pp. 317–339.

[272] ______, *A posteriori error estimation for nonlinear variational problems,* Comptes Rendus de l'Académie des Sciences, Mathématique 324 (1997), pp. 1169–1174.

[273] W.C. Rheinboldt, *On a theory of mesh-refinement processes,* SIAM J. Numer. Anal. 17 (1980), pp. 766–778.

[274] W. Ritz, *Über eine neue Methode zur Lösung gewisser Variationsprobleme der mathematischen Physik,* J. Reine Angew. Math. 135 (1908), pp. 1–61.

[275] J.E. Roberts and J.-M. Thomas, *Mixed and hybrid methods,* Elsevier, 1991.

[276] R. Rodríguez, *Some remarks on Zienkiewicz–Zhu estimator,* Numer. Methods PDE 10 (1994), pp. 625–635.

[277] O. Scherzer, H.W. Engl, and K. Kunisch, *Optimal a posteriori parameter choice for Tikhonov regularization for solving nonlinear ill-posed problems,* SIAM J. Numer. Anal. 30 (1993), pp. 1796–1838.

[278] C. Schwab, *A-posteriori modelling error estimation for hierarchic plate models,* Numer. Math. 74 (1996), pp. 221–259.

[279] M. Schultz and O. Steinbach, *A new a posteriori error estimator in adaptive direct boundary element methods: the Dirichlet problem,* Calcolo 37 (2000), pp. 79–96.

[280] M. Schulz and W.L. Wendland, *A general approach to a posteriori error estimates for strictly monotone and Lipschitz continuous nonlinear operators illustrated in elasto-plasticity,* Proc. 2nd European Conference on Numerical Mathematics and Advanced Applications 1997 (Heidelberg), World Sci, Publishing, River Edge, New York, 1998, pp. 572–579.

[281] J.L. Segerlind, *Applied finite element analysis*, Wiley and Sons, 1976.

[282] S. Shaw and J.R. Whiteman, *Negative norm error control for second–kind convolution Volterra equations,* Numer. Math. 85 (2000), pp. 329–341.

[283] K.G. Siebert, *An a posteriori error estimation for anisotropic refinement,* Numer. Math. 73 (1996), pp. 373–398.

[284] E. Stein and S. Ohnimus, *Coupled model- and solution-adaptivity in the finite element method,* Comput. Methods Appl. Mech. Engrg. 150 (1997), pp. 327–350.

[285] E. Stein, F.J. Bartold, S. Ohnimus, and M. Schmidt, *Adaptive finite elements in elastoplasticity with mechanical error indicators and Neumann-type estimators,* Advances in Adaptive Computational Methods in Mechanics, P. Ladevèze and J.T. Oden (eds.), Elsevier, New York, 1998, pp. 81–100.

[286] R. Stenberg, *Some new families of finite elements for the Stokes equations,* Numer. Math. 56 (1990), pp. 827–838.

[287] ———, *Error analysis of some finite element methods for the Stokes problem,* Math. Comp. 54 (1990), pp. 495–508.

[288] G. Stoyan, *Towards Discrete Velte Decompositions and Narrow Bownds for Inf-Sup Constants,* Computer and Mathematics with Applications 38 (1999), pp. 243–261.

[289] G. Strang and G. Fix, *An analysis of the finite element method,* Prentice Hall, Englewood Cliffs, 1973.

[290] T. Strouboulis and J.T. Oden, *A posteriori error estimation of finite element approximations in fluid mechanics,* Comput. Meth. Appl. Mech. Engrg. 78 (1990), pp. 201–242.

[291] S. Sun and M.F. Wheeler, *$L^2(H^1)$-norm a posteriori error estimation for Discontinuous Galerkin approximations of reactive transport problems,* Journal of Scientific Computing 22/23 (2005), pp. 501–530.

[292] J.L. Synge, *The hypercircle method,* Studies in numerical analysis (papers in honour of Cornelius Lanczos on the occasion of his 80th birthday), Academic Press, London, 1974, pp. 201–217.

[293] R. Temam, *Navier–Stokes equations. Theory and numerical analysis,* Studies in Mathematics and its Applications 2, North-Holland, Amsterdam, 1979.

[294] K. Tsuruta and K. Ohmori, *A posteriori error estimation for Volterra integro-differential equations,* Mem. Numer. Math. 3 (1976), 33–47.

[295] S. Turek, *Efficient solvers for incompressible flow problems: An algorythmic and computational approach*, Springer, Berlin, 1999.

[296] R. Verfürth, *A posteriori error estimators for the Stokes equations,* Numer. Math. 55 (1989), pp. 309–326.

[297] ______, *A posteriori error estimates for nonlinear problems. Finite element discretisations of elliptic equations,* Math. Comp. 62 (1994), pp. 445–475.

[298] ______, *A review of a posteriori error estimation and adaptive mesh-refinement techniques* Wiley and Sons, Teubner, New York, 1996.

[299] ______, *A posteriori error estimates for nonlinear problems.* $L^r(0,T;L^p(\Omega))$*-error estimates for finite element discretizations of parabolic equations,* Math. Comp. 67 (1998), pp. 1335–1360.

[300] ______, *A posteriori error estimators for convection-diffusion problems,* Numer. Math. 80 (1998), pp. 641–663.

[301] ______, *Robust a posteriori error estimates for sationary convection-diffusion equations,* SIAM J. Numer. Anal. 43 (2005), pp. 1766–1782.

[302] ______, *A posteriori error estimates for finite element discretizations of the heat equation,* Calcolo 40 (2003), pp. 195–212.

[303] A. Vesser, *Efficient and reliable a posteriori error estimators for elliptic obstacle problems,* SIAM J. Numer. Anal. 39 (2001), pp. 146–167.

[304] M. Vishik, *La methode des projections orthogonales pour les equations differentielles conjuguees a ellesmemes,* Doklady Acad. Sci. URSS (N.S.) 56 (1947), pp. 115–118.

[305] M. Vogelius and I. Babuška, *On a dimensional reduction method I. The optimal selection of basis functions,* Math. Comp. 37 (1981), pp. 31–46.

[306] ______, *On a dimensional reduction method III. A posteriori error estimation and an adaptive approach,* Math. Comp. 37 (1981), pp. 361–384.

[307] L.B. Wahlbin, *Superconvergence in Galerkin Finite Element Methods*, Lecture Notes in Mathematics 1605, Springer, Berlin, 1995.

[308] J. Wang, *Superconvergence analysis of finite element solutions by the least-squares surface fitting on irregular meshes for smooth problems,* J. Math. Study 33 (2000), pp. 229–243.

[309] J. Wang and X. Ye, *Superconvergence analysis for the NavierStokes equations,* Applied Numerical Mathematics 41 (2002), pp. 515–527.

[310] H. Weil, *The method of orthogonal projections in potential theory,* Duke Math. J. 7 (1940), pp. 411–444.

[311] J.R. Whiteman and G. Goodsell, *A survey of gradient superconvergence for finite approximations to second order elliptic problems on triangular and tetrahedral meshes,* The Mathematics of Finite Elements and Applications VII, J.R. Whiteman (ed.), pp. 55–74, Academic Press, 1991.

[312] B.I. Wohlmuth and R.H.W. Hoppe, *A comparison of a posteriori error estimators for mixed finite element discretizations by Raviart-Thomas elements,* Math. Comp. 68 (1999), pp. 1347–1378.

[313] T. Yamamoto, *A note on a posteriori error bound of Zabrejko and Nguen for Zincenko's iteration,* Numer. Funct. Anal. Optim. 9 (1987), pp. 987–994.

[314] K. Yosida, *Functional Analysis,* Springer, New York, 1996.

[315] S. Zaremba, *Sur un probleme toujours possible comprenant, a titre de cas particuliers, le probleme de Dirichlet et celui de Neumann,* J. math. pures et appl. 6 (1927), pp. 127–163.

[316] E. Zeidler, *Nonlinear functional analysis and its applications. I. Fixed-point theorems,* Springer, New York, 1986.

[317] O.C. Zienkiewicz, *Achievements and some unsolved problems of the finite element method,* Int. J. Numer. Meth. Engrg. 47 (2000), pp. 9–28.

[318] O.C. Zienkiewicz, B. Boroomand, and J.Z. Zhu, *Recovery procedures in error estimation and adaptivity: Adaptivity in linear problems,* Advances in Adaptive Computational Methods in Mechanics, P. Ladevèze and J.T. Oden (eds.), Elsevier, New York, 1998, pp. 3–23.

[319] O.C. Zienkiewicz and K. Morgan, *Finite elements and approximation*, Wiley and Sons, New York, 1983.

[320] O.C. Zienkiewicz and J.Z. Zhu, *A simple error estimator and adaptive procedure for practical engineering analysis,* Internat. J. Numer. Meth. Engrg. 24 (1987), pp. 337–357.

[321] ______, *Adaptive techniques in the finite element method,* Commun. Appl. Numer. Methods 4 (1988), pp. 197–204.

[322] ______, *The superconvergent patch recovery and a posteriori error estimates. Part 1: The recovery technique,* Int. J. Numer. Meth. Engrg. 33 (1992), pp. 1331–1364.

[323] J.Z. Zhu, *A posteriori error estimation – the relationship between different procedures,* Comput. Methods Appl. Mech. Engrg. 150 (1997), pp. 411–422.

[324] M. Zlámal, *Some superconvergence results in the finite element method. Mathematical aspects of finite element methods,* Lecture Notes in Math. 606, Springer, Berlin, 1977, pp. 353–362.

[325] ______, *Superconvergence and reduced integration in the finite element method,* Math. Comp. 32 (1978), pp. 663–685.

Author information

Sergey Repin, V.A. Steklov Institute of Mathematics in St.-Petersburg, Russia.
Email: `repin@pdmi.ras.ru`

Bei Fragen zur Produktsicherheit wenden Sie sich bitte an:
If you have any questions regarding product safety,
please contact:

Walter de Gruyter GmbH
Genthiner Straße 13
10785 Berlin
productsafety@degruyterbrill.com